Controlling Environmental Pollution

An Introduction to the Technologies, History, and Ethics

P. Aarne Vesilind and Thomas D. DiStefano
Bucknell University

DEStech Publications, Inc.

Controlling Environmental Pollution

DEStech Publications, Inc.
1148 Elizabeth Avenue #2
Lancaster, Pennsylvania 17601 U.S.A.

Printed in the United States of America
10   9   8   7   6   5   4   3   2   1

Main entry under title:
   Controlling Environmental Pollution: An Introduction to the Technologies, History, and Ethics

A DEStech Publications book
Bibliography: p.
Includes index p. 409

ISBN No. 1-932078-39-8

HOW TO ORDER THIS BOOK

BY PHONE: 866-401-4337 or 717-290-1660, 9AM–5PM Eastern Time
BY FAX: 717-509-6100
BY MAIL: Order Department
DEStech Publications, Inc.
1148 Elizabeth Avenue #2
Lancaster, PA 17601, U.S.A.
BY CREDIT CARD: American Express, VISA, MasterCard
BY WWW SITE: http://www.destechpub.com

To Ann

Table of Contents

Preface

OVER the past decades we have seen some stunning victories in our fight against environmental pollution. Our lakes and rivers are cleaner than they have been in a century, and our air, even in the dirtiest cities, is breathable. The solid waste situation seems to be under control, and we have made a lot of progress in attacking the legacy of neglect in the management of hazardous waste.

But much work remains to be done, and we cannot revel complacently in our achievements. Sustained commitment of human and financial resources will be needed for the foreseeable future to address the management and cleanup of hazardous waste sites. An increasing population with an ever-growing demand for consumer goods will generate waste at record levels, not just in the United States but around the world. The free-market system continues to sell products that may either emit as-yet-unimagined pollutants or have unknown health effects.

The work of managing environmental pollutants will never cease. It will only change. But to understand where we are going, we need to know where we have been, and thus this book looks both forward and backward. The book describes state-of-the-art technology used in pollution control, and at the same time relates stories of environmental catastrophe, technological breakthroughs, and people who have been most influential in the development of this field.

This book is intended for a non-engineering audience. A knowledge of calculus, physics, or fluid mechanics is not required, whereas a high-school level chemistry course is necessary for full understanding. The text is based on a one-semester Bucknell University course designed for students in environmental studies. The course consists of three class hours plus a two-hour laboratory per week. Our experience in teaching the course has shown that students who have not had any training in technology are able to learn the concepts presented in this book. Even most BA candidates who have not taken college-level science do quite well.

As with any undertaking, the writing of a book requires the help and assistance of many people. Most importantly, we want to acknowledge the assistance and participation of our colleague, Matthew Higgins, who wrote part of the chapter on water treatment, including the discussion on bottled vs. tap water and contributed his ideas and insights during the planning phase of this book.

Finally, we want to thank the many students who have been through the ENVR 211 course and have provided us with useful feedback on how the topics addressed in the following pages can best be presented.

P. AARNE VESILIND
THOMAS D. DISTEFANO
Lewisburg, Pennsylvania
May, 2005

CHAPTER 1

The Pollution of the Environment

WE begin this book on an optimistic note. We want you, the reader, to recognize that there have been tremendous achievements in controlling pollution and that our environmental quality is vastly improved from what it used to be only a few decades ago.

Pollution problems are as old as human history. Imagine the living conditions back in pre-historic times when smoke filled caves and huts, and human waste transmitted disease. And then fast-forward to the beginning of the Industrial Revolution in the 18th century. As bad as things might be today with human health and environmental quality, they were a far sight worse then.

One of the unsolved problems during the 19th century in industrial cities all over the world was the lack of toilet facilities for all but the wealthy, and the total absence of wastewater treatment for all. Sewage from those homes with water simply flowed into the nearest watercourse, while the less wealthy used toilets in outhouses that overhung rivers. The very poor often lived around crowded courtyards where the only means of relieving oneself was to make a pile in the courtyard. The worst part about all this was the diseases that were transmitted by such unsanitary conditions. In addition, the smell was overwhelming.

We know that sensitive gentlemen often stuffed cloves into a pomegranate and walked with this stuck under their nose in order to ward off the odors. The River Thames in London stank so badly that the Parliament had to stuff rags soaked with lye into the cracks in the windows to try to keep out the smell. Contemporary accounts claim that people chatting in the parlor would keel over in mid-sentence. The servants got the worst of human waste disposal because they so often lived in the basements. The waste from above would be stored in leaking tanks, and this would ooze through the walls into their rooms.

Air pollution in some cities was unimaginable by today's standards. The popular seaside resort Swansea, now a tourist mecca in Wales, was at one time one of the most polluted cities in the world. In the early part of the 19th century,

1

iron and coal were discovered in the hills surrounding this old fishing community, and the discovery was followed in rapid succession by copper smelters and tinplate factories all along the shore. A travel writer described Swansea this way:

> Swansea in the 1860s, say when its metal exchange was the copper center of the whole world, seemed to have been visited by some horrific plague. Visitors approaching it by train from the east, seeing for the first time the green and sulfurous glow of its smelters, finding their carriages darkening by the black of its atmosphere, above all perhaps smelling its chemical fumes seeping and swirling all around, were sometime terrified by the experience, so absolutely of another world did the place seem, and so poisoned by its own exhalations.
>
> It was the works on the river bends that everyone remembered, for they cast a shroud-like spell. Everything was dead along that shore. All the grass was blistered off, all the water was fouled, not a tree lived, not a flower blossomed. A pall of vapour lay low over the valley: no bird flew through it, only the chimneys contributed to their filth, and the flicker of the furnaces was reflected eerily in its clouds. [Quote is courtesy of Richard Hummel, University of Toronto]

But we cannot assume that the advances made in environmental pollution control over the last decades have solved the problem. Many of the grossest forms or pollution from the 19th century have been eliminated, but quite a few new ones—ones that are not nearly as obvious as a belching smokestack—have been substituted. Problems with endocrine disruptors, acid rain, global warming, pesticides, disinfection byproducts, heavy metals—the list can go on and on—are just now being identified, and these problems are a great deal more frightening than a stream polluted with paper mill waste or air contaminated with wood smoke. So not all of the problems in controlling environmental pollution have been solved, and new ones that threaten our health and well-being appear every day. Nevertheless, we have come a long way in the last centuries in making our world a cleaner place to live.

If factual evidence is presented that would lead one to conclude that certain forms of pollution are causing health effects or are creating conditions that will

Figure 1.1. The Swansea copper smelter during the 19th century.

affect the welfare of distant or future populations, there generally are two responses. The first response is to agree with the data and analysis and strive to do something about the problem. The second response is to a) disagree with the data, or b) argue that there is no need to do anything about it. We see this latter approach most visibly in the debate on global warming. Many people simply will not believe the data, and insist that the whole thing is a conspiracy intended to destroy our economy and even our country. We can approach these people rationally, and we hope eventually to convince them that the data are accurate. A far more difficult group to convince are those who believe all the data, but simply do not care about what the pollution will do. These people argue it is the destiny of every generation to just take what we can for our own benefit, use up what is available, and throw away what we do not need. They believe that this approach will maximize our material happiness. To them the rest of the world is not of concern, and neither are future generations. Let them fend for themselves, they argue. Future generations are going to have many advantages we don't have, so we can take what we want right now and let these future people find their own way. Besides, how could we possibly know what the world would be like in the future? These people believe that the idea that one single individual can change the world is preposterous, so why not cruise along in a 25-foot powerboat, eat hamburgers, and throw aluminum beer cans into the lake? Life is good.

But fortunately, most people do not see the problems with environmental quality in that light. Most people *do* recognize our commitment to future generations, and they also understand that the actions of a single person *do* matter, when magnified by the thousands and millions of like-minded people. The Earth is worth saving, they argue, and it is the responsibility of all of us to do something to make this happen. If you are one of these caring optimists, then this book is for you!

1.1 WHAT ARE THE ROOT CAUSES OF ENVIRONMENTAL POLLUTION?

It can be argued that there was no environmental movement before the 1960s. Sure, there were activists like John Muir who went on to found the Sierra Club, President Teddy Roosevelt who oversaw the creation of the national park system, and Gifford Pinchot who advocated the wise use of our resources. Before them we had Henry David Thoreau and Ralph Waldo Emerson who eloquently praised the wonder of the natural world. But there was no welling of public opinion. Environmental quality simply was not a topic of conversation, and there were no activists marching or lobbying to support environmental legislation.

But small signs were there, and the willingness of the public to force action by the government was beginning to take hold, fueled in great part by other political issues, including the Vietnam War and civil rights. The situation was like a supersaturated solution, waiting for a single crystal that would cause it to solidify.

That crystal was the publication of Rachel Carson's book *Silent Spring*. While *Silent Spring* did not *cause* the environmental revolution any more than *Uncle Tom's Cabin* caused the Civil War, it nevertheless was hugely influential in promoting the cause of environmentalism. The vocal negative reaction from the chemical/agricultural industry had a lot to do with the success of the book. Many people, reading the reactions of the industry, wondered what nerve Carson had touched, and perhaps there was something to what she was talking about after all.

RACHEL CARSON

After graduating from the Pennsylvania College for Women (now Chatham College) in 1929 Rachel Carson (1907–1964) worked at Woods Hole Marine Biological Laboratory and continued her education with a master's degree in zoology from Johns Hopkins University in 1932. She was a writer, scientist, and in 1962 wrote perhaps the most influential book ever published in the environmental field, *Silent Spring*. The title comes from what she foresaw as the death and destruction of birds due to the extensive use at that time of chlorinated pesticides and she called for an end to their indiscriminate use. The reaction by the chemical industry to Carson's book was immediate and vitriolic. She was branded as everything from a flake to a Communist sympathizer. Here is a sample of the emotional tirades her book unleashed:

> The great fight in the world today is between Godless Communism on the one hand and Christian Democracy on the other. Two of the biggest battles in this war are the battle against starvation and the battle against disease. No two things make people more ripe for Communism. The most effective tool in the hands of the farmer and in the hands of the public health official as they fight these battles is pesticides. [Parke C. Brinkley, President of the National Agricultural Chemicals Association]

But in the end, it was clear that her cause was just, and as an increasing amount of evidence piled up on how the upper food chain was being affected by these non-biodegradable pesticides, she become a hero in the environmental movement. She did not, unfortunately, live to see her life and work honored by the world.

Quote from: Graham, F. "The Mississippi River Kill" in *Environmental Problems*, W. Mason and G. Fokerts (ed) William C. Brown Co. 1973

The other two very important environmental personages in the 1970s as the environmental movement was gathering steam were Paul Ehrlich and Barry Commoner. Ehrlich was at the time a respected scientist at Stanford, and Commoner was at Washington University in St. Louis. They both spoke out eloquently for the movement and became its de facto leaders.

Unfortunately, they had a serious disagreement that eventually led to their professional falling out. Ehrlich claimed that the core problem we were facing was overpopulation, while Commoner blamed the sorry state of the environment on over-consumption and the unwise use of non-replenishable resources. The answer, of course, is that both were right, but at that time their egos would not allow them to accept this obvious conclusion. The root causes of our environmental pollution are twofold: overpopulation and unwise use of resources.

1.1.1 Human Population Growth

All species reproduce in larger numbers than is necessary for the preservation of the species. The rate at which a population or any species *could* increase is known as the *biotic potential,* which is influenced by the age at which reproduction begins, the percentage of the life span during which the organism is capable of reproducing, and the number of offspring produced during each reproductive cycle. Larger animals such as whales have a very low biotic potential, while bacteria can divide every 20 minutes. At this rate, a single bacterium can produce 1,000,000 bacteria in only 10 hours! This growth is called *exponential growth*, as illustrated by the J-curve in Figure 1.2.

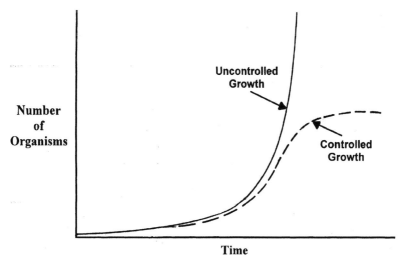

Figure 1.2. Growth curves for uncontrolled populations (the J curve) and populations that are controlled by environmental constraints (the S curve).

Exponential growth can best be illustrated by a story said to have originated in Persia. It tells of a clever courtier who presented a beautiful chess set to his king and in return asked only that the king give him one grain of rice for the first square, two grains, or double the amount, for the second square, four grains (or double again) for the third, and so forth. The king, being mathematically challenged, agreed and ordered the rice to be brought from storage. The eighth square required 128 grains, the 12th took more than one pound. Long before reaching the 64th square, every grain of rice in the kingdom had been used. Even today, the total world rice production would not be enough to meet the amount required for the final square of the chessboard. The secret to understanding the arithmetic is that the rate of growth (doubling for each square) applies to an ever-expanding amount of rice, so the number of grains added with each doubling increases.

Although some organisms can experience exponential growth for a short period of time, this rate of increase can not continue forever. What prevents the world from being overrun by various organisms is the *carrying capacity* of the environment, or the ability of the environment to support the population. At some point, as the population becomes overly large, there is *environmental resistance* to further growth due to predation, or lack of food and living space. The growth curve that occurs is thus more like the S-curve shown in Figure 1.2. At first the growth is exponential, but then various environmental resistance factors begin to limit the number of organisms and the number levels off to a constant number at the carrying capacity.

The global human population numbers over the past few thousand years

---

The Caribou on St. Matthew's Island, Alaska

St. Matthew's Island, located in the Bering Sea Wildlife Refuge, is roughly 128 square miles in area and supports a poorly developed land fauna. In 1944 a U. S. Coast Guard station was established on the island, and they decided to introduce a small herd of caribou to the island for recreational purposes. The Coast Guard station was abandoned shortly afterwards, and since that time the island has been uninhabited by humans. Because there were no predators on the island, and because there were no other large animals competing for the food supply, the small herd prospered. The herd's population increased exponentially from the original 29 to over 6,000 by 1963. This was far too many caribou than the vegetation of the island could support and the caribou overgrazed the island, nearly wiping out everything edible. Lichens, which were a staple of their winter diet, were completely eliminated, with predictable consequences. In the winter of 1964 nearly the entire herd of caribou starved, with only 8 females surviving to the spring. The caribou herd had experienced exponential, uncontrolled growth and exceeded the carrying capacity of the island. The only population control was death due to starvation.

eerily resemble the J-curve in Figure 1.2. The incredible increase in the number of humans has occurred because the controls that used to exists, such as famine, disease, and war, are being systematically eliminated.

Populations are governed by both birth rates and death rates. The growth of human populations is expressed as the difference between the births and deaths in a given population in a given time such as a year. In equation form,

$$R = B - D \qquad (1.1)$$

where

R = annual population change
B = annual number of births
D = annual number of deaths

If population changes are to be compared, these numbers have to be normalized by dividing each term in the above equation by the population. The *birth rate* is then the number of births per year per capita, and the *death rate* is the number of deaths annually per capita. The rate of change, or the *growth rate*, is therefore the birth rate minus the death rate, or

$$r = (b - d) \qquad (1.2)$$

where

r = the growth rate, or the rate of population change, number of people per year per capita (note that r can be negative if there is a net drop in population).
b = birth rate, number of births annually per capita
d = death rate, number of deaths annually per capita

To convert r to percent, multiply by 100.

Example 1.1

Ebolia has 150,000 people, and during the past year, there were 9,000 births and 3,000 deaths. What is the growth rate in Ebolia?

The birth rate, $b = 9,000/150,000 = 0.060$

The death rate, $d = 3,000/150,000 = 0.020$

The growth rate, $r = (0.060 - 0.020) = 0.040$, or $\times 100$, 4%

That is, the population of Ebolia is increasing by 4% a year.

Another way of expressing the growth of populations is the *doubling time*, or the time it takes for a population to double. An approximation of doubling time is

$$t_d = 0.693/r \tag{1.3}$$

where

t_d = doubling time, years
r = growth rate, number of people per year per capita

Example 1.2

What is the doubling time for the population in Ebolia in the above example?

$$t_d = 0.693/0.04 = 17.3 \text{ years.}$$

At this rate of growth, the population of Ebolia will be roughly 300,000 people in 17.3 years.

Within a geographical area, there are two more ways the population can change: *immigration* and *emigration*, moving into the geographical area or out of it. The population growth rate for a certain geographical area is then expressed as

$$r = (b - d) + (i - e) \tag{1.4}$$

where

i = rate of immigration, number of people per year per capita
e = rate of emigration, number people per year per capita

Example 1.3

In the Ebolia situation in the above example, the rate of emigration from Ebolia is 8,000 people per year, whereas the rate of immigration is 500 people per year. What is the growth rate of the population in Ebolia?

Rate if immigration $i = (500/150,000) = 0.0035$ people per year per capita

Rate of emigration $e = (8000/150,000) = 0.0533$ people per year per capita

From Example 1.1, the birth rate is 0.060 and the death rate is 0.020 people per year per capita. Hence the Ebolian growth rate is

$$r = (0.060 - 0.020) + (0.0035 - 0.0533) = -0.0098 \text{ people per year per capita}$$

That is, the population in Ebolia is actually decreasing about 1% per year due to the high rate of emigration.

On a worldwide basis, both the birth and death rates are decreasing, as shown in Figure 1.3, but the world population continues to increase. Because the number of births has decreased slightly more rapidly than the number of deaths, the worldwide rate of increase has also declined. That is, the growth rate r is decreasing with time. But despite this decline in the global growth rate we will not see a stable population ($r = 0$) until about the year 2090, at which point the population will level off at about 10.5 billion, or about twice the present population. (Figure 1.4)

The growth of populations is uneven across the world, with the rate in more developed countries being much lower than in poorer countries. Almost all the population increase during the next half century will occur in the less developed countries in Africa, Asia, and Latin America, whose population growth rates are much higher than those in richer countries. The populations in the less developed regions will most likely experience the greatest growth, with Asia's share of world population stabilizing at about 55 percent of the world's population. Europe's portion has declined sharply and could drop even more during the 21st century. The more developed countries in Europe and North America, as well as Japan, Australia, and New Zealand, are growing by less than 1 percent annually. Population growth rates are negative in many European countries, including Russia (–0.6%), Estonia (–0.5%), Hungary (–0.4%), and Ukraine (–0.4%). If the growth rates in these countries continue to fall below zero, the populations will slowly decline (Figure 1.5).

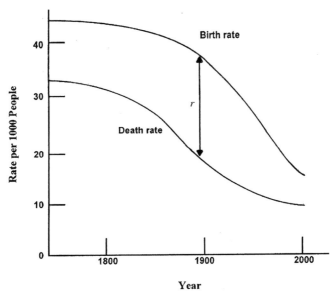

Figure 1.3. The decrease in both global birth rate and death rate. The difference is the growth rate, r.

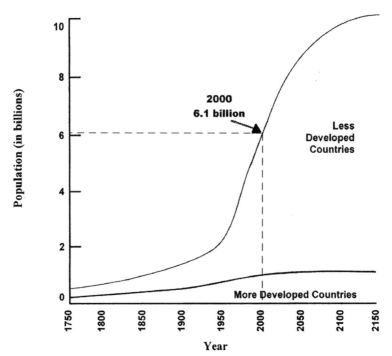

Figure 1.4. Global population prediction.

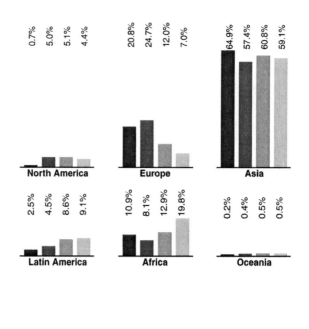

Figure 1.5. Populations for the six continents. The graphic shows the percent of the world's population. (*Source:* The United Nations Population Division, 1998).

10

Making such predictions presumes there will be no catastrophic changes to the population such as new viruses that defy treatment. We also do not know what the carrying capacity of the earth is and when this will be reached. It is possible that the carrying capacity is considerably less than 10.5 billion people, and that at some point there will occur a catastrophic population decline due to lack of food or the prevalence of disease. Given the past population growth, and the apparent inability of humans to understand the "tragedy of the commons," the Earth might well be another St. Matthew's Island.

GARRETT HARDIN

Garrett Hardin (1915–2003) received a BS in zoology from the University of Chicago and then went to Stanford University where he received his Ph.D. in microbiology. He is best remembered as a curmudgeon—a person who is not afraid to speak the truth, however unpopular the truth may be. In 1968 he wrote a hugely influential article entitled "The Tragedy of the Commons" which as become a must-read in every ecology course. In this article Hardin imagines an English village with a common area where everyone's cow may graze. The common is able to sustain the cows and village life is stable, until one of the villagers figures out that if he gets two cows instead of one, the cost of the extra cow will be shared by everyone, while the profit will be his alone. So he gets two cows and prospers, but others see this and similarly want two cows. If two, why not three—and so on—until the village common is no longer able to support the large number of cows, and everyone suffers. Harding applies this to the problem of birth control, arguing that the environmental cost of a child born today (air, water, and other resources) is shared by everyone, while the benefit is only to the parents. Hence every family will want more children, until the ability of the earth to support these populations is exceeded, resulting in global disaster.

A similar argument can be made for the use of non-renewable resources. If we treat diminishing resources such as oil and minerals as capital gains we will soon find ourselves in the "common" predicament of having too many people and not enough resources.

A thread running all through Hardin's books is that ethics has to be based on rational argument and not on emotion. He argues that for ethics to be useful people have to be literate, they must use words correctly, and they must appreciate the power of numbers. His most interesting book is *Stalking the Wild Taboo* in which he takes on any number of social misconceptions that demand rational reasoning.

1.1.2 Unwise Use of Resources

The resources necessary to allow humans to survive on this planet include air to breathe, water to drink, food to eat, and materials with which to build civilization. The objective is to use such resources in a manner that the Earth will be able to sustain the human population. This objective has been given the name *sustainability*.

Without doubt, human use of resources as presently practiced is unsustainable. That is, our present generations are using resources and polluting the earth at a rate that should make life more difficult (if not impossible) for future generations.

The *precautionary principle* states that if a problem is sufficiently severe and the consequences sufficiently serious, one would not need *proof* before action is taken to alleviate the potential damage. This recognition led the World Commission on Environment and Development, sponsored by the United Nations, to conduct a study of the world's resources. Also known as the Brundtland Commission, their 1987 report, *Our Common Future*, introduced the term *sustainable development* and defines it as "development that meets the needs of the present without compromising the ability of future generations to meet their own needs." The United Nations Conference on Environment and Development, i.e., the Earth Summit held in Rio de Janeiro in 1992 communicated the idea that sustainable development is both a scientific concept and a philosophical ideal. The document, *Agenda 21*, was endorsed by 178 governments and hailed as a blueprint for sustainable development. The administration of the first President Bush objected to some of the provisions that would require the United States to voluntarily cooperate with the rest of the world and refused to sign the document. Fortunately, the intransigence of the United States government has not prevented most other countries from adopting the central principles of this accord.

The underlying purpose of sustainable development is to help developing nations manage their resources such as rain forests without depleting these resources and making them unusable for future generations, and to prevent the collapse of global ecosystems. The Brundtland report presumes that we have a core ethic of intergenerational equity, and that future generations should have an equal opportunity to achieve a high quality of life. (The report is silent, however, on just *why* we should embrace the ideal of intergenerational equity, or why one should be concerned with the survival of the human species.)

One of the central problems with achieving sustainability is that consumption of resources is highly uneven. People in less developed countries consume considerably less resources than people in wealthier countries. Figure 1.6 is a dramatic illustration of the unevenness of the consumption of energy on a per capita basis. If the unwise consumption of resources is the root cause of

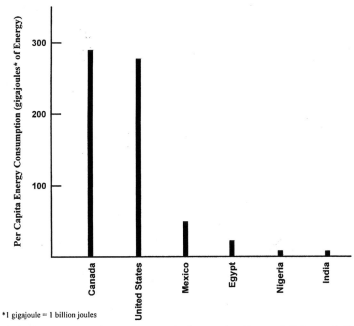

*1 gigajoule = 1 billion joules

Figure 1.6. The uneven use of energy resources in six countries. (*Source:* P. H. Raven, L. R. Berg, and G. B. Johnson, *Environment* Saunders College Publishing, 1993).

our environmental problems, then certainly the rich countries have to take the vast majority of the blame.

1.1.3 Population vs. Resources

So who is right? Ehrlich or Commoner? Is our problem one of overpopulation, or is it a problem of the unwise use of resources? Of course, they are both right. A country or geographic area is overpopulated if it has more people that consume more resources than it can supply, and a country can be overpopulated in two ways:

1. If there are too many people and the environment is being damaged because of the sheer number of people, even though their consumption is at a minimal level then this is called *people overpopulation.*

2. Even though there are few people, if their consumption is so high that they are not able to attain sustainability, then this is called *consumption overpopulation.*

It is the congruence of unwise use and consumption of resources by too many people that results in environmental pollution.

1.2 WHAT DETERMINES HOW WE CONTROL POLLUTION?

There are many variables that determine how any nation controls pollution, but some of them are the availability of technology, the wealth of the nation, its political system and governance, its laws and regulations, and most importantly, the prevailing ethical climate.

1.2.1 Technology

The most obvious approach to controlling pollution is the application of technology, usually using "end-of-pipe" solutions—accepting a waste in whatever form or quantity and treating it to reduce its adverse effect on health or the environment. There is no doubt that such technology has been successful in reducing pollution and enhancing the quality of life.

Consider the automobile as a source of pollution and a challenge to engineers. The internal combustion engine burns gasoline, and does so relatively inefficiently. Instead of the carbon and hydrogen in the gasoline burning to make only carbon dioxide and water,

$$(CH)_x + O_2 \rightarrow CO_2 + H_2O + (CH)_y$$

the combustion is inefficient and a whole slew of unburned hydrocarbons are produced. In addition, the high-temperature/high-pressure combustion, using nitrogen from the air, produces nitrogen oxides:

$$(CH)_x + O_2 + N_2 \rightarrow (CH)_y + CO + CO_2 + NO_x + H_2O$$

The unburned hydrocarbons $(CH)_y$, combined with the nitrogen oxides, NO_x, are responsible for the formation of photochemical smog in cities. The carbon monoxide is itself a problem in producing various physiological effects such as drowsiness (and death at high concentrations).

The internal combustion engine therefore presented a challenge to the automotive engineers, and they responded by designing catalytic converters that are installed in the exhaust pipe under the car. The oxidizing catalytic converter oxidizes most of the unburned hydrocarbons and the carbon monoxide to CO_2 and water, and the reducing catalytic converter reduces most of the nitrogen oxides to nitrogen gas. Problem solved.

But the problem is not each car, but the total effect of the cars on urban air. Even though the cars are cleaner, the total number of cars is rapidly increasing, so that the increased effectiveness of the catalytic converters is reduced due to the sheer number of cars. The effect of cleaner engines on photochemical smog in Los Angeles ought to have been noticeable by now, but the problem is that there now are many more cars and the total tonnage of emissions continues to increase every year. Something else needs to be done.

The next step would be to work "up-the-pipeline" to see what can reduce the emissions. One solution would be to reduce the amount of gasoline burned by vehicles, or increase the gasoline mileage achieved by cars. This is not a technical solution, however, since mandating the increase in mileage requires either government action or a sharp increase in gasoline prices. Neither seems to be likely in the near future.

Technology now looks even further up the pipeline and asks if the fuel might not be changed from gasoline to some other combustible material. This will require retrofitting cars and/or building newer engines that can use these fuels. Many cars in Southern California, for example, already use natural gas as a fuel, and its combustion is far cleaner than burning gasoline. Problems still exist with refueling such cars, and the production of CO_2, a global warming gas, is still a problem.

Moving even further up the pipeline, technology might develop a car that runs on hydrogen instead of some form of hydrocarbon. This combustion is exceptionally clean, since there are no waste hydrocarbons produced, and the nitrogen oxides can be controlled by lower temperatures and better catalytic converters if only reducing converters are needed. But the problem is the distribution of hydrogen, which is gaseous and highly explosive. Finally, the production of hydrogen requires energy, and this would have to come from the combustion of fossil fuels, thus negating the benefits of the cleaner burning cars.

The cleanest car is the electric vehicle, which burns nothing and emits nothing. But again, the electricity must be produced somewhere, often by burning fossil fuels, and transported long distances, experiencing large losses in efficiency.

Our technology has been effective in controlling environmental pollution, and there is every expectation that advances in treatment systems and the reduction of pollutants from manufacturing and everyday life will continue to be reduced thanks to an increased understanding of how to put knowledge to work in enhancing the quality of our environment.

1.2.2 Wealth

But it is one thing to have technology available to control pollution, and it is quite another to have the financial resources to purchase this technology. The poorer nations of the world are having pollution problems that are now only a dim historical memory for those of us fortunate to live in wealthy nations. An example of the problems created by economic transition from a poor country to one that has wealth can be found in today's China. One of the requisites of development is the availability of fuel, and China is fortunate to have huge deposits of coal, which is now being mined in increasing amounts and burned for energy in the cities. As a result, the air pollution in many Chinese cities rivals the worst days of the Industrial Revolution in Great Britain and the United States. There currently is not enough wealth in China either to convert to cleaner fuels, or to install pollution control equipment.

1.2.3 Politics and Governance

One of the reasons China is now developing so rapidly is its unfortunate recent history. During the past half century much of the world witnessed one of the most devastating experiments in the application of an economic theory. On paper, communism looks good. Why have duplication of facilities, products, and services if this is clearly unnecessary? If there already is a Coke machine in the basement of the dorm, why also have a Pepsi machine? If there is one manufacturer that makes useful tractors, why invest any more resources in the production of a different tractor? If economies of scale are important for efficiency, why not centralize production? For example, there is no need to have a lot of factories that manufacture shoes. Why not centralize all shoe manufacturing in Peoria or somewhere? Think of the money we would save. There would be plenty of shoes for everyone, and at very reasonable prices.

In such a system the management of a facility is under pressure to meet its production goals. If the centralized planners believe that the country will need 5 million shoes next year, it is up to the one and only shoe manufacturer to produce 5 million shoes, or else people will not have enough shoes.

The first tactic to make sure production goals are met is to trim the quality. If a shoe should have 200 stitches to mount the sole on to the shoe, then reducing this number to 150 stitches will save a lot of time. The shoe still looks like a shoe, and it behaves like a shoe, until the sole falls off, but that is someone else's problem. So the end result is that quality suffers in order that production goals can be met.

The second way the plant manager can make sure the production goals can be met is to get his people to work hard, but this is a problem. Since the wages are flat and there is no incentive to work hard (if you work hard you get paid the same as the guy who loafs), the reasonable choice is to do as little work as possible. This forces the government to take two actions: (1) try to convince the workers that they are on a great crusade and that they ought to be working hard for the good of the country. If this does not work, (2) threaten them with punishment if they don't produce. In communist countries it was a crime punishable by many years in jail (or worse) for criticizing the system and for not participating in the great national experiment. Troublemakers were systematically removed from society for that reason. Stalin killed 30 million of his own people in order to get rid of dissidents.

The other way production can be increased within the communist economic system is to eliminate everything in the manufacturing process that does not directly lead to increased production. This includes most importantly every means of pollution control. Only now, when the Eastern Bloc countries are rejoining the European commonwealth, are the environmental tragedies of this policy coming to light. The incredible air pollution of Romania, the unconscionable contamination of the Baltic Sea, the destruction of the groundwater resources in Poland—all are the victims of the last 50 years of communist rule in eastern Europe.

As Winston Churchill said: "Democracy is the worst form of government; except for all the rest." In democracies we have the luxury of public opinion, and the ability of the public to express its opinion leads to a form of government where human rights are upheld. One of those human rights is the right to live in an unpolluted environment.

In our democracy, an issue in every political clash is the environment. Every politician is for protecting the environment and controlling pollution, but few of them will actually make the difficult decisions when in office. Why is that?

First, defending the environment is without question a popular stance. Public opinion always lists environmental quality as one of the top issues. The public is, however, woefully ignorant of what the true issues are. Typically, the prevailing public opinion is that hazardous wastes are the number one health problem, while in fact hazardous wastes and their disposal is far down the list of environmental problems that affect human health. (One of the top problems, if it is included in the list, is smoking, but since this is a voluntary activity, the public will not perceive it as particularly dangerous, and many are willing to take the risk of health problems occurring.) The truth is that all in all our environment in the United States is very clean. We have made tremendous progress over the past decades to the point where most surface waters are clean enough for water contact sports and fishing, and where the air is clean enough to no longer present a health risk. There are local spots, of course, and we are working on them. And there is the global problem, but our government does not seem to care too much about that. So while environmental concern is a political issue, the great improvements in controlling pollution have resulted in some politicians being able to safely ignore environmental quality in their political campaigns, and when in office, crassly despoil the land in order to enrich themselves and their friends with little concern for public backlash.

The second reason some people are not concerned enough about the environment to make it an issue around which elections are decided is misinformation about the environment. Part of the problem is with people who intentionally write and broadcast untruthful and skewed information, but some of it is because the so-called environmentalists are often careless with their accusations and conclusions. Some alleged facts have been repeated so often that they assume an aura of truth, but when challenged, prove to be fallacious. One such "fact" is that if one walks the streets of New York, breathing in the city air, the pollutants are so bad that one might as well have been smoking 38 cigarettes a day. The origin of this number is difficult to trace, but it apparently was calculated based on the level of atmospheric particulate matter at street level, which would include dust from the traffic, and these particulate concentrations were compared to the particulates inhaled from cigarettes. The chemical and physical dissimilarity of the two types of particulates is ignored, as are the health effects. If an "environmentalist" repeats this claim in public, he or she can be easily and quickly refuted by factual evidence, which can bring other more relevant and factually correct matters into question. It is necessary, therefore, to maintain credibility when arguing with those for whom the environment is not important and who firmly believe that the entire environmental movement is a cynical plot to destroy our society.

The Death of the Oceans

An example of the kind of nonsense that some environmentalists publish, and which then gets used and quoted by political opponents to illustrate how all environmentalists are irrational is from a pamphlet published in 1969 by Paul Ehrlich, the noted Stanford professor:

> The end of the ocean came late in the summer of 1979, and it came even more rapidly that the biologists had expected. There had been signs for more than a decade, commencing with the discovery of 1968 that DDT slows down photosynthesis in marine plant life. It was announced in a short paper in the technical journal, *Science* but to ecologists it smacked of doomsday. They knew that all life in the sea depends on photosynthesis, the chemical process by which green plants bind the sun's energy and make it available to living things. And they know that DDT and similar chlorinated hydrocarbons had polluted the entire surface of the earth, including the sea.

The "short paper" alluded to by Ehrlich was a study in which mixed phytoplankton cultures were dosed with high concentrations of DDT, up to 500 parts per billion (ppb). The cultures were then incubated and the photosynthetic rate measured. The results showed indeed that at higher concentrations of DDT, the rate dropped. What Ehrlich did not know, or chose not to report, was that DDT is very poorly soluble in water, in the range of 1.2 ppb (parts per billion, or one part DDT per billion parts of water). In order to get the DDT into solution at 500 ppb, the experimenters had to dissolve the DDT in alcohol before it was used to dose the flasks, an unlikely situation in nature to say the least. Subsequent experiments by others showed that in that range of DDT solubility, its presence has no effect on photosynthesis.

But with the publication of the pamphlet the damage was done, and anti-environmentalists gleefully quoted Ehrlich as an example of the nonsense espoused by all environmentalists.

1.2.4 Laws and Regulations

It is not enough, however, for the technology to be available and to have the financial resources to purchase it. In free societies it is necessary for the people, through their government, to force the polluters to invest in pollution controls and to produce products that will not cause new problems.

Law is necessary because, at least in countries with a Western tradition, there is a primary conflict of interest between ownership and care for other citizens. A deep-seeded personal right and freedom that we cherish is the right to do what we want with our property. Nobody, especially the government, is to tell us what we can own and what we can do with what we own. This principle is especially important in the ownership of land. An owner can do what he or she wishes with land, and it is not the function of government or other people to tell the owner what can be done with land.

The second fundamental principle is that none of us is allowed, intentionally and maliciously, to hurt others. We cannot go around beating people over the head with baseball bats, for example, even if we own the baseball bats. This principle also applies to the use of land. An owner of a hog farm, for example, is not allowed to let waste from the land run into the neighbor's land, or the stench from the hog operations to hurt the sensibilities of neighbors. And thus we have a conflict that requires the passage of laws and the establishment of regulations.

These principles were first developed in what became *common law* in England, and was then brought over to the colonies. Common law is derived from the principle of fair play, or justice—a purely British invention. Under common law, if a person is wronged the perpetrator is convicted and sentenced on the basis of *precedence*. That is, if a similar wrongful act occurred previously, then the only right and fair thing to do is to treat the next person in a similar manner. Common law is not written down, except as cases that define the precedent for new cases.

When relying on common law, an individual or groups of individuals injured by pollution may cite general principles in two branches of that law: *tort law* and *property law*. The harmed party, the plaintiff, can enter a courtroom and seek remedies from the defendant for damaged personal well-being or damaged property. A tort is an injury incurred by one or more individuals. Careless accidents and exposure to harmful chemicals are the types of wrongs included under this branch of common law.

A polluter could be held responsible for the damage to human health under three broad categories of tort law: intentional liability, negligence, and strict liability.

Intentional liability requires proof that somebody did a wrong to another party on purpose. This proof is especially complicated in the case of damages from pollution. The fact that a "wrong" actually occurred must first be established, a process that may rely on direct statistical evidence or strong inference, such as the results of laboratory tests on rats. Additionally, intent to do the wrong must be established, and that involves producing evidence in the form of written documents or direct testimony from the accused individual or group of individuals. Such evidence is not easily obtained. If intentional liability can be proven to the satisfaction of the courts, actual damages as well as punitive damages can be awarded to the injured plaintiff.

Negligence may involve mere inattention by the polluter who allowed the injury to occur. Proof in the courtroom focuses on the lack of reasonable care taken on the defendant's part. Examples of such neglect in pollution include failure to inspect the operation and maintenance of pollution control devices, or the failure to design and size an adequate abatement technology. Again, damages can be awarded to the plaintiff.

Strict liability does not consider the fault or state of mind of the defendant. Under certain extreme cases, a court of common law has held that some acts are abnormally dangerous and that individuals conducting those acts are strictly liable if injury occurs. The court does tend to balance the danger of an act against the public utility associated with the act. An example could be the emission of a radioactive or highly toxic gas from an industrial smoke stack.

Again, if personal damage is caused by pollution from a known source, the damaged party may enter a court of common law and argue for monetary damages to be paid by the defendant or an injunction to stop the polluter from polluting or both. But sufficient proof and precedent are often difficult, if not impossible, to muster, and in many cases, tort law has been found to be inadequate in controlling pollution and awarding damages.

Property law, on the other hand, is based on interference with the use or enjoyment of property, and property rights are based on actual invasion of the property. Property law is founded on ancient actions between land owners and involves such considerations as property damage and trespassing. A plaintiff basing a case on property law takes chances, rolls the dice, and hopes the court will rule favorably as it balances social utility against individual property rights. The two legal theories based on property rights most applicable to pollution control are nuisance and trespass.

Nuisance is the most widely used form of common law action concerning the environment. Public nuisance involves unreasonable interference with a right, such as the "right to clean air," common to the general public. A public official must bring the case to the courtroom and represent the public that is harmed by the pollution. Private nuisance, on the other hand, is based on unreasonable interference with the use and enjoyment of private property. The key to a nuisance action is how the courts define "unreasonable" interference. Based on precedents and the arguments of the parties involved, the common law court balances the equities, hardships, and injuries in the particular case, and rules in favor of either the plaintiff or the defendant.

Trespass is closely related to the theory of nuisance. The major difference is that some physical invasion, no matter how minor, is technically a trespass. Recall that nuisance theory demands an unreasonable interference with land and the outcome of a particular case depends on how a court defines "unreasonable." Trespass is relatively uncomplicated. Examples of trespass include physical walks, vibrations from nearby surface or subsurface strata, and possibly gases and microscopic particles emitted from an individual smoke stack.

Both of these theories, nuisance and trespass, are difficult to use with regard to pollution. An unbearable odor to one person might simply be a fragrance to another, and it would not be considered a nuisance or a trespass.

In conclusion, common law has generally proven inadequate in dealing with problems of pollution. The strict burdens of proof required in the courtroom often result in decisions that favor the defendant and lead to continued pollution. Additionally, the technicality and complexity of individual cases often limit the ability of a court to act; complicated tests and hard-to-find experts often leave a court and a plaintiff with their hands tied. Furthermore, the absence of standing in a common law courtroom often prevents private individuals from bringing a case before the judge and jury unless the individual actually suffers material or bodily harm from the air pollution. *Standing* means that one has actually suffered a significant harm and the cause of this harm is in question. If, for example, you owe your friend 25 cents and refuse to pay it, your friend cannot go to court to collect it. He would be wasting the court's time by such a trivial matter, and thus he would not have standing in court.

One key aspect of these common law principles is their degree of variation. Each state has its own body of common law, and individuals relying on the

---

Water Pollution Before Water Pollution Laws

In the 1960s some of the rivers in our industrial areas were incredibly polluted. The states were unable or unwilling to clamp down on the polluters, fearing loss of industry and jobs. Two of the worst rivers were the Cayuga River in Cleveland and the Houston Ship Canal in Houston. The Cayuga has the distinction of being the only river to actually catch on fire and burn down a pier. The Houston Ship Canal was so polluted that some of the ships entering the harbor were experiencing stripping paint from the hulls and subsequent corrosion that left indelible marks attesting to their visit to Houston.

Common law was pretty much worthless is preventing such pollution. Since the Houston Ship Canal entered the Gulf of Mexico, and if the industries that lined the shores of the canal were content with the obnoxious water, then there was little that could be done unless the state mandated the cleanup. Unfortunately, the industries had such political power that state politicians were unwilling to mandate change. Federal laws were necessary if water pollution was to be controlled.

court system are generally confined to using the common laws of the applicable state.

The genius of common law, nevertheless, is that it tries to be fair to all, and at the same time is able to change as the needs and values of the people change. It is no longer a crime, for example, to be a witch. But common law changes very slowly, since most courts are loathe to make new law. Common law is also ineffective in protecting the environment and in correcting environmental ills because under common law a person (not a forest or a river or any other non-human entity) has to be wronged in order that relief can be sought in the

CHRISTOPHER D. STONE

Just because an animal, or a tree, or even a place cannot hire a lawyer and argue its case in court, does this non-human then have an absence of standing? Why can't a human who can argue the case do so on behalf of an animal, or tree, or other non-human?

This was the question that came before the Supreme Court in 1967 in a famous case, *Storm King vs. Federal Power Commission*. The Federal Power Commission wanted to lease out some federally owned land for a ski area, and the Sierra Club objected, believing that the development would harm the forests. But the problem was that no single member of the Sierra Club could prove that he or she was being directly harmed by the construction of the ski slopes. The harm to each person would be small, and collectively the harm would be great, but there was no one person who would have standing in the court. This is the problem of the 25 cent debt. The damage to a single individual is just not great enough to warrant taking up the court's time.

One of the lawyers on the case was Christopher Stone (1937–), a professor on the faculty of the University of Southern California law school. Stone, with an undergraduate degree from Harvard and a law degree from Yale, had been active in the Sierra Club and helped them on numerous occasions. In support of the Sierra Club's case, Stone wrote an article cleverly entitled "Should Trees Have Standing?" in which he argued that natural objects have every right to be represented in court if the damage is sufficiently great. The article appeared in the law review and became part of the brief submitted to the Supreme Court. Sitting on the court at that time was Justice William O. Douglas who was an avid outdoorsman and was quite sympathetic to the destruction of natural habitats. Using Stone's arguments, Douglas was able to sway the court in favor of the plaintiffs, and to stop the development of the ski area. While it is unlikely that this common law precedent will be widely used in the control of environmental pollution in the future, its successful application in the 1960s fueled and encouraged the environmental movement.

courts. The assumption is that all of nature is owned by humans and it is only the wealth and welfare of humans that common law protects. In legal terms, non-humans have no standing in court.

Common law is simply not very useful in quieting environmental ills. Rather, public opinion has caused the Congress of the United States to pass environmental legislation. Such *legislated law*, or *statutory law*, in effect establishes new common law by setting artificial and immediate precedents. Statutory law has several layers, where the lower layers of the law are subservient to the higher ones.

In the United States, the most basic statutory law is the U.S. Constitution on which all of other statutory laws depend and to which they are subservient. If the Congress, for example, passes a law that says that all suspects in certain criminal situations are to be held without bail and without counsel for years and years, the Supreme Court will step in and declare such a law unconstitutional since each citizen has the right, guaranteed by the Constitution, to a speedy trial.

The next most important set of laws are federal laws passed by the Congress and signed into law by the President. Within these laws the Congress can give various agencies of the United States the right to set regulations that have the strength of law. For example, Congress might say that the U.S. EPA has the power to pass regulations for controlling air pollution emissions from solid waste incinerators. The regulations thus established by the U.S. EPA are not really law, but have the force of law since Congress has delegated this duty to the U.S. EPA, reserving of course the right to rescind these regulations if it feels that the U.S. EPA has not done a proper job.

In addition to federal laws there are state laws, and each of the fifty states is busy making laws on issues not covered by the federal government. For example, each state has the right to decide what its public school system should be like and how it should be funded. While the federal government contributes money to education, the individual states are the ones that carry out the function of organizing public education. If at some time the federal government decides that it wants to control education, it will have a constitutional problem, since the Constitution specifically gives the states the right to oversee local issues not national in scope, such as education. Congress has, however, gotten around this sticky issue by saying that federal standards (such as highway speed limits and ages for alcohol consumption—clearly state-wide of even local issues) are voluntary, but that if a state refuses to abide by such regulations the federal government will cut off funding to state programs. This is too big of stick, and states meekly go along. Theoretically, state legislators are supposed to make all laws, and delegate some issues such as national defense to the federal government, but this is no longer what is actually happening. The Congress and executive agencies have the true power in the country.

Finally, on an even more local level, the county and municipality have the

---

Early Smoke Ordinances

The history of smoke ordinances in this country illustrates the problem of using common law in pollution control. In 1881, the City of Chicago passed the first smoke ordinance in the United States, and other cities tried to follow but often ran into legal problems. For example, in 1893, St.Louis passed the following ordinance:

> The emission into the open air of dense black or thick gray smoke within the corporate limits of the City of St.Louis is hereby declared to be a nuisance. The owners, occupants, managers, or agents of any establishment, locomotives, or premises from which dense black or thick gray smoke is emitted or discharged, shall be deemed guilty of a misdemeanor, and, upon conviction therefore, shall pay a fine of not less than ten nor more than fifty dollars.

When the city sued an obdurate offender, the Heitzberg Packing and Provision Company, however, the Missouri Supreme Court declared the law invalid because "while it is entirely competent for the city to pass a reasonable ordinance looking at the suppression of smoke when it becomes a nuisance to property or health or annoying to the public at large, this ordinance must be held void because it exceeds the powers of the city under its charter to declare and abate nuisances." In other words, it was acceptable for a community to pass such an ordinance if it had the power to do so, but the state had not given St.Louis such a power.

---

right to pass ordinances to allow them to govern. Local ordinances banning backyard burning of trash are examples of such ordinances. If, however, the state decides that it wants to pass a statewide law banning the burning of trash on private property, it can do so, and local ordnances are pre-empted. A community cannot declare that it does not want to follow the state laws.

On the other hand, if the federal government wants to pass a law banning backyard burning of trash, it can do so and all state laws are then irrelevant. A state cannot decide it does not want to abide by the law. Of course, the state may decide to challenge the federal law in court, claiming jurisdiction, but it is unlikely to win. Besides, all the federal government has to do is to tie the ban to some appropriate funding package, and the states will comply.

All law- and regulatory-making bodies are juggling the problem of preserving the maximum freedom of the individual in what can be done with private property, while trying to protect the health, safety, and welfare of the public from the use of such property. Banning backyard burning is an infringement of private rights, but the ban will result in great reduction in air pollution and odor which benefits everyone.

An example of how statutory law was used to clean up the environment is the Clean Water Act of 1972. This ground-breaking legislation set a goal of bringing the water quality of all the streams and rivers in the United States to levels where fishing and swimming would be safe. Since that time we have made tremendous progress, even in light of increased population and complexity in industrial discharges, and the success of the law is evident in its duplication in nearly every industrialized country in the world.

The procedure was to first allow each state to set up water quality goals, *stream standards* defined on the basis of such characteristics as dissolved oxygen, pH, bacteriological quality, and temperature. Then each stretch of stream or river is tested and labeled according to these standards, so all surface waters in each state are identified by a *water quality classification*. These classifications are different for each state since some parameters are important in some states but not others. For example, North Carolina standards include temperature as one of the parameters because trout would not be able to spawn in warmer water and the state wanted to protect trout streams in the mountains. This would not be a concern in Louisiana, for example.

Once the streams are all classified, the law then requires each state to issue permits to all municipalities and industries that discharge into the watercourses. These permits are conditional in that the discharge water quality is to be under some set parameters. For example, a permit might be written for a discharge to a small stream that would limit the temperature of the discharged water to no more than 30°C. If an industry changes processes and produce warmer water, it has to provide some means of cooling it down so that the temperature of the discharge does not exceed 30°C.

The initial classification of streams does not result in improvement of water quality in those streams. It simply confirms the status quo for water quality because industries and municipalities are discharging exactly what they have always discharged, and now they have a permit to do so. But the law had a kicker. When the discharge permits come up for renewal, the states can begin to ratchet up the water quality levels. For example, if the 30°C industrial discharge results in a stream temperature that prevents trout from spawning, the state can change the discharge permit and force the industry to lower the discharge temperature to say 20°C. Now the industry, in order to legally discharge its water to the watercourse, has to provide some means of cooling the water prior to discharge. Using this technique, one notch at a time, the states are able to bring most of the rivers and streams to an acceptable level of water quality.

We come to the conclusion that while all of the four approaches to controlling environmental pollution—technology, wealth, political will, and laws and regulations—are effective, they are not adequate to solve environmental problems. But all of them are necessary, and they are glued together by our sense of what is the right thing to do—by ethics.

1.2.5 Ethics

Citizens can pass laws and develop technology to control pollution only if there is a will to do so, and this will is based on ethical principles. While the scholarly field of ethics is both interesting and substantive, most of us do not even think of ethics in our everyday life—we simply carry on with what we are doing. Our decisions that involve ethical questions are instinctive and not made by reverting to some rule or principle. Our ethical behavior is very much like our use of grammar in speaking. When we want to be understood, we do not compose a sentence in our head and check if it has a verb or if there are dangling participles. We simply know how to talk in complete understandable sentences. Ethics is like that as well. We are conditioned by our environment and experiences to behave in a certain way. If that way is to take into account the welfare of others, we call such activity virtuous behavior, and the person a *virtuous* person.

Being virtuous is not difficult. Crudely put, a virtuous person uses common sense morality in his or her interactions with others. Common sense morality is basically doing what would be expected of any of us if we had a sense of caring and kindness toward others. Common sense morality suggests that we would help others as long as this action does not ask some unreasonable cost from us. For example, if you see a person in trouble in the water, and you have a chance to throw him a rope, the action shows kindness to the person in trouble at almost no cost to you. You would agree, we hope, that this is what all of us ought to do if we would like to be thought of as good persons. But suppose the person is drowning in a raging flood, and the only way to help him would be to jump in, with the most likely outcome that both of you would drown. This is simply too much to ask of anyone, and anyone who does that would be considered foolhardy.

In all cases having to do with commons sense ethics, the question is how human beings ought to treat each other. In short, the *moral community*, or those individuals with whom we would need to interact ethically, includes only humans and the only *moral agents* within the moral community are human beings. Moral agency requires *reciprocity* in that each person agrees to treat one another in a mutually acceptable manner.

If there is one moral principle that people value above all others it is truth. A truthful person is one who can be believed and who will not lead you astray. A truthful person is one who will not knowingly deceive you, and one you can trust. In many environmental matters, being truthful about the effects of pollutants, especially on human health and welfare, is the highest moral requirement. And yet, there are numerous examples where this has not occurred. Four such examples—the unconscionable actions of the asbestos industry, the deceitfulness of a major corporation, the unwillingness of our government to tell the truth about air quality, and the broken promises to a small rural community—are described in the discussion boxes.

Morality is, to repeat, the treatment of other people in a way that is beneficial to all. Morality demands that the promise made by one town administration to a

---

Orange County Landfill

The siting of landfills is one of the most contentious activities engaged in by local government. Since solid waste disposal is considered a local problem, few communities have the foresight to enter into regional solid waste management programs and seek instead to take care of their own solid waste within their own borders.

Chapel Hill, North Carolina and Orange County had established a landfill in the early 1970s in a rural section of the county outside the city limits of Chapel Hill. Very close to the proposed landfill was a long-established rural African-American neighborhood called the Rogers Road Community. When the landfill was first proposed, the members of this community complained that they were asked to shoulder the burden of the landfill and that they would gain no direct benefit from it. In all probability their property values would drop and the effect of the landfill (odor, noise, and littered roads) would be detrimental to their community.

The members of the Chapel Hill Town Council responded by promising the Rogers Road Community that the new landfill would last only 20 years and that when it was full the area would be made into a park, which they could then enjoy.

In the mid 1990s when the landfill was filling up, the town and county set out to find a new landfill site. They hired an engineering firm to study the situation, and the engineering report designated several potential sites for the new facility—one of which was the expansion of the existing landfill. The representatives of the Rogers Road Community reminded the Chapel Hill Town Council that 25 years ago the Council had promised to the community that the landfill would be closed and a new one sited elsewhere. The other three neighborhoods close to the alternative sites also reacted and since in every case these areas were much wealthier and had greater political clout, the Town Council decided that the best (political) alternative was to expand the existing landfill and to ignore the promise made to the residents of Rogers Road Community. As one council member rationalized, "The promise to the Rogers Road Community was made by another Council. None of those Council members are now on the Council. Since we, the present Council members, did not promise anything, we are not beholden to keep the old promises."

---

neighborhood be kept. Morality demands that a company that knows that it is severely harming its workers at the very least tell the workers what is happening. We want to behave morally, and expect others to do likewise, knowing that doing so benefits all of us. This sense of *reciprocity* is the rational basis for morality, a concept originally developed by Thomas Hobbs, who called it a *social contract*. We have an agreement, or a contract, with our fellow humans to treat each other with respect, and if we do, we all benefit. The truth of this idea can be powerfully demonstrated by examples in our contemporary world where the social contract has broken down, where people are not acting with regard to welfare of others and where all social order has disintegrated.

None of us would want to live in such communities, and so we have a special responsibility to continually work toward maintaining a civil society.

The concept of a social contract and the need for reciprocity, however, do not explain our treatment of non-humans. We obviously aren't the only inhabitants on earth, and is it not also important how we treat non-human animals? Or plants? Or places? Should the moral community be extended to include other animals? . . . plants? . . . inanimate objects such as rocks, mountains, and even places? If so, should we also extend the moral community to our progeny who are destined to live in the environment we will to them?

The Asbestos Conspiracy

Asbestos is a mineral and a fiber, and has the dual characteristics of being very strong and non-combustible. It has found wide use in the insulation of boilers and other heating components, as well as fireproofing and insulating buildings. It was widely used in automobile brakes and in asphalt tiles and shingles.

Asbestos is not a health problem unless it gets into your lungs, where the fibers get caught in the alveoli (the small sacs where gas transfer takes place) and once there irritate the membranes, eventually producing a form of lung cancer called asbestosis.

Although the detrimental effects of asbestos became widely known only in the past few decades, insurance companies stopped insuring asbestos workers as early as 1918. In 1930, Johns Manville, the largest manufacturer of asbestos products, distributed an internal report outlining the negative health effects of asbestos, and in 1933 settled a confidential lawsuit brought by eleven workers with the understanding that the proceedings would be forever sealed, suggesting that the testimony would have been damaging to the company if it had been made public. In 1935 Johns Manville executives told the editor of the trade journal *Asbestos* to not publish anything detrimental about asbestosis. The conspiracy was most evident when the president of Johns Manville was quoted in 1942 to have called "a bunch of fools" another asbestos-manufacturer who shared health information with its workers. In 1951 the Johns Manville management insisted that any reference to cancer be removed from any scientific publication resulting from research sponsored by the company.

When a class action suit was brought in 1977 against the company by workers who were exposed to asbestos in the 1940s, the documents showing the conspiracy became public. Since it takes decades for the effects of asbestos to be manifested, these former workers were only then experiencing health problems. A former Johns Manville manager testified that there was a "hush-hush policy" of not revealing what the company knew of the health dangers of asbestos.

The extended lawsuit, which eventually ended in favor of the plaintiffs, caused Johns Manville to go into bankruptcy. The trial judge concluded that there had been a "conscious effort by the (asbestos) industry from the 1930s to downplay or arguably suppress the dissemination of information to employees and the public for fear of the promotion of lawsuits."

Air Pollution Following the Destruction of the World Trade Center

Following the collapse of the World Trade Center towers on 11 September 2001, for most New Yorkers the sudden increase in respiratory discomfort or infection did not seem high on the list of concerns. Only months after the tragedy did the facts surrounding the air pollution caused by the collapse of the towers become clear. Perhaps 10,000 people at or near the site suffered from various forms of asthma, sinusitis, or bronchitis. The long-term effects such as emphysema, cancer, and other pulmonary diseases are still to be realized.

When the towers collapsed, an immense amount of dust was discharged into the air. The quantity of asbestos released is estimated at between 300 and 400 tons! The towers had been constructed before the use of asbestos for fireproofing and insulation was banned and it had been deemed much too expensive to try to remove from the towers. The transformers in the electrical system had over 100,000 gallons of polychlorinated biphenyl (PCB) and the remains of the towers smoldered for months afterward, releasing untold quantities of particulates and toxic chemicals.

Despite the knowledge that the air around ground zero was highly contaminated and that rescue teams ought to wear protective devices, a few days after the tragedy the U. S. EPA proclaimed that the there was no significant contamination at the site. As a result, about 12,000 people who lived in the neighborhood returned to their homes. In actual fact, the measurements the EPA conducted showed that the levels of fine metal particles, sulfuric acid, fine undissolvable glass particles, and high-temperature carcinogenic matter were *higher than any values ever measured by the EPA test team.* The pressure to release the "all clear" signal from the EPA apparently came from the White House, which wanted to reduce the anxiety of the citizens of New York.

How could this have happened, and why did the EPA scientists succumb to the political pressure? Part of the explanation is in the kind of studies they chose to report. The particulate matter, for example, was done for particles less than 2.5 microns in size. While it is true that these particles are the most important in long-range chronic illnesses associated with air pollution, this was not the problem at ground zero. What they should have reported was total particulates in the air since this was what was causing the respiratory problems. The levels of total particulates in the air soon after the attack have been estimated as high as 100,000 micrograms per cubic meter ($\mu g/m^3$). By comparison, the acceptable level of particulates in a coal mine is 15 $\mu g/m^3$. Similar selective use of data apparently allowed the EPA to justify its "all clear" signal and disclaim that it was due to pressure from the White House. The long-range health effects of this government untruthfulness remain unknown.

The Potlatch Incident

It cost us a bundle but the Clearwater River still runs clear

Potlatch

"It cost us a bundle but the Clearwater River still runs clear" read the headline beneath the picture. And sure enough, it was a scene of breath-taking natural beauty—hills, shaggy with evergreens, framing a stretch of clear, blue water flecked with white foam where it raced over hidden rocks. But what Potlatch Forests, Inc. neglected to mention in its national ad campaign was that the picture had been snapped 50 miles upstream from the company's pulp and paper plant in Lewiston, Idaho, where it had its wastewater discharge.

When an enterprising local college newspaper editor pointed out the discrepancy between ad copy and reality, the company responded by canceling all corporate advertising, and its president Benton R. Cancell was quoted as saying: "We tried out best. You just can't say anything right any more—so to hell with it."

Quote from: Newsweek "Pollution: Puffery or Progress" December 28, 1970

Such questions are being debated and argued in a continuing search for what has become known as *environmental ethics*, a framework that is intended to allow us to make decisions *within* our environment, decisions that will concern not only ourselves but the rest of the world as well.

One approach to the formulation of an ethic that incorporates the environment is to consider environmental values as *instrumental values*, values that can be measured in dollars and/or the support nature provides for our survival (e.g., production of oxygen by green plants). The instrumental value of nature view holds that the environment is useful and valuable to people, just as other desirable commodities such as freedom, health and opportunity. This *anthropocentric* view of environmental ethics, the idea that nature is here only for the benefit of people, is of course an old one. Aristotle states that "Plants exist to give food to animals, and animals to give food to men. . . . Since nature makes nothing purposeless or in vain, all animals must have been made by nature for the sake of men". Thus by this reasoning, the value of non-human animals can be calculated as their value to people. We would not want to kill off all the Plains buffalo, for example, because they are beautiful and interesting creatures, and we enjoy looking at them. To exterminate the buffalo would mean that we are causing harm to other humans.

We can also agree that it is necessary to live in a healthy environment in order to be able to enjoy the pleasures of life, and therefore we include a concern for non-human animals on our list of moral goods. Likewise, other aspects of the environment also have instrumental value. One could argue that to contaminate the water, or pollute the air, or destroy natural beauty is taking something that doesn't belong only to a single person. Such pollution is stealing from others, plain and simple.

Not only would we not want to kill useful animals, but we would not want to exterminate species since there is the possibility that they will somehow be useful in the future. An obscure plant or microbe might be essential in the future for medical research, and we should not deprive others of that benefit. And it would be unethical to destroy the natural environment because so many people enjoy hiking in the woods or canoeing down rivers, and we should preserve these for our benefit.

While the instrumental value of nature as the main basis for an environmental ethics has merit, it also has a serious problem. If we accept this argument, it would not prevent us from killing or torturing individual animals as long as this does not cause harm to other people. But the mandate is not compatible with our attitudes about animals. We would condemn a person who causes unnecessary harm to any animal that could feel pain, and many of us do what we can to prevent this.

Given these problems with the concept of instrumental value of nature as a basis for our attitudes toward the environment, there has ensued a search for some other basis for including non-human animals, plants and even things within our sphere of ethical concern. The basic thrust in this development is the attempt to attribute *intrinsic* value to nature, and to incorporate non-human nature within our moral community.

This is known as *extensionist* ethical thinking, since it attempts to extend the moral community to include other creatures. Such a concept is perhaps as revolutionary as the recognition only a century ago that slaves are also humans and must be included in the moral community. Aristotle, for example, did not apply ethics to slaves since they were not, in his opinion, intellectual equals. We now recognize that this was a hollow argument, and today slavery is considered morally repugnant. It is possible that in the not too distant future, the moral community will include the remainder of nature as well, and we will afford nature the same moral protection.

The extensionist environmental ethic was initially publicized not by a philosopher, but a forester. Aldo Leopold (1887–1948) defined the environmental ethic, or as he called it, a land ethic, as an ethic which " . . . simply enlarges the boundaries of the community to include soils, waters, plants and animals, or collectively the land."

He recognized that both our religious as well as secular training had created a conflict between humans and the rest of nature, where nature had to be subdued and conquered, and nature was something powerful and dangerous against

ALDO LEOPOLD

Aldo Leopold (1887–1948) was born in Burlington, Iowa, and spent his younger years roaming the woods near his home. He went east to go to school, first Lawrenceville School in New Jersey and then Yale, where he received a graduate degree in forestry. He joined the United States Forest Service in 1909 where he worked to preserve wilderness areas. He joined the Forest Products Lab in Madison, Wisconsin, but soon became anxious to be back in the woods. His knowledge of timber and game management soon made him an acknowledged expert, and he was offered a faculty position at the University of Wisconsin. He bought a poor farm in a county where the topsoil had been badly eroded (hence the "Sand County") and started to husband it back to health. He died fighting a forest fire on his neighbor's farm.

His writings, especially his book, *The Sand County Almanac*, had enormous influence on the environmental movement, and he is acknowledged as the founder of environmental ethics. He had a very simple and ecologically rational view of what was right: "A thing is right when it tends to preserve the integrity, stability, and beauty of the biotic community. It is wrong when it tends otherwise."

which we had to continually fight, but that a rational view of nature would lead us to an environmental ethic which " . . . changes the role of *Homo Sapiens* from conqueror of the land community to plain member and citizen of it."

Leopold was, in fact, questioning the ageold belief that humans are special, that somehow we are not a part of nature, but pitted against nature in a constant combat for survival, and that we have a Godgiven role of dominating nature, as specified in Genesis. Much as later philosophers (and people in general) began to see slavery as an untenable institution and recognized that slaves belonged within our moral community, succeeding generations may recognize that the rest of nature is equally important in the sense of having rights.

The question of admitting non-humans to the moral community is a contentious one, and centers on the fact that ethics and morality only exist among people who are willing and able to reciprocate. The reason I do not want to lie to you is that I expect you to not lie to me. This reciprocity does not exist with non-humans. Some would argue that it therefore is not reasonable to extend our moral community to include anyone other than humans because of the requirement of reciprocity.

But is reciprocity a proper criterion for admission to the moral community? Do we not already include within our moral community human beings that cannot reciprocate—infants, the senile, the comatose, our ancestors, and even future people? Maybe we are making Aristotle's mistake again by our exclusionary practices. Perhaps being human is not a necessary condition for inclusion in the moral community, and other beings have rights similar to the rights that humans have. These rights may not be something that we necessary give them, but the rights they possess by virtue of being.

If this is true, then there is nothing to prevent non-human animals from having "inalienable rights" simply by virtue of their being, just as humans do. They have rights to exist and to live and to prosper in their own environment, and not to have humans deny them these rights unnecessarily or wantonly. If we agree that humans have rights to life, liberty and absence of pain, then it seems only reasonable that animals, who can feels similar sensations, should have similar rights.

With these rights comes moral agency, independent of the requirement of reciprocity. The entire construct of reciprocity is of course an anthropocentric concept that serves well in keeping others out of our private club. If we abandon this requirement, it is possible to admit more than humans into the moral community.

But if we crack open the door, what are we going to let in? What can legitimately be included in our moral community, or, to put it more crassly, where should we draw the line? If the moral community is to be enlarged, many people agree that it should be on the basis of sentience, or the ability to feel pain. This argument suggests that all sentient animals have rights that demand our concern.

To include animal suffering within our circle of concern however opens up a Pandora's box of problems. While we might be able to argue with some vigor that suffering is an evil, and that we do not wish to inflict evil on any living being that suffers, we do not know for sure which animals (or plants) feel pain, and therefore are unsure about who should be included. We can presume with fair certainty that higher vertebrate animals can feel pain because their reactions to pain resemble ours. A dog howls, and a cat screams, and tries to stop the source of the pain. Anyone who has put a worm on a hook can attest to the fact that the worm probably feels pain. But how about creatures that cannot show us in unambiguous ways that they are feeling pain? Does a butterfly feel pain when a pin is put through its body? An even more difficult problem is the plant kingdom. There are those people who insist that plants feel pain when they are hurt, and that we are just too insensitive to recognize it.

If we do *not* focus only on the pain and pleasure suffered by animals, then it is necessary to either recognize that the rights of the animals to avoid pain are equal to those of humans, or to somehow list and rank the animals in order to specify which rights animals have under what circumstances. In the first instance, trapping animals and torturing prisoners would have equal moral significance. In the second, it would be necessary to decide that the life of a chickenhawk is less important than the life of a chicken, and an infinite number of other comparisons.

Finally, if this is the extent of our environmental ethics, we are not able to argue for the preservation of places and natural environments, except with regard to how they might affect the welfare of sentient creatures. Damming up the Grand Canyon would be quite acceptable if we adopted the sentient animal criterion as our sole environmental ethic.

It seems, therefore, that it is not possible to draw the line at *sentience*, and the next logical step is to simply incorporate all of life within the folds of the moral community. This is not as outrageous as it seems, the idea having been developed by Albert Schweitzer who calls his ethic a "reverence for life." He concludes that limiting ethics to only human interactions is a mistake, and that a person is ethical "only when life, as such, is sacred to him, that of plants and animals as that of his fellow men." Schweitzer believes that an ethical person would not maliciously harm anything that grows, but would exist in harmony with nature. He recognizes of course that in order to eat, humans must kill other organisms, but he holds that this should be done with compassion and a sense of sacredness toward all of life. To Schweitzer, human beings are simply a part of the natural system.

This approach to environmental ethics has a lot of appeal, and quite a few proponents. Unfortunately, it fails to convince on several accounts. First, there is no way to determine where the line between living and non-living really should be drawn. Second, the problem of how to weigh the value of non-human animal life relative to the life of humans is unresolved. Should the life of all creatures be equal, and thus a human life be equal to that of any other creature? If so, the squashing of a cockroach would be of equal moral significance to the murder of a human being. If this is implausible, then there must again be some scale of values, and each living creature must have a slot in the hierarchy of values as placed on them by humans. If such a hierarchy is to be constructed, how would the value of the life of various organisms be determined? Are microorganisms of equal value to polar bears? Is lettuce of the same order of importance as a gazelle?

Such ranking will also introduce impossible difficulties in determining what is and is not deserving of moral protection. "You, the amoeba, you're in. You, the paramecium, you're out. Sorry," just doesn't compute. Drawing the line for inclusion in the moral community at "all of life" therefore seems to be indefensible.

Some ethicists suggest that there exists a system of concentric rings, with the most important moral entities in the middle, and the rings extending outward, incorporating others within the moral community, but at decreasing levels of moral protection. The question of how the various creatures and places on the earth are to be graded in terms of their moral worth is not resolved, and indeed is up to the people doing the valuing. This process is human centered, and ecocentric environmental ethics is a form of the anthropocentric environmental ethic with fuzzy boundaries, since it does not know where to draw the lines. Because we don't want to value everything equally, we have to have some hierarchy. We would all agree that humans have greater moral value than rocks, but what standards do we use to make that decision?

It is not at all obvious why we should have protective, caring attitudes for some thing only if these attitudes can be reciprocated. Perhaps we are hung up on the idea of acceptance into the *moral community*, and the logjam can be broken by thinking of it as the inclusion of all things in a universal community. In this community, reciprocity is not necessary, and what matters is love and caring for others, simply because they exist, and by so doing, deserve our respect. The amount of love and care we give to others, people and things, should be proportional to the ability to give it, and demands nothing in return.

1.3 PROBLEMS

1.1 You have been given the responsibility of designing a large trunk sewer. The sewer alignment is to follow a creek that is widely used as a recreational facility. It is a popular place for picnics; nature walks have been constructed by volunteers along its banks; and the local community wants to eventually make it a part of the state park system. The trunk sewer will badly disrupt the creek, destroy its ecosystem and make it unattractive for recreation. What thoughts would you have on this assignment? Write a journal entry as if you were keeping a personal journal (diary).

1.2 Would you intentionally run over a box turtle trying to cross a road? Why or why not? Present an argument for convincing others that your view is correct.

1.3 Often environmental decisions affect people of the next generation, and even several generations to come. A major bridge, such as the Brooklyn Bridge, is over 150 years old and it looks like it will stand for another century. This bridge is a mere piker compared to some of the Roman engineering feats, of course.

Societal decisions such as nuclear waste disposal also affect future generations, but in a negative way. We are willing this problem to our kids, their kids, and untold generations down the line. But should this be our concern? What, after all, have future generations ever done for us?

Philosophers have been wrestling with the problem of concern for future generations, and without much success. Listed below are some arguments often found in literature about why we should *not* worry about future people. Think about these arguments, and construct opposing arguments if you disagree. Be prepared to discuss these in class.

We do not have to bother worrying about future generations because:

1. These people do not even exist, so they have no claim to any rights such as moral consideration. We cannot allow moral claims by people who do not exist, and may not ever exist.

2. We have no idea what the future generations may be like, and what their problems will be, so it makes no sense to plan for them. To conserve (not use) non-replenishable resources just in case they may need them makes no sense whatever.

3. If there will be future people, we have no idea how many, so how can we plan for them?

4. Future generations will be better off than each previous generation, given the advance of technology, so we have no moral obligation to them.

5. If we discount the future, we find that reserves of resources are of little use to future generations. For example, if a barrel of oil will cost say $100 fifty years from now, today it is worth only pennies. That is, a person would have to invest only pennies to have it grow to $100 in fifty years. So it makes no sense to conserve a barrel of oil today, since relative to the distant future it is worth only a few pennies today.

1.4 Suppose all the women in the world will suddenly have a maximum of two children. Will the global population stabilize at the present level? Why or why not?

1.5 Diane works for a large international consulting firm which has been retained by a federal agency to assist in the construction of a natural gas pipeline in Arizona. Her job is to lay out the centerline of the pipeline according to the plans developed in Washington.

After a few weeks on the job, she is approached by the leaders of a local Navajo village and told that it seems that the gas line will traverse an ancient sacred Navajo burial ground. She looks on the map and explains to the Navajo leaders that the initial land survey did not identify any such burial grounds.

"Yes, although the burial ground has not been used recently, our people believe that in ancient times this was a burial ground, even though we cannot prove it. What matters is not that we can show by archeological digs that this was indeed a burial ground, but rather that the people believe that it was. We therefore would like to change the alignment of the pipeline to avoid the mountain".

"I can't do that by myself. I have to get approval from Washington. And whatever is done would cost a great deal of money. I suggest that you not pursue this further." replies Diane.

"We already have talked to the people in Washington, and as you say, they insist that in the absence of archeological proof, they cannot accept the presence of the burial ground. Yet to our people the land is sacred. We would like you to try one more time to divert the pipeline."

"But the pipeline will be buried. Once construction is complete, the vegetation will be restored, and you'll never know the pipeline is there" suggests Diane.

"Oh, yes. We will know it is there. And our ancestors will know it is there."

The next day Diane is on the telephone with Tom, her boss in Washington and she tells him about the Navajo visit.

"Ignore them," advises Tom.

"I can't ignore them. They truly *feel* violated by the pipeline on their sacred land." replies Diane.

"If you are going to be so sensitive to every whim and wish of every pressure group, maybe you shouldn't be on this job," suggests Tom.

Should Diane simply tell the Navajo leaders that she has called Washington and that they are sorry, but the alignment cannot be changed? If she feels that the Navajo people have been wronged, what courses of action does she have? How far should she stick her neck out? Respond by writing a two page paper analyzing the ethics of this case.

1.6 Can the population of a country have a positive growth rate when its birth rate is declining? If so, how?

1.7 A baby born in the United States is expected to live for 70 years, consuming 5000 calories a day. If a cheeseburger is equal to 100 calories, how many "hamburger equivalents" would the American consumer consume during a lifetime?

1.8 What common law legal theories might be used to fight the siting of a large incinerator for automobile tires directly adjacent to your campus?

1.9 One of the sometimes contentious fights among environmentalists have pitted *conservationists* against *preservationists*. The two giants in this battle have been John Muir and Gifford Pinchot.

For this problem, choose one aspect of the environment (e.g. a local lake, or the woods near your home) and discuss the pressures on this environmental resource. Which philosophy, protection or conservation, do you feel most exemplifies your attitudes? In a two-page paper, explain your reasons for holding these views.

JOHN MUIR

John Muir (1838–1913) came to America when he was 11 and his family settled in Wisconsin. He suffered an eye injury that made him temporarily blind in 1867. When he recovered he decided to turn his eyes toward the natural environment. He walked from Wisconsin to the Gulf of Mexico, then sailed the Caribbean and the West Coast of North America, landing in San Francisco in 1868. He started to write about his travels, eventually publishing over 300 articles and 10 books. He became an environmental activist (before there were such people) and partly due to his efforts and the assistance of President Teddy Roosevelt with whom he had become fiends, the Sequoia, Mount Rainier, Petrified Forest, and Grand Canyon national parks were established. His single greatest defeat was the flooding of the Hetch Hetchy canyon in California, but he was able to save the area that later become Yosemite Park. In 1892 he and his friends established the Sierra Club, still the most influential environmental organization in the United States.

GIFFORD PINCHOT

In contrast to John Muir's humble beginnings, Gifford Pinchot (1865–1946) graduated from Yale University in 1889 and studied at L'Ecole Nationale Forestiere in Nancy, France. He eventually became the Chief Forester of President Teddy Roosevelt's newly-formed U.S. Forest Service. He believed that resources such as forest should not be destroyed but should be used wisely. He was a conservationist, in contrast to John Muir, who was a preservationist. Under Pinchot's leadership, the number of national forests increased from 32 in 1898 to 149 in 1910 for a total of 193 million acres.

Pinchot was an able politician, being elected governor of Pennsylvania for two non-consecutive terms, and campaigning for women's right to vote, prohibition of the sale and use of alcoholic beverages, and a graduated income tax, and other progressive ideas. In the 1930s Pinchot ran for the U. S. Senate on the Republican ticket, but was unable to get other Republicans to back him because of his support of the reforms advocated by President Franklin D. Roosevelt.

1.10 This is what Garrett Hardin had to say about Christopher Stone's essay, "Should Trees Have Standing?":

> Laws, to be stable, must be based on ethics. In evoking a new ethic to protect land and other natural amenities, [Aldo] Leopold implicitly called for concomitant changes in the philosophy of the law. Now, less than a generation after the publication of Leopold's classic essay, Professor Christopher D. Stone has laid the foundation for just such a philosophy in a graceful essay that itself bids fair to become a classic.

What was so important about this essay? Get a copy of it (it appears in numerous anthologies and a book by the same name) and see what Stone has to say. Just read the introduction, which summarizes his arguments. Do you agree with him? Why or why not?

CHAPTER 2

Tools for Pollution Control

S OLVING pollution problems first requires the application of risk analysis, or the quantitative understanding of just how the potential effect of pollution can affect humans or the environment. Risks can be weighed against benefits gained using both benefit/cost analysis and the estimation of environmental impact. Most importantly, the pollution problem has to be measured quantitatively so it can be understood, and the effect of control measures estimated. In this chapter we present some tools that are useful in the analysis of pollution problems and in the implementation of pollution control measures.

2.1 RISK ANALYSIS

Risk is defined as the probability that something undesirable will happen. For humans, the undesirable event is usually death or illness, although other undesirable events, such as being jilted by a lover or losing your money in a casino can also be considered risks. While these are legitimate risks and the probability of such events occurring can be estimated, in this discussion we limit the definition of risk to the probability of death.

2.1.1 Expressions of Risk

If death is the undesirable event, then it is possible to analyze the risk of dying. This calculation can be done in several ways:

Deaths Per 100,000 Persons

This expression of risk calculates the chances of dying as a result of a given cause. The risk is expressed as

$$R_1 = \frac{D}{P}$$

41

where

R_1 = risk of death
D = number of deaths due to a given cause, people per year
P = population

Example 2.1

In the United States about 260,000 smokers die every year as a result of lung cancer and chronic obstructive pulmonary disease. If the population of the United States is 280 million, what is the risk of death (from these two factors) associated with habitual smoking?

$$\frac{260,000}{280,00,00} = 0.00093 \text{ smoking related deaths per person in the USA}$$

Because the answer is such a small number, typically the risk of death is expressed as per 100,000 population, or

$$\frac{260,000}{280,000,000 / 100,000}$$

= 93 smoking related deaths per 100,000 people in the USA

In other words, a habitual smoker in the U.S. has an annual risk of 93 in 100,000, or about one in a thousand, of dying of lung cancer or other smoking-related diseases during any given year.

Deaths Per 1,000 Deaths

Another way of expressing the risk of dying is to calculate the deaths from a specific cause relative to all deaths regardless of cause. Table 2.1 shows approximate numbers of deaths in the United States during any given year.

TABLE 2.1. Approximate Annual Deaths in the United States.

Cause of Death	Annual Deaths
Cardiovascular disease	918,000
Cancer (all)	65,600
Chronic obstructive pulmonary disease	112,000
Motor vehicle accidents	45,400
Alcohol-related disease	52,200
Other causes	1,663,000
All causes	3,351,600

This risk is calculated as

$$R_2 = \frac{D}{T}$$

where

R_2 = risk of dying of a certain cause relative to total deaths
D = number of deaths due to a given cause per year
T = total number of deaths from all causes during a given year

---

Example 2.2

Of the 3,351,600 deaths in the United States every year, 918,000 of these are due to cardiovascular disease. If a person dies, what is the chance (probability) that he or she will die of cardiovascular disease?

$$\frac{918,000}{3,351,000} = 0.273$$

That is, about 27% of the deaths in the United States are due to cardiovascular disease.

Relative Risk

A third way of expressing risk is by comparing the rate of death to that in other populations. For example, the risk of fatal lung cancer in smokers can be compared to the risk of dying of the same disease for non-smokers. The risk for smokers and non-smokers are calculated and then compared.

$$R_3 = \frac{[R_1]_+}{[R_1]_o}$$

where

R_3 = relative risk
$[R_1]_+$ = risk of dying of a given cause for a population having some attribute or characteristic
$[R_1]_o$ = risk of dying of a given cause for a population not having some attribute or characteristic

---

Example 2.3

If in a population of 280,000,000, there are 260,000 deaths annually from smoking-related diseases, not all of the people dying of these diseases are smokers. Of the 260,000 annual deaths, about 21,000 people who die of lung cancer and cardiovascular problems are non-smokers. This means that the relative risk of fatal lung cancer in smokers

$$R_3 = \frac{[R_1]_+}{[R_1]_o} = \frac{[D/P]_+}{[D/P]_o} = \frac{[D]_+}{[D]_o} = \frac{260,000 - 21,000}{21,000} = 11.4$$

That is, the chances of a person dying of lung cancer or cardiovascular problems is over 11 times higher if the person smokes cigarettes.

Relative risk of death is also called the *standard mortality ratio* (SMR), defined as the ratio of the observed deaths due to a certain disease in a given population that shares a certain characteristic (e.g., smokes cigarettes) divided by the expected deaths in a population of the same size that does not have that characteristic (e.g. non-smokers). In the above example, the standard mortality ratio is 11.4.

Three important characteristics of epidemiological reasoning are illustrated by the above example:

- Everyone who smokes heavily will not die of lung cancer.
- Some non-smokers die of lung cancer.
- Therefore, one cannot unequivocally relate any given individual lung cancer death to cigarette smoking.

Thus when the Tobacco Institute claims that there still is no smoking gun, they are right. We do not have *proof* that smoking causes lung cancer. All we can say is that if persons really wanted to get lung cancer, the best thing they can do is to smoke cigarettes. This will result in a high probability of contracting that disease.

Example 2.4

A butadiene plastics manufacturing plant is located in Beaverville, and the atmosphere is contaminated by butadiene, a suspected carcinogen. The cancer death rate in the community of 8000 residents is 36 people per year, and the total death rate is 106 people per year. Does Beaverville appear to be a healthy place to live, or is the cancer risk unusually high?

From Table 2.1, we see that the number of cancer deaths in the United States is 65,600 annually for a population of 280,000,000, or 23.4 deaths per 100,000 people. In Beaverville, however, there are 36 cancer deaths in a population of 8000, or 36/8000, or 450 deaths per 100,000 people. The standard morality ratio (SMR) is thus

$$SMR = \frac{450}{23.4} = 19.2$$

This suggests that a person living in Beaverville will have more than 19 times greater chance of dying of cancer than the average citizen of the United States.

The probability of dying of all causes in the United States, as shown in Table 2.1, is 3,351,600/280,000,000 = 0.012, or 1200 deaths per 100,000 people. In Beaverville, the probability of dying is 106/8000 = 0.013 or 1300 deaths per 100,000. This suggests that a Beaverville resident is only a little more likely to die in any given year from any cause as the average resident of the U.S. Another way of looking at the data is to ask if the cancer death rate per 1000 deaths is higher in Beaverville than in the USA. In Beaverville, of the 106 deaths annually, 36 are due to cancer, or

$$\frac{36}{106} = 0.34$$

or 34% of all the deaths are due to cancer. In the United States, however, the risk of cancer being the cause of death is

$$\frac{65,600}{3,351,600} = 0.19$$

or 19%. We may thus conclude further that a death in Beaverville in any given year is almost twice as likely to be a cancer death than is the case in the U. S. as a whole. Something is clearly wrong in Beaverville. Perhaps butadiene really is carcinogenic? Or perhaps it isn't, and the real problem is the lead smelter across the river from Beaverville?

Estimates of Risk Based on Epidemiological Evidence or Animal Models

There are two basic techniques for estimating the risk of any environmental toxin or harmful agent. The first is to use epidemiological data, and the second is to use animals.

The problems with calculating risk from epidemiological data are immense. *Epidemiology* is the science of finding both risks and causative agents for human health problems. For example, suppose there is concern with the effect of fluoride on public health. If the effect of fluoridation of drinking water on childhood cavities is to be compared between two communities, one with and one without fluoridation, we have to be certain that the two towns are indeed so similar that the only difference is the fluoridation of water. Perhaps one of the towns is wealthy and can afford excellent health care, while the other lacks such care. Or suppose the kids in one community eat a sugary diet. These and an infinite number of other variables may make a simplistic comparison based on epidemiological data almost impossible.

One way around that problem is to use pairs of people and follow them longitudinally. For example, suppose you found two people who were alike in all measurable ways—age, gender, occupation, ethnicity, smoking, dietary habits, etc.—and set up pairs with one person in each town. You would then argue that the only difference between them would be that one drank water with

What Causes Cholera?

In the 1850s many theories of why epidemics occurred were advanced. Edwin Chadwick, a champion of public health during the Victorian era in England, believed that odor was to blame. He put it succinctly: "All smell, if it be intense, initiate acute disease." William Farr, one of the greatest public health physicians, stoutly believed that cholera was contracted through the atmosphere, with something he called cholerine, a zymotic material of cholera. He was an excellent epidemiologist and was one of the first to bring statistics to the assistance of disease prevention. But sometimes his statistics were misleading. He plotted the incidence of cholera in London as a function of elevation above the River Thames and concluded that cholera must be contracted though the air, the miasma evaporating from the river and carrying these cholerine particles with it.

Another explanation for cholera was advanced by John Snow who suggested that cholera was a waterborne disease. Snow believed that contaminated water must contain zymotic material that found its way from the intestines of the diseased persons to the digestive tracts of others. Chadwick did not buy this idea at all. Bad water was only a predisposing cause of cholera: it was the smell that caused it.

The 1853 cholera outbreak in London provided a classical opportunity to test Snow's theories. Most of the water companies serving water to the city drew their water from the Thames at the most convenient location. The Metropolitan Water Act of 1852 required the companies to change their source to upstream locations, away from major contamination. By 1853 only one water company had done so, but with this change came an immediate and dramatic reduction in the incidence of cholera in that section of London served by the water company.

When the disease returned to London in 1854, one of the water companies was still providing contaminated water. John Snow plotted the incidence of cholera on a city map, thus creating the first spot map in public health history, and showed that the incidence was clearly related to the contaminated water coming from that pump. The pump handle was removed and the epidemic subsided.

The idea that contamination can cause disease was also emerging in the medical field where infection was a horror in all hospitals. In the maternity ward in Vienna, for example, one in ten women contracted birthing fever and died. Ignaz Semmelweis, a Hungarian physician working in the maternity ward, demonstrated in 1854 that washing hands in chloride of lime between examining patients greatly reduced infections. He did not, however, understand that the infections were caused by microorganisms, but rather blamed them on the transmission of organic particles from one patient to the next. Meanwhile, Joseph Lister in England showed that infections could be greatly reduced by applying carbolic acid to a wound. Lister had a better idea of what he was actually doing because he was aware of the work on microorganisms conducted by Louis Pasteur in France. Lister accepted the germ theory and concluded that microorganisms were dangerous and deadly and that killing microorganisms (disinfecting) was beneficial.

The Mystery of Legionnaire's Disease

The bicentennial convention of the American Legion was held in Philadelphia in 1976 at the elegant but aging Bellevue Stratford Hotel. Toward the end of the conference, a number of the attendees were suffering from flu-like symptoms, and four days after leaving Philadelphia, an Air Force veteran died from pneumonia. During the next week, 30 additional Legionnaires who had been at the conference died from what became known as "Legionnaire's Disease."

The Center for Disease Control in Atlanta was called in to find out what was going on. Theories ranged from viral pneumonia to terrorism (before this became a popular alternative) to a communist plot (after all, these were mostly veterans who died). The first step was to find out if there was some common organism that might explain the disease. Using tissues from lung biopsies and samples of sputum, microbiologists discovered a Gram-negative rod-shaped bacterium that was present in all of the samples. The next step was to try to grow this organism, now called *Legionella pneumophila*, in the laboratory, and then to use it to infect an animal that would exhibit symptoms similar to pneumonia. Through trial and error, microbiologists were able to grow *L. pneumophila* and use it to infect guinea pigs. There was now little doubt that this organism was the cause of Legionnaire's disease, but where did it come from and why and how did it infect the Legionnaires in Philadelphia?

Microbiological detective work suggested that the organism is thermophilic, and thus if it is to be found in nature, it should exist in warm places. Sure enough, *L. pneumophila* turned out to be similar to organisms cultured from warm springs and geysers. Further work showed that the organism liked warm, dark places, and grew in a biofilm associated with specific species of algae. The next step was to find how the Legionnaires came into contact with the pathogen. Recognizing that this must be an airborne infection, the epidemiologists focused on the air quality in the hotel, and discovered that biofilms were growing in the air ducts of the hotel heating/ventilating system. Within these biofilms they isolated *L. pneumophila*, and the mystery was solved.

a high concentration of fluoride and one drank water with low fluoride, and any differences should then be attributed to the effect of the drinking water. Theoretically, this is a useful study, but we will never know if we have accounted for all variables. More importantly, the likelihood of finding a statistically significant number of twins in the two towns is very unlikely. Thus epidemiological data are at best crude guides to the relative toxicity of various pollutants.

Epidemiology can, however, be useful in determining if certain chemicals are or are not toxic to humans and in raising questions that can then be resolved by biochemistry. At times epidemiologists are asked to define the causative agent after a disease is recognized. Historically one of the most daunting problems faced by early epidemiologists was finding the causative agent for cholera, a dreaded disease in the 1800s. Such challenges have not abated, as demonstrated by the case of Legionnaire's disease, a modern epidemiological detective story.

The second way of estimating unit annual risk is with animal studies, or animals as models of human health. The most important problem in using animal models is that lethal diseases, such as cancer often have long gestation times, sometimes 20 or more years. Mice only live a couple of years, so in order to induce cancer in mice, an outrageously high dose is necessary. Is this in any way reasonable when applied to humans? Also, mice have a very different metabolism, and one chemical might be very toxic to mice but only mildly dangerous to humans. Dioxin is such an example. For many years we thought that dioxin was one of the most deadly chemicals known. But industrial accidents such as the accident in Seveso, Italy have shown that dioxin is not a very potent acute toxin. Humans seem to have a way of expelling the chemical without causing much short-term harm. Subsequent studies have suggested that dioxin is an endocrine disruptor, causing enzymatic damage in humans, and such damage is a long-term problem.

The Accident in Seveso, Italy

The accident was blamed on a broken valve. On a Saturday afternoon in 1984, the plant, owned by the Industrie Chimiche Meda Societa Azionaria, started releasing a toxic cloud into the atmosphere. At first there was not alarm, and little concern as the 3000 kg of various chemicals including 2,4,5 trichlorophenol used in the manufacture of herbicides wafted toward the municipality of Seveso. The emission, however, also carried with it between 10 and 20 kg of dioxin, which had been found to be an incredibly potent toxin, acting in the parts per trillion range.

After the toxic cloud passed through the city, the medical personnel were on the scene estimating the effect of the toxin on the people of Seveso. The first sign of human health problems was a high incidence of chloracne, a burn-like skin lesion, that appeared on children a few hours after the accident. The problems were thus minimized and the residents assured that there was no risk, but then the presence of dioxin was identified, two weeks after the accident. This caused the evacuation of nearly 400 people from the most severely affected areas, which was, or course, much too late to prevent exposure to the dioxin. Eventually, about 37,000 people are believed to have been exposed to the chemical. The hardest hit were the animals, which either died from the exposure or were killed to prevent the transmission of dioxin up the food chain.

(continued)

(continued)

The exposure to dioxin suffered by these people ought to have killed them outright, or at the very least, caused a very high rate of cancer. Nothing like this has happened, however, proving that the extrapolation of toxins from mouse to human is not simply a case of multiplying by ratio of the body weights. What has happened is that during the first seven years after the accident, a statistically unlikely number of female babies have been born to the exposed population. The cause of this is still a mystery. Other effects reported by the people who were exposed to the toxins have been immune system and neurological disorders as well as spontaneous abortions, but these problems do not seem to be linked to dioxin.

The highly contaminated area was dug up and the soil stored in two massive concrete tanks, along with the slaughtered animals, and the area has been covered with clean soil that is now a park for the people of Seveso.

Voluntary vs. Involuntary Risks

Some risks are imposed from without, however, and these we can do little about. For example, it has been shown that the life expectancy of people living in a dirty urban atmosphere is considerably shorter than people living identical lives but breathing clean air. This is a risk we can do little about (except to move our lives), and it is this type of risk that most people resent the most. In fact, studies have shown that the acceptability on an involuntary risk is on the order of 1000 times less than our acceptability of a voluntary risk. Consider, for example, the problem of trichloroethylene (TCE), a known carcinogen and a common hazardous waste polluting groundwater. We acknowledge the risk of TCE in drinking water and strive to reduce it, but at the same time, if a person drank 0.005 mg/L of TCE in the drinking water (the EPA standard), the chance of dying from cancer induced by the TCE is only 1 in a million, whereas the equivalent mortality rates are 50 miles of car travel, working in a factory for 2 weeks, or 1.5 minutes of rock climbing! We accept the latter voluntary risks without much concern, but worry about the involuntary ones, such as chemicals in our drinking water. Such human behavior can explain why people who smoke cigarettes still get upset about poor outdoor air quality, or why people will drive while intoxicated to a public hearing protesting the siting of an airport, fearing the crash of airplanes.

Much of our environmental pollution control is driven by the fear of health problems, and thus the analysis and calculation of risk is a very important topic which we barely introduce in the above discussion. There is more on risk in the chapter on hazardous waste, since the risk of toxins from uncontrolled disposal is often the number one environmental problem as perceived by the public.

2.2 BENEFIT/COST ANALYSIS

In the 1940s, the Bureau of Reclamation and the U.S. Army Corps of

The Disaster in Bhopal, India

Union Carbide, with the cooperation of the government of India, built a large fertilizer manufacturing plant in Bhopal, India, providing jobs for the people and inexpensive fertilizers for the local farmers. During the early hours of the thirteenth of December, 1984, an accident occurred in the plant, and a toxic cloud of methyl isocyanate swept down into a sleeping city, killing thousands of people. When the extent of the disaster was finally tallied, about 3,800 people had died, with thousands more permanently disabled.

The plant originally imported the methyl isocyanate used in the manufacturing of the pesticide with a trade name of Sevin. With the objective of being more independent of imports, the Indian government encouraged the plant to start the manufacture of its own methyl isocyanate. Because of the involvement of the Indian government, a number of restrictions were placed on the company including a requirement that only Indians be employed at the plant. As American engineers and other workers left the plant they were replaced by Indians, many of whom built homes all around the plant. These homes were, unfortunately, merely shanties and there was no governmental infrastructure to support the influx of people to the area. The fact that the workers and their families were packed close to the plant in largely open structures contributed to the severity of the accident.

When the gas escaped, there was no way to call for help. The city of nearly 750,000 people had only two phone trunks which could handle only a few calls at a time. The news of the disaster was slow in reaching the government and Union Carbide and as a result the initial response was slow and uncertain. The Indian govenment was not fully cooperative at first and made it difficult to send relief, thus exacerbating the situation.

The cause of the accident was uncertain until Union Carbide did a thorough study and discovered that large volumes of water had been put into the methyl isocyanate holding tanks which created an uncontrolled chemical reaction. The question still existed as to why this was done. Union Carbide's answer was sabotage. A "disgruntled employee" intentionally filled the tank with water. This is possible, but somewhat improbable since the employee would have known that the resulting reaction would be deadly to thousands of people, including most likely his own friends and family. A more likely scenario is that it was poor plant management and/or incompetence, for which both the Indian govenment and Union Carbide would be responsible. As of this writing, the Indian govenment has spent over $75 million dollars in rehabilitating and reconstucting Bhopal, including the construction of 12 hospitals. Union Carbide, which was required by the courts to pay $470 million, has only contributed $3.1 million toward disaster aid.

Could a disaster like Bhopal happen in the United States? Within the year of the tragedy in India another Union Carbide plant in Industry, West Virginia, had an accidental release of methyl isocyanate into the Kanawha Valley, eerily mirroring the events in Bhopal. Fortunately, nobody died in this accident, but the episode brought home to Americans the risks and dangers of living close to plants that produce or use toxic chemicals.

Engineers battled for public dollars in their drive to dam up all the free-flowing rivers in America. In order to convince the Congress of the need for major water storage projects, a technique called *benefit/cost analysis* was developed. At face value, this is both useful and uncomplicated. If a project is contemplated, an estimate of the benefits derived is compared in ratio form to the cost incurred. Should this ratio be more than 1.0, the project is clearly worthwhile, and the projects with the highest benefit/cost ratios should be constructed first because these will provide the greatest returns on the investment. By submitting their projects to such an analysis, the Bureau and the Corps could argue for increased expenditure of public funds, and could rank the proposed projects in order of priority. As is the case with cost effectiveness analyses, the calculations in benefit/cost analyses are in dollars, with each benefit and each cost expressed in monetary terms. For example, the benefits of a canal would be calculated as monetary savings in transportation costs. But some benefits and costs such as clean air, flowers, whitewater canoeing, foul odors, polluted groundwater and littered streets cannot be easily expressed in monetary terms. Yet these benefits and costs are very real and should somehow be included in benefit/cost analyses. One solution is simply to force monetary values on these benefits. In estimating the benefits for artificial lakes, for example, recreational benefits are calculated by predicting what people would be willing to pay to use such a facility. There are, of course, many difficulties with this. The value of a dollar varies substantially from person to person, and some persons benefit a great deal more from a public project than others, yet all may share in the cost. Because of the problems involved in estimating such benefits, recreational benefits can be bloated in order to increase the benefit/cost ratio. It is thus possible to justify almost any project since the benefits can be adjusted as needed. In the example below, monetary values are placed on subjective benefits and costs to illustrate how such an analysis is conducted. The reader should recognize that the benefit/cost analysis is a simple arithmetic calculation, and that even though the final value can be calculated to many decimal places, it is only as valuable as the weakest estimate used in the calculation.

Example 2.5

   A small dam is to be constructed on a tributary, and the question is whether the savings in flood damage will justify the cost of the dam. Here is some information:

Expected life of the dam	50 years
Expected cost of the dam	$4,000,000
Land cost	$1,000,000

Flood damage over 50 years expected without the dam:

Homes	$1,000,000
Roads	$2,000,000
Farms	$3,000,000

The total cost is therefore $5,000,000 while the total benefits in damage not done by the floods is $6,000,000. The benefit/cost ratio is then calculated as

$$B/C = \frac{6,000,000}{5,000,000} = 1.2$$

If B/C is greater than 1.0, then building the dam makes sense.

Note that in the above example only some of the items could be quantified, such as the costs of construction. The benefits from flood damage are highly subjective. If the dam is to last for 50 years, who knows what the area will be like then? There may not be any houses or farms to protect, or there may be a city, built partly because the dam protects them from flood damage. Also note that there is no cost even attempted for the loss of a river. If such a cost were included, the calculation might have resulted in a different conclusion. Benefit/cost analysis is useful in pollution control when we know and can quantify both the benefits and the costs. The problem is that we seldom can find a proper common denominator. We might be able to calculate the flood damage, and we might be able to estimate what the recreational benefit of some facility might be (often estimated as how much people would be willing to pay to use it), but we have great difficulty with intrinsic benefits like free space, beautiful vistas, rushing water, and human health. Even more problematical is including the benefits to non-human creatures. During the past years the United States federal government has been pressured by industrial groups to take into account economic considerations when calculating the cost of pollution control. Industries can readily calculate the costs of pollution control, and they insist that the benefits of a clean environment or greater human health also be calculated so that a benefit/cost calculation can be performed. Their hope, of course, is that it would be shown to be uneconomical to install further pollution control (the benefit/cost result would be less then one), resulting in lower production costs for the industry. Suffice it to say that such calculations are not possible to make, and the insistence on a benefit/cost calculation when intrinsic values are involved is simply impossible. The benefit/cost analysis is an inappropriate tool for making pollution control decisions.

2.3 ENVIRONMENTAL IMPACT ANALYSIS

On 1 January 1970, President Nixon signed into law the National Environmental Policy Act (NEPA) which was intended to ". . . encourage productive and enjoyable harmony between man and his environment."

The law proved to be groundbreaking legislation and provided the model for environmental legislation in most of the western world. The president also, by executive order, created the Environmental Protection Agency (U.S. EPA) and

WILLIAM RUCKELSHAUS

The strength and legitimacy of the U. S. Environmental Protection Agency owes much to the leadership of its first director, William D. Ruckelshaus (1932–). Ruckelshaus is a graduate of Princeton University with a law degree from Harvard. After graduation he was a Deputy Attorney General in Indiana and was then elected to the Indiana House of Representatives. In 1970 he was asked by President Nixon to head the nascent U. S. EPA. During the EPA's formative years he was able to blend the various federal agencies that oversaw pollution and environmental health into one cohesive structure, took action against severely polluted cities and industrial polluters, oversaw the setting of health-based standards for both air and water pollution, and developed the first regulations controlling emissions from automobiles (amid general anguish from the automobile and petroleum industries who claimed it could not be done). He worked with the states to develop both water quality standards and ambient air quality plans, and he worked to ban the use of some pesticides such as DDT. Almost all of the environmental legislation we presently enjoy in the United States was guided through the Congress during the years William Ruckelshaus was the first head of the U. S. EPA.

In 1973 he stepped down from the directorship to become at first the acting director of the FBI and then briefly as the Deputy Attorney General in the Justice Department. He distinguished himself in this post by refusing to fire the special prosecutor investigating the Watergate break-in and instead resigned his post.

Following the disastrous tenure of Ann Gorsuch as the director of the U. S. EPA during the first Reagan administration in which she was apparently charged with scuttling the agency (a Republican objective), William Ruckelshaus was once again asked to take over. He worked to rekindle both the work and the morale of the agency employees and developed widely accepted principles of risk-based decision-making in environmental controls. Restoring and protecting the water quality in the Chesapeake Bay, developing the processes for cleaning up hazardous waste sites, and the banning of many chlorinated pesticides were significant accomplishments during his second tenure as the chief of the U. S. EPA.

appointed William Ruckelshaus as its first director. The U. S. EPA was to have executive oversight over environmental matters including implementing environmental legislation.

The National Environmental Policy Act set up the Council on Environmental Quality (CEQ), which was to be a watchdog on federal activities as they influence the environment, and the CEQ reported directly to the President. The vehicle by which the CEQ would monitor significant federal activity impacting the environment was to be a report called the *Environmental Impact Statement* (EIS). This lightly regarded provision in NEPA, tucked away in Section 102, stipulates that the EIS is to be an inventory, analysis, and evaluation of the effect of a planned project on environmental quality. The EIS is to be written first in draft form by the federal agency in question for each significant project that is expected to affect the environment, and then this draft is to be submitted for public comment. Finally the report is rewritten taking into account public sentiment and comments from other governmental agencies. When complete, the EIS is to be submitted to the CEQ, who then makes a recommendation to the President as to the wisdom of undertaking that project.

The impact of Section 102 of NEPA on federal agencies was traumatic, since they were not geared up in manpower or in training, nor were they psychologically able to accept this new (what they viewed as a) restriction on their activities. Thus the first few years of the EIS were tumultuous, with many environmental impact statements being judged for adequacy in courts of law.

Conflict of course arises when the cost-effective alternative, or the one with the highest benefit/cost ratio, also results in the greatest adverse environmental impact. Decisions have to be made, and quite often the B/C wins out over the EI. It is nevertheless significant that since 1970, the effect of a planned project on the environment must be considered, whereas before 1970 concerns about the environment were never even acknowledged, much less included in the decision-making process.

In practice, governmental agencies tend to conduct internal EI studies and propose only those projects that have both a high B/C ratio and a low adverse environmental impact. Most EI statements are thus written as a justification for an alternative that has already been selected by the agency.

Recent reorganization in the White House has resulted in the abolishment of the Council on Environmental Quality and the establishment of a White House Office on Environmental Policy. This office performs the functions of the Council on Environmental Quality, as well as establishes environmental policy at the highest level. Importantly, the abolishment of the CEQ and the creation of the new office does not change the need for reviewing draft environmental impact statements, and the NEPA requirements still apply.

Although the old CEQ developed some fairly complete guidelines for environmental impact statements, the form of the EIS is still variable and considerable judgment and qualitative information (some say prejudice) goes into every EIS. Each agency seems to have developed its own methodology, within the constraints of the CEQ guidelines, and it is difficult to argue that any one format is superior to another. Since there is no standard EIS, the following discussion is a description of several alternatives within a general framework. It is suggested that the EIS should be in three parts: inventory, assessment, and evaluation.

Inventory

The first duty in the writing of any EIS is the gathering of data, such as hydrological, meteorological, biological etc. information. A listing of the species of plants and animals in the area of concern for example is included in the inventory. There are no decisions made at this stage, since everything properly belongs in the inventory.

Assessment

The second stage is the analysis part commonly called the *assessment*. This is the mechanical part of the EIS in that the data gathered in the inventory are fed to the assessment mechanism and the numbers are crunched accordingly. Numerous assessment methodologies have been suggested; but only the most widely used method, the quantified check list, is described below for illustrative purposes.

The *quantified check list* is possibly the simplest quantitative method of comparing alternatives. It involves first the listing of those areas of the environment that might be affected by the proposed project, and then an estimation of

1. the importance of the impact
2. the magnitude of the impact
3. the nature of the impact (whether negative or positive)

Commonly the importance is given numbers such a 0 to 5, where 0 means no importance whatever, while 5 implies extreme importance. A similar scale is used for magnitude, while the nature of the impact is expressed as simply −1 as negative (adverse) and +1 as positive (beneficial) impact. The Environmental Impact (EI) is then calculated as

$$EI = \sum_{i=1}^{n} (I_i \times M_i \times N_i) \qquad (2.4)$$

where

I_i = importance of i-th impact
M_i = magnitude of i-th impact
N_i = nature of i-th impact, so that $N = +1$ if beneficial and $N = -1$ if detrimental
n = total number of impacts.

The following example illustrates the use of the quantitative check list.

Example 2.6

A community agrees that their refuse collection is inadequate and something has to change. The two alternatives being discussed are to increase the refuse collection frequency from 1 to 2 times per week, or to allow for the burning of rubbish on premises. Analyze these two alternatives using a quantified check list.

First, the areas of environmental impact are listed. In the interest of brevity, only a few areas are shown below, while recognizing that a thorough assessment would include many other concerns. Following this, values for importance and magnitude are assigned (0 to 5) and the nature of the impact (+/−) is indicated. The three columns are then multiplied.

Alternative 1: Increasing collection frequency

Area of concern	Importance (I)	Magnitude (M)	Nature (N)	$(I \times M \times N)$
Air pollution (trucks)	4	2	−1	−8
Noise	3	3	−1	−9
Litter in streets	2	2	−1	−4
Odor	2	3	−1	−6
Traffic congestion	3	3	−1	−9
Groundwater pollution	4	0	−1	−0

(*Note:* no new refuse will be landfilled)
EI = −36

Alternative 2: Burning on premises

Area of concern	Importance (I)	Magnitude (M)	Nature (N)	$(I \times M \times N)$
Air pollution (burning)	4	4	−1	−16
Noise	0	0	−1	0
Litter (present system of collection causes litter)	2	1	+1	+2
Odor	2	4	−1	−8
Traffic congestion	0	0	−1	0
Groundwater pollution	4	1	+1	+4

(*Note:* less refuse will be landfilled)
EI = −18

On the basis of this analysis, burning the refuse would result in the lower adverse environmental impact.

It should be clear that any such analysis is wide open to individual modifications and interpretations, and the example above should not be considered in any way a standard method. It does, however, provide a numerical answer to the question of environmental impact, and when this is done for several alternatives, the numbers can be compared. This process of comparison and evaluation represents the third part of an EIS.

Evaluation

The comparison of the results of the assessment procedure and the development of the final conclusions are all covered under evaluation. It's important to recognize that the previous two steps, inventory and assessment, are simple and straightforward procedures compared to the final step, which requires judgment and common sense. During this step in the writing of the EIS the conclusions are drawn up and presented. Often the reader of the EIS sees only the conclusions and never bothers to review all the assumptions that went into the assessment calculations, so it is important to include in the evaluation the flavor of these calculations and to emphasize the level of uncertainty in the assessment step.

But even when the EIS is as complete as possible, and the data have been gathered and evaluated as carefully as possible, conclusions concerning the use of the analysis are open to severe differences of opinion. For example, the EIS written for the Alaska oil pipeline, when all the volumes are placed into a single pile, represents 14 *feet* of work. And at the end of all that effort, good people on both sides drew diametrically opposite conclusions on the effect of the pipeline. The trouble was, they were arguing over the *wrong thing*. They thought they were arguing about how many caribou would be affected by the pipeline, but their real disagreement was how deeply they cared that the caribou were affected by the pipeline in the first place. For a person who does not care one twit about caribou, the impact is zero, while those who are concerned about the herds and the long-range effects on the sensitive tundra ecology, care very much. What then is the solution? How can environmental decisions be made in the face of conflicting values? This is the challenge, and we have no easy answer. Suffice it to say that the EIS has made a tremendous difference in how federal agencies do their work, and flawed though it may be, it is a positive step toward understanding our commitment to environmental quality.

2.4 INFORMATION ANALYSIS

In working with environmental problems, we seldom know anything for sure. Although there may never before have been a flood in August, as soon as we want to install a new stormwater catchment system that depends on dry weather, it will rain cats and dogs. But if a flood has not occurred in August in over 100 years, chances are fairly good it won't happen next year.

If we have 100 years of data, we feel pretty confident about predicting, but what if the only information we have was that there was no flood the last 5 years? How likely would it be that a flood will occur? Such questions are answered by a form of risk analysis, but in this case we are talking about the probability of a natural event such as a flood occurring during a specified time.

Such *probability* calculations are central to many decisions involving the environment. Related to probability is the analysis of incomplete data using *statistics*. A piece of data (e.g., streamflow for one day) is valuable in itself, but when combined with hundreds of other daily streamflow data points, the information becomes even more useful, but only if it can be somehow manipulated and reduced. For example, if we want to impound water from a stream for a water supply, the individual daily flow rates would be averaged (a statistic), and this would be one useful number in deciding just how much water would be available.

Environmental data are often analyzed using a *frequency analysis* where the ordinate (vertical axis) is the cumulative fraction of observations. These curves are commonly used in hydrology, soils engineering, resource recovery, and many other applications. The curves are constructed by first ranking the available data. In the case of floods, the data are ranked according to the flow rate in the stream or river and the probabilities of that flow occurring are calculated. If the data are assembled for n years and the rank is designated by m, with $m = 1$ for the largest flow rate, the probability that any give size flood will occur is given by

$$P = \frac{m}{n} \tag{2.6}$$

where

P = probability of an event occurring
m = rank of the event
n = total number of events in data set

The probability of some event such as a flood occurring is expressed in terms of a the *return period*, or once in how many years is the event expected to recur. If the annual probability of an event occurring is 5% ($m/n = 0.05$), then the event can be expected to recur once in 20 years, or have a return period of 20 years. Stated in equation form:

$$\text{Return period} = (1/P) \tag{2.7}$$

The construction of a frequency distribution diagram is shown in Example 2.7.

Example 2.7

One year of flow data are available for a river.

Month	Flow rate, m^3/min
J	50
F	60
M	70
A	40
M	32
J	20
J	50
A	80
S	10
O	55
N	65
D	80

We need to know what chance there will be that the river flow will exceed 75 m^3/min, since this would flood out a construction site. That is, how often, on average, can we expect a month with such high flows?

The available data are ranked. Also tabulated is the frequency of occurrence of the flood, or $P = m/n$, where m is the rank and n is the number of observations, in this case 12.

Rank, m	Flow rate (m^3/min)	Frequency of occurrence, m/n
1	80	0.08
2	80	0.17
3	70	0.25
4	65	0.33
5	60	0.42
6	55	0.50
7	50	0.58
8	50	0.67
9	40	0.75
10	32	0.83
11	20	0.92
12	10	1.00

On Figure 2.1, the flow rate (on the horizontal axis) is plotted against the frequency of occurrence, m/n and a best-fit line is drawn. From this line, it is possible to estimate the expected occurrence of any size flood. For example, the flow that is expected once every other month ($P = 6/12 = 0.5$) is about 55 m^3/min, since this is flow that is most likely to occur 6 months out of 12. The return period is 1/0.5, or two months. Once every two months, on average, the river flow will be at least 55 m^3/min.

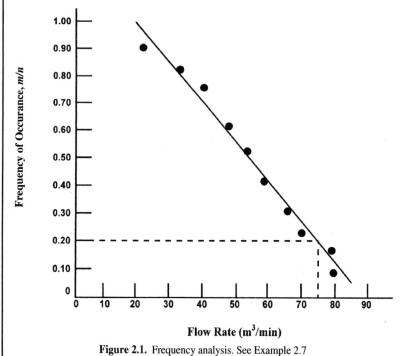

Flow Rate (m³/min)

Figure 2.1. Frequency analysis. See Example 2.7

If the objective is to estimate the return period of a flood of 75 m³/min, we enter the plot at flow = 75 m³/min and find that the frequency of occurrence is 0.2. That is, for any given month, there is a 20% chance that the flow will be 75 m³/mm or greater. This is the same as saying that the return period is 1/0.2 = 5 months. We would have to wait 5 months, on average, to see a flow of 75 m³/min or greater.

In the above example, the return period is 5 months, but that does not mean that we have to wait 5 months before we see that flow. This flow might occur next month, and there may be 3 months in a row when the flow is greater than 75 m³/min. Nature is fickle. All we can say is that on average, such a flow will occur once every 5 months. Not very comforting when the entire project depends on not having a flood!

2.5 MEASUREMENT OF POLLUTION

The solution of environmental pollution problems almost always involves the use of calculations. Such calculations make it possible to describe the physical world in commonly accepted and relatively unambiguous terms.

Some of the most important measurements are density, concentration, flow rate, and residence time.

2.5.1 Density

The density of any material is defined as its mass divided by its volume, or

$$\rho = \frac{M_A}{V_A} \tag{2.6}$$

where

ρ = density, kg/m^3
M_A = mass of material A, kg
V_A = volume of material A, m^3

In the International System (SI) the base unit for density is kg/m^3, while in the American system density is commonly expressed as lbM/ft^2. (lbM = pounds(mass)). Water in the SI system has a density of 1×10^3 kg/m^3, which is equal to 1 g/cm^3. In the American engineering system, water has a density of 62.4 lbM/ft^3.

2.5.2 Concentration

Whenever we talk about pollutants in the environment, we almost always describe their concentrations as the amount of the substance (in mass or volume) per unit volume or per unit mass of the surrounding medium. For example, we may be interested in the amount of dissolved oxygen in water, and we typically express the concentration as a mass per unit volume, such as mg O_2 per liter, or more simply, mg/L. When dealing with water, the derived dimension of concentration is usually expressed gravimetrically as the mass of a material A in a unit volume of A plus B. The concentration of some solid material A in a fluid B is expressed:

$$C_A = \frac{M_A}{V_A + V_B} \tag{2.7}$$

where

C_A = concentration of some solid A, (for example kg/m^3)
M_A = mass of solid material A, kg
V_A = volume of A in the mixture, m^3
V_B = volume of B, the liquid in which A is suspended, m^3

In the SI system, the basic unit for concentration is kg/m^3, although the most widely used concentration term in environmental engineering is milligrams per liter, written as mg/L.

Example 2.8

Plastic beads with a volume of 0.04 m^3 and a mass of 0.48 kg are placed in a container and 100 liters of water are poured into the container. What is the concentration of plastic beads, in mg/L?

Using the above equation,

$$C_A = \frac{M_A}{V_A + V_B} = \frac{0.48 \text{ kg}}{(0.04 \text{ m}^3)(10^3 \text{ L / m}^3) + 100L} = 0.00343 \text{ kg / L} = 3,430 \text{ mg / L}$$

where A represents the beads and B the water. Remember that 1 kg = 1,000,000 mg and there are 1000 liters in every cubic meter.

Note that in the above example the volume of water is added to the volume of the beads. If the plastic beads with a volume of 0.04 m^3 are placed in a 100 liter container and the container filled to the brim with water, the total volume is:

$$V_A + V_B = 100 \text{ L}$$

and the concentration of beads is:

$$C_A = 4,800 \text{ mg/L}$$

The concentration of beads is higher since the total volume is lower.

Another measure of concentration is *parts per million* (ppm). This unit means one part of a substance exists in one million parts of the surrounding media. For example, in water the "part" is typically taken as a mass, and therefore, we could give the amount of oxygen in water as 7 ppm. This would be the same as saying we have 7 grams of oxygen per 1 million grams of liquid (the liquid in this case is the water with all the other materials in it). The unit of ppm (on a mass basis) is numerically equivalent to mg/L if the fluid in question is water having a density of 1.0 g/cm^3. Remember that ppm in this case is a measure of mass over mass. The conversion from ppm to mg/L is demonstrated in Example 2.8.

Some material concentrations are most conveniently expressed as a fraction or percentage, usually in terms of mass.

$$\Phi_A = \frac{M_A}{M_T} \tag{2.8}$$

where

Φ_A = fraction of material A
M_A = mass of material A, g
M_T = total mass of all materials, g

To obtain a percentage, the fraction would need to simply be multiplied by 100. Φ_A can, of course, also be expressed as a ratio of volumes.

If the concentration is in parts per million (ppm), then

$$\Phi_A = \frac{M_A}{M_T} \times 10^6$$

where Φ_A = concentration of A in ppm

Example 2.9

A wastewater sludge has a solids concentration of 10,000 ppm (remember, this is equivalent to saying we have 10,000 g of solids per 1 million grams of sludge). Sludge is a name given to a combination of solids suspended in water. Express this as both percent solids (mass basis) and mg/L assuming that the density of the solids is 1 g/cm^3.

1. Expressed on a percent mass basis

$$10,000 \text{ ppm} = \frac{1 \times 10^4 \text{ g solids}}{1 \times 10^6 \text{ g sludge}} = 0.01 \text{ g/g} = 1\%$$

2. Express as mg/L assuming density is 1 g/cm^3

$$\frac{[1 \times 10^4 \text{ g solids}]}{[1 \times 10^6 \text{ g sludge}]} \frac{[1 \text{ g sludge}]}{[\text{cm}^3]} \frac{[1 \text{ cm}^3]}{[1 \text{ mL}]} \frac{[1000 \text{ mL}]}{[1 \text{ L}]} \frac{[1000 \text{ mg solids}]}{[1 \text{ g solids}]}$$

$$= 10,000 \text{ mg/L}$$

The last calculation illustrates a useful relationship;

10,000 mg/L = 10,000 ppm (if density = 1 g/cm^3) = 1% (by weight)

Very often in water pollution work, concentrations of solids are expressed both as mg/L and as percent solids. It is useful to be able to quickly make the conversion.

In air pollution control, concentrations are generally expressed gravimetrically as mass of pollutant per volume of air at standard temperature and pressure. For example, the national air quality standard for sulfur dioxide is 0.05 μg/m^3. (One microgram = 10^{-6} gram.) Occasionally, air quality is expressed in ppmv, or parts per million on a volume basis, and the calculations are in terms of volume/volume, or one ppm = 1 volume of a pollutant per 1×10^6 volumes of air.

Conversion from mass/volume (μg/m^3) to volume/volume (ppm) requires knowledge of the molecular weight of the gas. At standard condition, 0°C and 1 atmosphere of pressure, one mole of a gas occupies a volume of 22.4 L. One mole is the amount of gas in grams numerically equal to its molecular weight.

Example 2.10

Assume we have 1 ppmv of methane CH_4 (in other words, we have 1 L of methane in 1 million liters of air) in the air, express its concentration in $\mu g/m^3$. The molecular weight of methane is 16 (12 for carbon and 4×1 for the hydrogen). The conversion from ppmv to $\mu g/m^3$ is:

$$\frac{[1 \text{ L CH}_4]}{[1 \times 10^6 \text{ L Air}]} \quad \frac{[1 \text{ mole of CH}_4]}{[22.4 \text{ L CH}_4]} \quad \frac{[16 \text{ g CH}_4]}{[1 \text{ mole of CH}_4]} \quad \frac{[1 \times 10^6 \text{ }\mu g]}{[1 \text{ g}]} \quad \frac{[1000 \text{ L}]}{[1 \text{ m}^3]}$$

$$= 714 \text{ }\mu g/m^3$$

Based on this conversion, a general relationship can be written to covert ppmv to $\mu g/m^3$:

$\mu g/m^3$ = 1000 (ppm × molecular weight)/22.4 at 0°C and 1 atmosphere.

If the gas is at 25°C at one atmosphere, as is common in air quality standards calculations, the conversion is

$\mu g/m^3$ = 1000 (ppm × molecular weight)/24.45 at 25°C and 1 atmosphere.

A summary of typical concentration units used for expression concentration in water, air, and soil samples is given in Table 2.2.

2.6 PROCESS ANALYSIS

Many environmentally significant problems involve processes. A process is an operation where something happens, like malt and hops fermenting into beer, or students learning to read, or phosphorus being recycled in a farm pond. Understanding processes is thus very important in environmental work. In this section we introduce five concepts: flow rate, continuity, residence time, materials balance, and reactors.

TABLE 2.2. Typical units for concentration in water, air, and soil samples.

Media	Unit	Dimensions
Water	mg/L ppm (parts per million)	mass/volume mass/mass
Air	$\mu g/m^3$ ppmv (parts per million by volume)	mass/volume volume/volume
Soil	mg/kg	mass/mass

2.6.1 Flow Rate

The flow rate can be either *gravimetric (mass) flow rate* or *volumetric (volume) flow rate*. The former is in kg/s, or lbM/s while the latter is expressed as m³/s or ft³/s. The mass and volumetric flow rates are not independent quantities, since the mass (*M*) of material passing a point in a flow during a unit time is related to the volume (*V*) of that material as

$$[\text{mass}] = [\text{density}] \times [\text{volume}]$$

Thus a volumetric flow rate (Q_V) can be converted to a mass flow rate (Q_M) by multiplying by the density of the material.

$$Q_M = Q_V \rho \qquad (2.9)$$

where

Q_M = mass flow rate, kg/min
Q_V = volume flow rate, m³/min
ρ = density, kg/m³

The symbol Q is almost universally used to denote flow rate. The relationship between mass flow, of some component *A*, concentration of *A*, and the total volume flow (*A* plus *B*) is

$$Q_{M_A} = C_A \times Q_{V_{A+B}} \qquad (2.10)$$

This equation states that the mass flow rate (kg/min) of some material *A* can be calculated if the concentration of *A* (in a mixture of *A* and *B*) and the volume flow rate (of *A* and *B* together) are known. Note that this is not the same as Equation 2.9, which is applicable to only one material or one component in a flow stream. Equation 2.10 relates to two *different* materials or components in a flow. For example, a mass flow rate of plastic balls moving along and suspended in a stream of water is expressed as kg of these balls per second passing some point, which is equal to the concentration (kg balls/m³ total volume, balls plus water) times the stream flow (m³/s of balls plus water).

Example 2.11

A wastewater treatment plant discharges a flow of 1.5 m³/s (water plus solids) at a solids concentration of 20 mg/L (20 mg solids per liter of flow, solids plus water). How much solids is the plant discharging each day?

Using the equation above,

$$[\text{Mass flow}] = [\text{Volume flow}] \times [\text{Concentration}]$$

$$Q_{MA} = \left[20\,\frac{mg}{L} \times \frac{1 \times 10^{-6}\,kg}{mg}\right] \times \left[1.5\,\frac{mg}{s} \times \frac{10^3\,L}{m^3} \times 86{,}400\,\frac{s}{day}\right] = 2592 \text{ kg / day}$$

2.6.2 Continuity

If water at a flow rate Q flows through any opening of area A it will have a velocity v so that $Q = A \times v$. The best way to understand this is to imagine a pipe of cross-sectional area A (say in square meters), carrying water at a velocity of v (say m/sec). The flow rate is therefore

$$m^2 \times m/s = m^3/s \qquad (2.11)$$

The equation $Q = Av$ is known as the *continuity equation.*

Example 2.12

A 0.2 m diameter pipe carries a flow of 0.4 m³/second.

1. What is the velocity of the water? (Recall that $A = \pi r^2$ where r = radius)

$$Q = Av, \text{ or } v = Q/A = (0.4)/([0.1]2 \times 3.14) = 12.7 \text{ m/s}$$

2. If the flow in this pipe passes into a larger pipe, say of 0.3 m in diameter, what will be velocity in this pipe?

The flow of course has to be the same, so the velocity in the larger pipe is

$$(0.4)/([0.15]2 \times 3.14 = 5.7 \text{ m/s}$$

As the area through which the flow occurs gets bigger, the velocity gets smaller.

2.6.3 Retention Time

Another important concept in treatment processes is *retention time,* (sometimes called *detention time* or even *residence time*) defined as the avergae time the fluid spends in the container through which the fluid flows. An alternate definition is the time it takes to fill up the container.

Mathematically, if the volume of a container such as a large holding tank, is V (m³), and the flow rate into the tank is Q (m³/min), then the retention time is

$$\bar{t} = \frac{V}{Q}$$

where

$\bar{t}$ = retention time, min
Q = flow rate into the container, m^3/min
V = volume of the container, m^3

The average retention time can be increased by reducing the flow rate Q or increasing the volume V, and decreased by doing the opposite.

Example 2.13

A lagoon has a volume of 1500 m^3 and the flow into the lagoon is 3 m^3/hour, what is the retention time in this lagoon

$$\bar{t} = (1500)/3 = 500 \text{ hours}$$

2.6.4 Materials Balance

One of the most powerful tools used in process analysis is *materials balance.* In approaching this method, we use the concept of a "black box" to represent any unit operation where a process occurs. It is not important to know what happens inside a black box, but only to know that there are flows of materials going into it and out of it. Figure 2.2 shows a black box into which some material is flowing. All flows into the box are called *influents* and represented by the letter X. If the flow is described as mass per unit time, X_0 is a mass per unit time of materials X flowing into the box. Similarly, X_1 is the outflow or *effluent.* If no processes are going on inside the box that will either make more of the material or destroy any, and if the flow is assumed not to vary with time (to be at *steady state*), then it is possible to write a material balance around the box as

(Mass/time) IN = (Mass/time) OUT

or using the symbols in Figure 2.2,

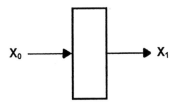

Figure 2.2. A simple black box with one input and one output.

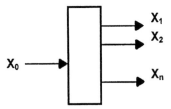

Figure 2.3. A simple black box with one input and several outputs.

$$X_0 = X_1$$

The black box can be used to establish a volume balance or a mass balance if the density does not change in the process. Because the definition of density is mass per unit volume, the conversion from a mass balance to a volume balance is achieved by dividing each term by the density (a constant). It is generally convenient to use the volume balance for liquids and the mass balance for solids.

Separating Flow Streams

Black boxes can be *separators*, or devices that split a flow into several parts. A black box shown in Figure 2.3 receives flow from one feed source and separates this into two or more flow streams. The flow into the box is labeled X_0 and the two flows out of the box are X_1 and X_2. If again it is assumed that steady-state conditions exist (the flow does not change) and that no material is being destroyed or produced, then the materials balance is

$$X_0 = X_1 + X_2$$

The material X can of course be separated into more than one fraction, so that the materials balance is

$$X_0 = X_1 + X_2 + X_3 \ldots X_n$$

where there are n effluents.

Example 2.14

A city generates 103 tons/day of refuse, all of which goes to a transfer station. At the transfer station the refuse is split into 4 flow streams headed for three incinerators and one landfill. If the capacity of the incinerators is 20, 50 and 22 tons/day, how much refuse must go to the landfill?

The problem is shown graphically in Figure 2.4. The four output streams are the incinerators of known capacities plus one landfill. The input stream is the solid waste delivered to the transfer station. Setting up the mass balance,

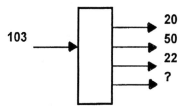

Figure 2.4. See Example 2.14.

[tons of refuse/day generated] = [tons of refuse/day to incinerator 1] +
[tons of refuse/day to incinerator 2] + [tons of refuse/day to incinerator 3] + [tons
of refuse/day to the landfill]

$$103 = 20 + 50 + 22 + M$$

where M = mass of refuse to the landfill. Solving for the unknown, M = 11 tons/day.

Combining Flow Streams

A black box can also be a blender, receiving numerous influents and discharging one effluent, as shown in Figure 2.5. If the influents are labeled X_1, X_2, ... X_m, the materials balance would yield

$$X_1 + X_2 + \ldots + X_m = X_0$$

Example 2.15

A trunk sewer that carries wastewater to a wastewater treatment plant has a flow capacity of 4.0 m³/sec. If the flows from the collecting sewers (smaller sewers from neighborhoods) exceed this amount, the water will back up and there will be an environmental (not to say olfactory) disaster. Currently three neighborhoods contribute to the sewer, and their maximum (peak) flows are 1.0, 0.5, and 2.7 m³/sec. A builder wants to construct a development that will contribute a maximum flow of 0.7 m³/sec to the trunk sewer. Would this cause the sewer to exceed its capacity?

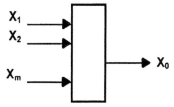

Figure 2.5. A black box with multiple inputs and one output (a blender).

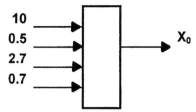

Figure 2.6. See Example 2.15.

Think of the sewer as a black box with flows going in and out, as shown in Figure 2.6. Setting up the materials balance in terms of volume,

(Volume/time of wastewater) IN = (Volume/time wastewater) OUT

$$1.0 + 0.5 + 2.7 + 0.7 = X_0$$

where X_0 is the flow in the trunk sewer. Solving, $X_0 = 4.9$ m^3/sec

It seems that the sewer would be overloaded if the new development is allowed to attach to the sewerage system. Even now, the system seems to be overloaded, and the only reason disaster has not struck is that not all of the neighborhoods produce the maximum flow at the same time of day.

Mixing Multiple Materials Flow Streams

Since the mass balance and volume balance equations are actually the same equation, it is not possible to develop more than one materials balance equation for a black box, unless there is more than one material involved in the flow, in which case materials balances can be written for both (or all in the case of more than two) materials. If the flow is a liquid in which a solid is suspended, materials balances can be written in terms of the liquid (volume balance) and the solid (mass balance). Consider the following example where silt flow in rivers (expressed as mass of solids per unit time) is analyzed.

Example 2.16

The Allegheny and Monongahela Rivers meet at Pittsburgh to form the mighty Ohio. The Allegheny, flowing south through forests and small towns runs at an average flow of 340 cubic feet per second (cfs) and has a low silt load, 250 mg/L. The Monongahela, on the other hand, flows north at a flow of 460 cfs through old steel towns and poor farm country, carrying a silt load of 1500 mg/L.

1. What is the average flow in the Ohio River?
2. What is the silt concentration in the Ohio?

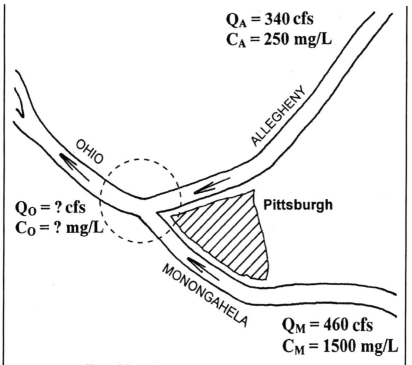

Figure 2.7. Confluence of two rivers. See Example 2.16.

Figure 2.7 shows the confluence of the rivers, and the variables are identified. All the available information is added to the sketch, including the unknown variables. The confluence of the rivers is the black box, as shown by the dotted line. Water flow is to be balanced first.

$$(\text{Water flow}) \text{ IN} = (\text{Water flow}) \text{ OUT}$$

There are two rivers flowing in, and one out, so the equation reads

$$340 + 460 = Q_O$$

where Q_O = flow in the Ohio. Solving,

$$Q_O = 800 \text{ cfs}$$

A mass balance must now be written in terms of the silt carried in the water. Recall that mass flow is calculated as concentration times volume, or

$$Q_{M_A} = C_A \times Q_{V_{A+B}}$$

Setting up the mass balance,

(Mass flow of sediment) IN = (Mass flow of sediment) OUT

$$(C_A Q_A) + (C_M Q_M) = C_O Q_O$$

where C = concentration of silt, and the subscripts A, M and O are for the three rivers. Substituting;

$$0 = (250 \times 340) + (1500 \times 460) - (C_O \times 800)$$

Note that the flow of the Ohio is 800 cfs as calculated from the volume balance. Solving,

C_O = 969 mg/L which is the concentration of silt in the Ohio River.

There is no need to convert the flow rate from ft^3/s to L/s since the conversion factor would be a constant that would appear in every term of the equation and simply cancel.

2.6.5 Reactors

Most of the processes in environmental engineering occur in reactors, or tanks designed to get something done. The process of making beer occurs in a reactor, for example. Reactor kinetics is a sophisticated science and we are able to only touch on it here, primarily to describe two ideal reactors that are referred to later in this book.

The first ideal reactor is called a *plug flow reactor*, and the best way to describe it is to think of a garden hose [Figure 2.8(a)]. Water goes into one end of the garden hose and comes out the other. If this is an ideal reactor, there is no mixing when the water is flowing through the hose. A good way to think of it is to imagine injecting a signal, such as a dye from a syringe, at the beginning of the hose. If this is an ideal reactor, there is no mixing within the reactor and the signal, or *plug* of colored dye, travels through the reactor without longitudinal mixing. The time the plug of dye spends in the reactor is the hydraulic residence time because this is the time needed to flow from one end of the reactor to the other. When the plug of dye comes out, it comes out all together because it has not (ideally) mixed with the water in the reactor, and if we plot the concentration C of the dye we will get a plot that looks like Figure 2.8(b). The dye comes out at residence time $\bar{t}$, and it comes out all at once at an infinitely high concentration. And the moment it is out, the concentration of dye in the flow returns to zero. This is impossible of course, but remember this is an ideal reactor.

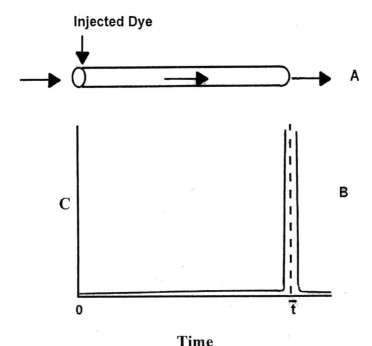

Injected Dye

A

C

B

0   t̄

Time

Figure 2.8. Plug flow reactor.

The plug flow reactor is characterized also by a condition that becomes important when we talk about some of the processes in water treatment—uniform flow. *Uniform flow* means that the flow velocity does not vary with respect to location. In an ideal plug flow reactor, the velocity at every point in the reactor is the same, and this is the reason there is no mixing. Mixing can only occur if there are different velocities within the reactor. But the uniform flow condition requires that no matter where we look within the reactor, we will always find that the water will have the same velocity.

Uniform flow, incidentally, is different from steady flow. With uniform flow, the velocity does not change with respect to location. With *steady flow*, the velocity does not change with respect to time.

A second ideal reactor is the *completely mixed flow reactor* shown in Figure 2.9(a). This reactor is totally and completely mixed, so that if a dye is injected into the incoming flow it is immediately (in zero time) mixed throughout the reactor, and the mixing is perfect in that the concentration of the dye in the reactor is the same at all places within the mixed reactor. Because the dye is mixed all at once, it also begins to flow out of the reactor at time zero. As it flows out, clean water begins to flow into the reactor which dilutes the concentration of the dye in the reactor. The concentration C of the dye in the reactor is the highest the moment the dye is injected into the reactor, and from then on the concentration decreases. This flushing out is shown by the curve in Figure 2.9(b).

Note on Figure 2.9(b) that at the hydraulic retention time, 64% of the dye has left the reactor, and 36% of the dye is still in the reactor, and that it takes a very long time for all of the dye to eventually be flushed out of the reactor.

The real world is not ideal, of course, and no reactor in real life is either an ideal plug flow reactor, or an ideal completely mixed flow reactor. The dye injected into the plug flow reactor will disperse somewhat in the garden hose as the water passes through it, and this will result in a dye concentration curve more like that shown in Figure 2.10. Some of the dye will come out early, most

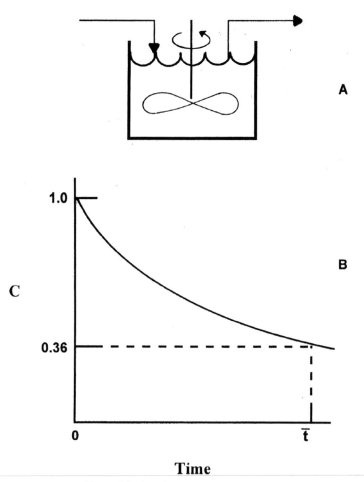

Figure 2.9. Completely mixed flow reactor.

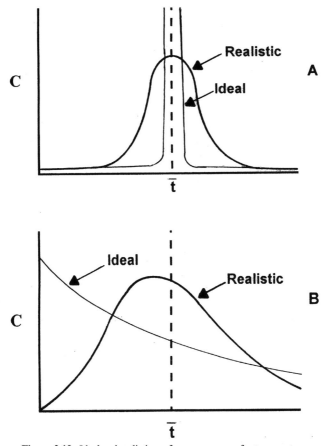

Figure 2.10. Ideal and realistic performance curves for two reactors.

will come out at the hydraulic retention time, and some will linger in the reactor and come out late.

Similarly, completely mixed flow reactors are also ideal. When the dye is injected into the reactor, there is no way it can mix instantaneously throughout the reactor, and there is no way some of it can start to come out of the reactor at time zero. Realistically, the water flowing out of the reactor will be clean at first, and then as the concentration of the dye mixes within the reactor, the concentration in the outflow increases, reaching a maximum, at which point it begins to decrease as the reactor is flushed out. The concentration curve for the injected dye for a real life completely mixed flow reactor looks something like Figure 2.10(b).

Note that curves for an inefficient plug flow reactor and an inefficient completely mixed flow reactor look fairly similar. Most reactors, in fact, have concentration curves that identify them as poorly operating plug flow or poorly operating completely mixed flow reactors. So much for ideal behavior.

2.7 SYMBOLS

A = area
B = benefit
C = cost
C = concentration, mass/volume
D = number of deaths per year
EI = environmental impact
I = importance of environmental impact
L = liters
M = magnitude of environmental impact
M = mass
m = rank of an event
n = number of events
P = probability that some event will occur
P = population
Q = flow, as either mass or volume per unit time
Q_V = volume flow rate, volume/unit time
Q_M = mass flow rate, mass/unit time
R = risk of death
SMR = standard mortality ratio
$\bar{t}$ = retention time
T = total number of deaths from all causes during one year
v = velocity, length/time
V = volume
ρ = density
Φ = concentration as percent (mass/mass)

2.8 PROBLEMS

2.1 Workers in a chemical plant producing a pesticide suffer from a rare form of cancer that is usually fatal. During the 20 years of the plant's operation, 20 employees out of 350—the total number of employees at the plant during those years—developed this cancer. Does working in the plant present an excess cancer risk? Why? What assumptions need to be made?

2.2 Previously unavailable data on this rare cancer incidence (Problem 2.1) indicates that, among people who have never worked in the pesticide industry, there are only 10 deaths per 100,000 persons per year from this disease. How does this change your answer to Problem 2.1?

2.3 The biochemical oxygen demand (BOD) of wastewater entering a wastewater treatment plant was measured for 8 days.

Date	BOD (mg/L)
12 June	110
13 June	130
14 June	150
15 June	120
16 June	180
17 June	210
18 June	230
19 June	190

What fraction of time will the BOD be expected to be more than 200 mg/L?

2.4 Find the won/lost records for your college basketball team for the past ten years. Calculate the winning percentage for each year. How often can the team be expected to have an equal number of wins and losses (play at least 500 ball)?

2.5 A lake has the following dissolved oxygen (DO) readings:

Month	DO (mg/L)
1	12
2	11
3	10
4	9
5	10
6	8
7	9
8	6
9	10
10	11
11	10
12	11
13	6
14	7
15	10

1. What fraction of time can the DO be expected to be above 10 mg/L?
2. What fraction of time will the DO be less than 8 mg/L?
3. What is the average DO?
4. What is the return period for a DO of 2 mg/L?

2.6 The return periods and flood levels for Buffalo Creek are shown below:

Return Period (years)	Flood level (m^3/s)
1	108
5	142
10	157
20	176
40	196
60	208
100	224

Estimate the flood level for a return period of 500 years.

2.7 The number of students taking a course for the past five years is as follows:

Year	No. of students
1	13
2	18
3	17
4	15
5	20

How many years in the next 20 years should the class be 10 students or smaller?

2.8 One gram of table salt is added to an 0.5 L glass and the glass is filled up with tap water. What is the concentration of salt in mg/L?

2.9 Metal concentrations in wastewater sludges are often expressed in terms of milligrams of metal per kg of total dry solids. A wet sludge has a solids concentration of 200,000 mg/L and 8000 mg/L of these solids is zinc. What is the concentration of zinc as (g Zn/g dry solids)?

2.10 One gram of pepper is placed in a 100 mL beaker and the beaker is filled up with water to the 100 mL mark. What is the concentration of pepper in mg/L?(What should you assume in this problem in order to be able to answer it using the given information?)

2.11 A wastewater treatment plant receives 10 million gallons per day of flow. This wastewater has a solids concentration of 192 mg/L. How many pounds of solids enter the plant every day?

2.12 Ten grams of plastic beads with a density of 1.2 g/cm^3 are added to 500 mL of an organic solvent with a density of 0.8 g/cm^3. What is the concentration of the plastic beads in mg/L? (This is a curve ball!)

2.13 A stream flowing at 600 L/min carries a sediment load of 2000 mg/L. What is the sediment load in kg/day?

2.14 A water treatment plant produces drinking water that has an arsenic concentration of 0.05 mg/L. If the average consumption of water is 2 gallons of water each day, how much arsenic is each person drinking each day?

2.15 A power plant emits 120 pounds of flyash per hour up the stack. The flow rate of the hot gases in the stack is 25 cubic feet per second. What is the concentration of the flyash in g/m^3?

2.16 A university has about 12,600 undergraduate students, and graduates 2,250 students each year. Suppose everyone enrolled at the university graduates eventually, what is the approximate retention time at the university?

2.17 A water treatment plant has six settling tanks that operate in parallel (the flow gets split into six equal flow streams), and each tank has a volume of 40 m^3. If the flow to the plant is 10 million gallons per day, what is the residence time in each of the settling tanks? If instead, the tanks operated in series (the entire flow goes first through one tank, then the second, and so on), what would be the retention time in each tank?

2.18 A solid waste processing facility receives loads of mixed cans, both aluminum and steel. It's annual report states that on average, they receive 1.5 tons of cans per day, and ship out 0.55 tons per day of steel cans and 0.85 tons per day of aluminum cans. Does this make sense?

2.19 A composting facility receives 2.0 tons per day of wet sludge at 25% solids. It produces finished compost at 10% moisture. How much finished compost, in wet tons per day, should be plant be able to produce? (Assume both sludge and compost are made up of solids and moisture, or water. Thus a 25% solids sludge is 25% solids and 75% water by weight.)

2.20 A settling tank in a wastewater treatment plant has a volume of 50,000 L, and it is designed for a retention time of 2 hours. How much flow can it receive without exceeding its design retention time?

2.21 The wind on a mountaintop is frequently quite strong. Why do you think this is? Consider the continuity equation.

2.22 A thickener used in a wastewater treatment plant is used to concentrate the solids in a flow stream. Suppose a thickener receives a flow of 2 m^3/min at a solids concentration of 0.2 percent solids. The effluent is discharged at 1.8 m^3/min at a solids concentration of 0.002 percent solids. What is the underflow solids concentration and flow rate?

2.23 A centrifuge in a wastewater treatment plant is used to dewater sludge, or to get as much water out of sludge as practicable. Suppose a centrifuge receives a flow of 5 gallons per minute at a solids concentration of 10,000 mg/L. The objective is to produce a sludge with a solids concentration of 200,000 mg/L solids, and to have zero solids in the centrate (clear liquid). What will be flow rate of the centrate if this is to be achieved?

2.24 If the volume of two reactors, an ideal plug flow reactor and an ideal completely mixed flow reactor, is the same, each starts out with the same concentration of a contaminant in the reactor, and the flow to each reactor is the same, which one will be flushed out first (to zero concentration)? Explain.

2.25 Suppose 5,000 students matriculate at a university, and in four years all of them graduate. Is the university more like a plug flow reactor, or a completely mixed flow reactor? Draw the curve showing first the ideal situation of students graduating vs. time, and then on the same curve show the realistic curve showing how students graduate from a typical university.

CHAPTER 3

Water Quality

A supply of water is critical to the survival of life. People need water to drink, animals need water to drink, and plants need water to drink. The basic functions of society require water: cleaning for public health, consumption for industrial processes, and cooling for electrical generation. Although there are areas of the Earth that are desperate for water, on average, the Earth is not running out of water. The challenge for water engineers and scientists is to provide sufficient quantities of water where it is required, at the desired water quality.

3.1 THE HYDROLOGIC CYCLE

The volume of water in oceans and seas is a staggering 324 million cubic miles, with about 6 million more cubic miles locked up in the ice caps, accounting for about 97% of all the water on Earth. Unfortunately, neither of these sources is readily available as fresh water for human use, and we must rely on the remaining 3% of the water in the atmosphere, water in the soil, and water on the surface of the ground. Many smaller communities rely on groundwater for their supplies, but much of the water in the ground is difficult to remove in large volumes and thus larger communities are left with surface water as the main source of water.

The *hydrologic cycle* is a useful starting point for the study of water supply. This cycle, illustrated in Figure 3.1, includes precipitation of water from clouds, infiltration into the ground or runoff into surface water, followed by evaporation and transpiration of the water back into the atmosphere. The rates of precipitation and evaporation/transpiration help define the baseline quantity of water available for human consumption. *Precipitation* is the term applied to all forms of moisture falling to the ground, and a range of instruments and techniques have been developed for measuring the amount and intensity of

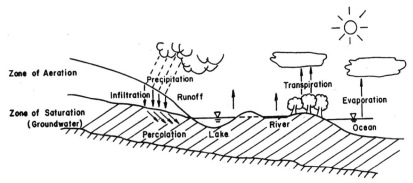

Figure 3.1. The hydrologic cycle.

rain, snow, sleet, and hail. The average depth of precipitation over a given region, on a storm, seasonal, or annual basis, is required in many water availability studies. Any open receptacle with vertical sides can be used as a rain gauge, but varying wind and splash effects must be considered if amounts collected by different gauges are to be compared.

Evaporation and *transpiration* are the movement of water back to the atmosphere from open water surfaces and from plant respiration. The same meteorological factors that influence evaporation are at work in the transpiration process: solar radiation, ambient air temperature, humidity and wind speed. The amount of soil moisture available to plants also affects the transpiration rate. Evaporation is measured by measuring water loss from a pan. Transpiration can be measured with a *phytometer*, a large vessel filled with soil and potted with selected plants. The soil surface is hermetically sealed to prevent evaporation; thus moisture can escape only through transpiration. Rate of moisture escape is determined by weighing the entire system at intervals up to the life of the plant. Phytometers cannot simulate natural conditions, so results have limited value. However, they can be used as an index of water demand by a crop under field conditions, and thus relate to calculations that help determine water supply requirements for that crop. Because it is often not necessary to distinguish between evaporation and transpiration, the two processes are often linked as *evapotranspiration*, or the total water loss to the atmosphere.

As shown on the schematic of the hydrologic cycle (Figure 3.1), water exists in many forms, and most of these can be used as supplies for human consumption. For example, icebergs are essentially fresh water since the freezing water expels the salt, and these can be towed from the north pole to water-starved regions. Rainwater can, and is, used in many rural areas. Some schools and other public buildings are now installing cisterns to capture

TABLE 3.1. Potential Use of Various Sources of Water
as Public Water Supply.

Sources of water	Quality	Quantity
Rainwater	Quite high unless rain falls through polluted atmosphere	Excellent in most regions of the world, but difficult to capture
Icebergs	Excellent	Need to be transported
Groundwater	Excellent unless contaminated by toxic wastes	Being depleted in many areas
Upstream rivers	Very good	Very good, with reservoirs
Downstream rivers	Often poor	Very dependable unless the river is is used for other consumptive purposes such as irrigation
Lakes	Often good	Very dependable
Estuaries	Problems with salinity intrusion into estuary	Very dependable
Ocean	High in salt, requires expensive desalinization	Most dependable
Wastewater	Requires extensive treatment	Very dependable

rainwater and use it to flush toilets and water the lawns. The Gibraltar community on the Mediterranean tip of Spain derives all of its water supply by capturing rainwater falling on the face of the huge rock.

The quality of these different sources of water varies widely, as shown in Table 3.1.

In this text we concentrate on two of the most common sources of water—groundwater and fresh surface water, particularly fresh surface water from streams and rivers.

3.2 GROUNDWATER

Groundwater is both an important direct source of supply that is tapped by wells and a significant indirect source, since surface streams are often supplied by subterranean water.

Near the surface of the earth, in the *zone of aeration*, soil pore spaces contain both air and water. This zone, which may have zero thickness in swamplands and be several hundred feet thick in mountainous regions, contains three types of moisture. After a storm, *gravity water* is in transit through the larger soil pore spaces. *Capillary water* is drawn through small pore spaces by capillary action and is available for plant uptake. *Hygroscopic moisture* is water held in place by molecular forces during all except the driest climatic conditions. Moisture from the zone of aeration cannot be tapped as a water supply source.

In the *zone of saturation*, located below the zone of aeration, the soil pores are filled with water, and this is what we call *groundwater*. A stratum that contains a substantial amount of groundwater is called an *aquifer*. The surface between the two zones is called the *water table*. The water table is evident when a well is dug and water fills it to some depth below the ground surface. An aquifer may extend to great depths, but because the weight of overburden material generally closes pore spaces, little water is found at depths greater than 600 m (2000 ft). When the water table intersects the surface of the ground we have a *spring*. Spring water provided the first useful public water supplies in colonial America, and some communities still rely on springs for their municipal supply.

The amount of water that can be stored in an aquifer is the volume of the void spaces between the soil grains. The fraction of voids volume to total volume of the soil is termed *porosity*, defined as the ratio of volume of voids to the total soil volume. Gravel has great porosity, for example, while solid granite has almost no porosity. The greater the porosity of the soil, the more water is available.

Some soils may have substantial porosity but not all of this water is available because it is tightly tied to the soil particles. Clay soils, for example, might have high porosity, but because the clay particles are so tightly packed with small interstitial spaces between the clay layers the water cannot flow out. The fraction of water that can be extracted from a given well is known as *specific yield*, defined as the percent of total volume of water in the aquifer that will drain freely from the aquifer. Typical values of porosity and specific yield are shown in Table 3.2.

The modern well is constructed by drilling into the ground until water is encountered. The depth of wells can be as little as 10 meters, or as much as 100 meters or more, depending on the geology. Once the well is drilled, a well casing or pipe is dropped into the hole, with a perforated pipe or well screen at the very bottom. An electric motor and pump are lowered to the bottom of the well, and they sit inside the perforated pipe. This is where the water is pumped out, and the location is known as the *well point*.

If a well is sunk into an aquifer, shown in Figure 3.2, and water is pumped out, the water in the aquifer will begin to flow toward the well. As the water

TABLE 3.2. Estimate of Porosity and Specific Yield for Selected Materials.

Material	% Porosity	% Specific Yield
Clay	45	3
Sand	35	25
Gravel	25	22
Sandstone	15	8
Granite	1	0.5

Source: Linsley,R.K. and J.B. Franzini, *Elements of Hydrology*. McGraw-Hill, Copyright © 1958. Used with permission of the McGraw-Hill Book Company.

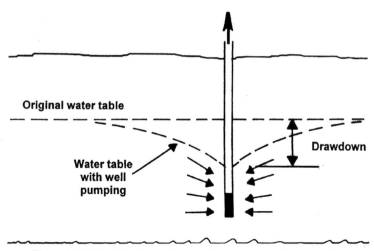

Figure 3.2. Typical drawdown when a well extracts water from an aquifer.

approaches the well, the area through which it flows gets progressively smaller and more energy is used by the water to flow through the soil. This results in a *cone of depression* and the lowering of the water table is known as a *drawdown*. If the rate of water flowing toward the well is equal to the rate of water being pumped out of the well, the condition is at equilibrium, and the drawdown remains constant. If we looked at the cone of depression over time it would not change. But if rate of water pumping is increased, the radial flow toward the well has to compensate, and this results in a deeper cone or drawdown. If the flow is increased, the rate of water flowing toward the well may be inadequate to supply the well and the drawdown will be depressed to the depth of the well head. At this point the well "goes dry." The most water a well can produce without lowering the drawdown too low and going dry is known as the *safe yield*.

Multiple wells in an aquifer can interfere with each other and cause excessive drawdown. Consider the situation in Figure 3.3 where first a single well creates a cone of depression. If a second extraction well is installed, the cones will overlap, causing greater drawdown at each well. If many wells are sunk into an aquifer, the combined effect of the wells could deplete the groundwater resources and all wells would "go dry."

The reverse is also true, of course. In an *injection well* water is injected into a well and this water flows toward other wells, building up the groundwater table and reducing the drawdown. The judicious use of extraction and injection wells is one way that the flow of contaminants from hazardous waste or refuse dumps can be controlled.

Finally, a lot of assumptions are made in the above discussion. First, we assume that the aquifer is homogeneous and infinite, that is, it sits on a level aquaclude and the permeability of the soil is the same at all places for an infinite distance in all directions. Second, steady state and uniform radial flow is assumed. The well is assumed to penetrate the entire aquifer, and is open for the entire depth of the aquifer. Finally, the pumping rate is assumed to be constant. Clearly, any of these assumptions may be unwarranted and cause the analysis to be wrong.

Groundwater supplies are generally the purest water available. Typically the only treatment required with groundwater is *softening,* or the removal of manganese and calcium. The presence of these elements makes water *hard*, a condition characterized by the difficulty of getting soap to lather. Soft water, on the other hand, feels slippery and excessive softness makes rinsing soap difficult.

Not all groundwater is pure and wholesome, however. Agricultural facilities such as feed lots can grossly contaminate aquifers, so as to render the groundwater essentially unusable. Even overland flow from small farms can end up in springs and make the water unfit to drink. Not every well and spring, although the water might look inviting, is safe to drink.

Another problem in rural areas is the presence of acid mine drainage that can contaminate groundwater. Acid mine drainage contamination is characterized by high concentrations of heavy metals such as copper, zinc, iron, manganese, and lead. The pH of such contaminated groundwater is also very low, resulting in corroded water pipes and pumps.

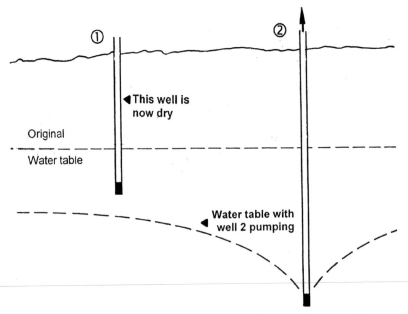

Figure 3.3. Possible scenario when too much water is extracted from an aquifer.

How to Kill a River

The Santa Cruz River is supposed to flow south from Tucson, Arizona, toward the Mexican border, but in reality the river no longer exists. What has happened to the Santa Cruz River?

A hundred years ago the river was bordered by huge mesquite, cottonwood and willows trees, with a rich wildlife of deer, coyote, and bobcats. Although the rainfall was limited, the river was fed by snowmelts from its upper reaches. Before the European invasion, the native populations had developed a stable farm-based economy along the river. Like many rivers, the Santa Cruz flooded in the spring, providing nutrients and water for planting. The early Europeans followed these practices in setting up prosperous farms.

The European settlers, however, were not satisfied with sustainable agriculture and in the late nineteenth century started to pump groundwater to irrigate the fields. The groundwater under the river might have looked like this:

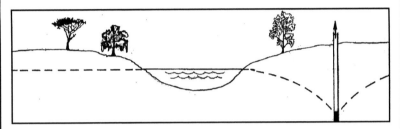

The farmers also cut the trees along the banks of the river, allowing the banks to cave in and forming a wider, shallower riverbed and enhancing evaporation. The growing city of Tucson needed increasingly more drinking water, and since the groundwater aquifer was the only source of water, the wells pumped at an increasing rate. Eventually the water table dropped to 200 feet below the ground, creating problems with land subsidence. As the water table dropped, the flow of groundwater, instead of replenishing the river, flowed away from the river, creating a dry river bed as shown below.

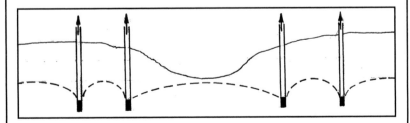

Since floods were still likely, the final insult to the river was the construction of impermeable banks along the river that effectively prevented the flood waters from recharging the aquifer. As the farmers pumped more and more water, the aquifer fell far below the riverbed and the groundwater no longer recharged the stream. Trees and other plants died off, taking with them the rich animal life that existed along the river. Today the Santa Cruz River is nothing more than a dry gulch.

The Iron Mountain Mine

Although the quality of groundwater is generally very high, if groundwater seeps to the surface under some circumstances, disastrous effects on surface water quality can result. One such problem is acid mine drainage, perhaps most vividly illustrated by the problems at the Iron Mountain Mine in California.

Beginning in the boom years of the 1860s through the 1960s, Iron Mountain was mined for gold, silver, copper, iron, and other minerals. The mine tailings from this huge mining site still litter the surface, and weathering has caused large fractures that exposed the remaining minerals in the mountain to surface water and oxygen. In such a condition, especially in the presence of pyrite, sulfuric acid forms and this flows from the mine portals and leaches out the metals in the piles of tailings. The pH of the runoff from the Iron Mountain Mine has been measured in negative numbers (less than zero!). This highly acidic runoff with high concentrations of various heavy metals flows into the Spring Creek Reservoir, from where the water for some years was released into another reservoir that eventually flowed into the Sacramento River. The release of this waste resulted in a virtual elimination of aquatic life in many of the creeks surrounding the Iron Mountain Mine site, and high levels of metals caused numerous fish kills in the Sacramento River. The Chinook salmon that use this river for spawning was declared an endangered species. The dissolved oxygen levels in the river, however, were high since there was no aquatic life left to use the available dissolved oxygen in the water.

A lime neutralization system has been installed to treat the worst of the acid mine drainage before it goes into the Spring Creek Reservoir. In addition, some of the worst areas have been capped to prevent seepage of water into the mine and some of the streams have been diverted to keep the water out of the reservoir. Work is continuing on what is possibly the largest and most difficult acid mine drainage problem in the United States.

In more urban areas, groundwater can be contaminated by industrial effluents. One of the worst pollutants is, surprisingly, nitrates. This generally non-toxic ion can cause *methylglobinemia*, or "blue baby disease." The nitrate enters the bloodstreams of small infants and ties up the oxygen in the hemoglobin, thus effectively suffocating the child and causing it to turn blue. Methylglobinemia was a dreaded disease during the last century before we understood what caused it.

Today the most problematic contaminants of groundwater are *nonaqueous phase liquids* (NAPL). Light NAPLs include such chemicals as gasoline, heating oil, and kerosene. Leaking underground storage tanks and surface spills contribute to this contamination that does not mix with groundwater and sits on top of it in large pools, contaminating the groundwater for a long time. Perhaps even more problematic are the dense NAPLs such as trichloroethane, carbon tetrachloride, pentachlorophenols, dichlorobenzenes, and creosote. These

Tampa Bay Water

The Tampa Bay area includes the city of Tampa on the land side and the venerable St. Petersburg on the bay side, connected by a long causeway. As the population of the region exploded during the past decades, the local governments needed to find water for the citizens. Desalinization is very expensive, and there are no large rivers to develop for surface supply, so the local water authority, Tampa Bay Water (TBW) decided that groundwater was the only option, and constructed well fields throughout the Tampa area. No further fresh water was available from the St. Petersburg side since salt water intrusion would have destroyed the remaining groundwater resource. With salt water intrusion, the salty sea water, being heavier than fresh water, seeps under the fresh water lens and eventually all the wells start pumping sea water, as shown below.

So TBW started to pump an increasing amount of fresh water from the land side. The effects of their pumping can best be demonstrated by what happened to one family, Steve and Cathy Monsees. In 1988 they moved to Pasco County and bought a property fronting a five-acre pond in the back and the 100-acre Prairie Lake in front of their house. To their horror, they watched Prairie Lake dry up and by the early 1990s it was completely gone. They questioned the TBW people about this and were told that the reason was a prolonged drought. In looking at the situation further, they found out that the past few years had been wetter than normal, so the dry lakebed could not have been caused by a drought. TBW continued to prevaricate, arguing variously that the cause was irrigation and then development. Finally, the Monsees found out that TBW had drilled a large well field about a mile from the lake. The field produced 45 million gallons per day, a huge amount of water that could not be sustained by rainwater percolation. The groundwater table dropped to below the lake level and the lake water all drained into the ground, and finally ended up in the taps of the increasingly thirsty citizens of Tampa and St. Petersburg. The damage to the groundwater resources were found to be massive, with over half of the lakes in the region dried up. Damage to streams and creeks in the area precipitated widespread loss of wildlife. Giant trees that had taken water from the aquifer died and fell, and huge sinkholes were formed as the cavities in the ground collapsed.

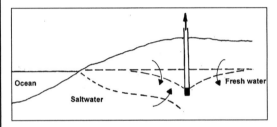

After many lawsuits, TBW offered an apology, but did not stop the damage to the groundwater resources and the natural environment. Incidentally, they eventually realized that they could not indefinitely lower the groundwater table and constructed a $120 million desalinization plant, but this has not been enough to meet the growing demand. At what point does the availability of water dictate limits on population growth?

Source: Glennon, Robert Water Follies Island Press, 2002

chemicals are quite mobile in groundwater and can contaminate large volumes of groundwater. When they are discharged into the ground they drop down through the aquifer until they reach an aquaclude where they will stay until the volume is sufficient to cause them to flow by gravity. An aquifer that has been contaminated by NAPLs often has to be abandoned if treatment is impractical.

3.3 AQUATIC ECOLOGY

The understanding of water quality available from surface water sources requires an understanding of aquatic ecology. *Ecology* is the study of plants, animals, and their physical environment; that is, the study of *ecosystems*, and how energy and materials behave in these ecosystems.

3.3.1 Energy and Materials Flows

Energy and materials both flow inside ecosystems, but with a fundamental difference: energy flow is only in one direction, while materials flow is cyclical. All energy on earth originates from the sun as light energy. Plants trap this energy through *photosynthesis* and using nutrients and carbon dioxide, convert the light energy to chemical energy by building high-energy molecules of starch, sugar, proteins, fats, and vitamins. In a crude way, photosynthesis can be pictured as:

$$[\text{Nutrients}] + CO_2 \xrightarrow{\text{sunlight}} O_2 + [\text{High energy molecules}]$$

All other organisms must use this energy for nourishment and growth; using a process called *respiration*:

$$[\text{High energy molecules}] + O_2 \rightarrow CO_2 + [\text{Nutrients}]$$

This conversion process is highly inefficient, with only about 1.6% of the total energy available being converted into high energy molecules through photosynthesis.

Plants, because they manufacture the high energy molecules, are called *producers*, and the animals using these molecules are called *consumers*. Both plants and animals produce wastes and eventually die. This material forms a pool of dead organic matter known as *detritus*, which still contains considerable energy (That's why we need wastewater treatment plants!). In the ecosystem, the organisms that use this detritus are known as *decomposers*.

In summary, there are three main groups of organisms within an ecosystem:

- producers receive energy from the sun and produce high energy molecules.
- consumers use these molecules as a source of energy.
- decomposers use the residual energy remaining in dead producers and consumers, as well as the residual energy in the waste products of producers and consumers.

This one-way flow is illustrated in Figure 3.4. Note that the rate at which this energy is extracted (symbolized by the slope of the line) slows considerably as the energy level decreases; a concept important in wastewater treatment.

Energy flow is one-way flow: from the sun to the plants, to be used by the consumers and decomposers for making new cellular material and for maintenance. Energy is not recycled within an ecosystem, as illustrated by the following argument.

Suppose a plant receives 1000 Joules (J) of energy from the sun. Of that amount, 760 J is rejected (not absorbed), and only 240 J absorbed. Most of this is released as heat, and only 12 J used for production, 7 of which must go for respiration (maintenance) and the remaining 5 J will go toward building new tissue. If the plant is eaten by a consumer, 90% of the 5 J will go toward the animal's maintenance, and only 10% (or 0.5 J) to new tissue. If this animal is in turn eaten, then again only 10% (or 0.05 J) will be used for new tissue, and the remaining energy is used for maintenance. If the second animal is a human being, then of the 1000 J coming from the sun, only 0.05 J or 0.005% is used for tissue building—a highly inefficient system.

While energy flow is in one direction only, nutrient flow through an ecosystem is cyclical, as represented by Figure 3.5. Starting with the dead organics, or the detritus, the initial decomposition by microorganisms is to

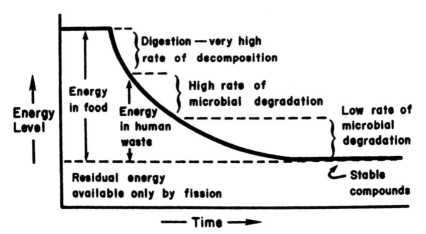

Figure 3.4. Energy flow in nature.

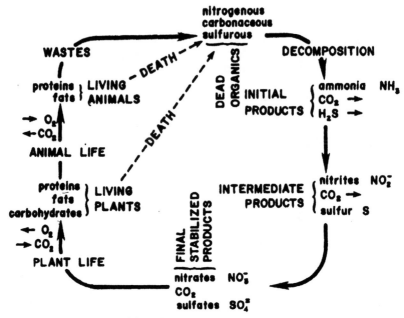

Figure 3.5. Nutrient flow in nature.

initial products such as ammonia (NH_3), carbon dioxide and hydrogen sulfide (H_2S) for nitrogenous, carbonaceous, and sulfurous matter respectively. These products are in turn decomposed further, until the final stabilized or fully oxidized forms are nitrate (NO_3^-), carbon dioxide (CO_2), sulfate (SO_4^{2-}), and phosphates (PO_4^{3-}). The carbon dioxide is used by the plants as a source of carbon, while the nitrates, phosphates and sulfates are used as nutrients, or the building blocks for the formation of new plant tissue.

 There are two types of microbial decomposers: aerobic and anaerobic microorganisms, classified according to whether or not they require molecular oxygen in their metabolic activity. The equation for aerobic and anaerobic decomposers are:

Aerobic: [Organic detritus] $+ O_2 \rightarrow CO_2 + H_2O +$ [Nutrients]

Anaerobic: [Organic detritus] $\rightarrow CO_2 + CH_4 + H_2S + NH_3 + \ldots +$ [Nutrients]

 The decomposition carried out by aerobic organisms is much more complete, since some of the end products of anaerobic decomposition (e.g. ammonia nitrogen) are not in their final fully oxidized state. Aerobic decomposition is necessary in order to oxidize ammonia nitrogen to the fully oxidized form, nitrate nitrogen. Both aerobic and anaerobic microorganisms are used in wastewater treatment, as discussed in Chapter 4. Decomposition in biochemistry is classified according to whether or not the microorganisms have the ability to use dissolved oxygen (O_2) as the *hydrogen acceptor* in the decomposition reaction. (The oxygen accepts hydrogen to form H_2O.) As the organic molecules decompose, the hydrogen from these molecules hooks on to some other chemical. *Obligate aerobes* are microorganisms that must have dissolved oxygen in order to survive, and they use oxygen as the hydrogen acceptor, so that in fairly simple terms, the hydrogen from the organic compounds ends up as water, as in the aerobic equation above. When no dissolved oxygen is available, some organisms, called *facultative* microorganisms will be able to use oxygen from other sources, such as nitrates and sulfates. The nitrates are converted to ammonia (NH_3) and the sulfates to hydrogen sulfide (H_2S), as shown in the above anaerobic equation. (The nitrogen and sulfur become the hydrogen acceptors.) The microorganisms find it easier obtaining the oxygen from nitrates, so this process occurs in preference to the use of sulfates. When oxygen is no longer available from nitrates, the next source of oxygen is sulfates. Some microorganisms, for which dissolved oxygen is in fact toxic, are called *obligate anaerobes*, and they also participate in anaerobic decomposition. One of the end-products of purely anaerobic decomposition is methane, CH_4, where the carbon is the hydrogen acceptor. Methane, of course, still has a lot of energy.

Plants form proteins, fats, and carbohydrates (the high-energy molecules), and in the process of photosynthesis, produce oxygen. Some of these plants die, forming the detritus or dead organic matter that must be decomposed by the microorganisms. Some plants are consumed by animals, and the nutrients are recycled back to detritus by the wastes produced by the animals or as a result of the death of the animals.

The process by which an ecosystem remains in a steady-state condition is called *homeostasis*. There are, of course, fluctuations within a system, but the overall effect is steady state. To illustrate this idea, consider a very simple ecosystem consisting of a grass, field mice, and owls, as pictorially represented in Figure 3.6.

The grass receives energy from the sun, the mouse eats the seeds from the grass, and the owl eats the mice. This progression is known as a *food chain*, and the interaction among the various organisms is a *food web*. Each organism is said to occupy a *trophic level*, depending on its proximity to the producers. Since the mouse eats the plants, it is at trophic level 1, the owl is at trophic level 2, etc. It's also possible that a grasshopper eats the grass (trophic level 1) and a praying mantis eats the grasshopper (trophic level 2) and a shrew eats the praying mantis (trophic level 3). If the owl now gobbles up the shrew, it is performing at trophic level 4.

Competition in an ecosystem occurs in niches. A *niche* is an animal's or plant's best accommodation to its environment, i.e., where it can best exist in the food web. Returning to the simple grass-mouse-owl example, each of the participants occupies a niche in the food web. If there is more than one kind of grass, the plants may occupy very similar niches. However, there will always be subtle, and in reality extremely important, differences. Two kinds of clover, for example, may seem at first to occupy the same niche, until it is recognized that one species blossoms early, and the other later in the summer, thus not competing directly.

The greater the number of organisms available to occupy various niches within the food web, the more stable is the system. If, in the above example, the grass dies due to a drought or disease, the mice would have no food and they as well as the owls would die of starvation. If, however, there were *two* grasses, each of which the mouse could use as a food source (they both fill almost the same niche relative to the mouse's needs), the death of one grass would not result in the collapse of the system. This system is therefore considered to be more stable because it can withstand perturbations without collapsing. Examples of very *stable ecosystems* are tropical forests and estuaries, while *unstable ecosystems* include the northern tundra and the deep oceans.

One of the most important effects humans have had on ecosystem stems from the use of pesticides. The large-scale use of pesticides for the control of undesired organisms began during World War II with the invention and wide use of the first effective organic pesticide, DDT. Before that time, arsenic and other chemicals had been employed as agricultural pesticides, but the high cost of these chemicals, and their toxicity to humans, limited their use. DDT however was cheap, lasting, effective, and did not seem to harm human beings.

Figure 3.6. A simple ecosystem.

The Kelp/Urchin/Sea Otter Ecological Homeostasis

A simplified example of a complex natural interaction is the restoration of the giant kelp forests along the Pacific coasts of Canada, United States, and Mexico. This valuable seaweed, *Macrocystes*, forms 200-foot-long streamers fastened on the ocean floor and rising to the surface. Twice each year the kelp dies and regrows, being one of the fastest growing plants. In addition to the wealth of marine life for which the kelp forests provide habitat, kelp is a source of algin, a chemical used in foods, paints, cosmetics, and pharmaceuticals. Commercial harvesting is controlled to insure the reproduction of the forests.

Some years ago, however, the destruction of these beds seemed inevitable. They were rapidly disappearing, leaving behind a barren ocean floor with no habitat for the marine life that had been so prevalent. The reason for the disappearance was disputed. Pollution? Over harvesting? Or was there some other more subtle cause?

The mystery was finally solved when it was discovered that sea urchins, which feed on the bottom, were eating the lower parts of the kelp plants weakening their hold on the ocean bottom, and allowing them to float away. The sea urchins are the main source of food for the sea otters, and the otters keep the population of sea urchins in check. However, the hunting of sea otters had sufficiently depleted their numbers and allowed for an explosion of the sea urchin population, which in turn resulted in the depletion of the kelp forests.

With the protection of the sea otters, the balance of life was restored, and the kelp forests are again growing in the Pacific.

Many years later it was discovered that DDT decomposes very slowly, is stored in the fatty tissues of animals and is readily transferred from one organism to another through the food chain. As it moves through the food chain, it is biomagnified, or concentrated, as the trophic levels increase. The concentration factor from the DDT level in water to the larger birds is about 500,000. Table 3.3 shows the magnification within a salt marsh.

As a result of these very high concentrations of DDT (and subsequently

TABLE 3.3. DDT Residues in Organisms Taken from a Land Island Salt Marsh.

Organism	DDT residues (ppm)
Water	0.00005
Plankton	0.04
Minnow	0.23
Pickerel (predatory fish)	1.33
Herring gull (scavenger)	6.00
Merganser (fish-eating duck)	22.8

Source: Woodwell, G. M., C. F. Wurster, and P. A. Isaacson, "DDT residues in east coast estuary: A case of biological concentration of a persistent insecticide" *Science*, v. 156, pp. 821–825, 1967.

A Pesticide Story

An interesting example of how DDT can cause a perturbation of an ecosystem occurred in a remote village in Borneo. A World Health Organization worker, attempting to enhance the health of the people, sprayed DDT in the inside of thatch huts in order to kill flies, which he feared would carry disease. The dying flies were easy pray for small lizards which live inside the thatch and feed on the flies. The large dose of DDT however made the lizards ill, and they in turn fell pray to the village cats. As the cats began to die, the village was invaded by rats, which were suspected of carrying bubonic plague. As a result, live cats were parachuted into the village in order to try to restore the balance upset by a well-meaning health official.

Source: Smith, G. J. C. *et. al.*, *Our Ecological Crisis* (New York: Macmillan Pub. Co., 1974)

---

from other chlorinated hydrocarbon pesticides as well), a number of birds are on the verge of extinction, because the DDT affects their calcium metabolism, resulting in the laying of eggs with very thin (and easily broken) shells.

People occupy the top of the food chain and would be expected to have very high levels of DDT. Because it is impossible to relate DDT directly to any acute human disease, the increase in human DDT levels for many years was not an area of public health concern. Public concern finally forced DDT to be banned

---

The Vestal Virgins

The Roman goddess of the hearth, Vesta, was worshiped in a temple that contained a perpetual fire, tended by six virgins. The Vestal Virgins were highly respected and took on demi-god status within the community. Mortals were not allowed to even touch them. Occasionally, one of the Vestal Virgins fell from grace and had a liaison. When this violation was discovered the offending Virgin was led to a deep underground cell, given some bread and water, and left to die. This punishment seemed to be a compromise between two strict rules: do not touch the Vestal Virgin and the Vestal Virgin must remain chaste or be put to death. By leaving her in the underground cell to die she was not touched by anyone and yet she paid for her transgression.

The lore of the Vestal Virgins illustrates that there are two ways to kill anything, be it a human or a planet: by harming it directly, or killing it indirectly by removing the sustaining environment.

Effect of Oil on the Ecology of Prince William Sound, Alaska

The *Exxon Valdez* left the Alyeska Pipeline Terminal at 9 pm on 23 March 1989, loaded with over 53 million gallons of crude oil, bound for Long Beach, California. Such tankers had navigated Prince William Sound over 8,700 times with no serious accidents, so there was no reason to expect any trouble. And yet, three hours later, the *Exxon Valdez* struck Bright Reef, rupturing its tanks and spilling 11 million gallons of oil into the sound.

The oil was loaded during the day under the direction of Chief Mate James Kunkel. The rest of the crew of 19 had the day off. The captain, Joseph Hazelwood, had spent the day doing business and socializing in various bars and returned to the ship at 8:30. Captain Hazelwood acknowledged having had several beers during the day, but denied being intoxicated, and this was corroborated by others.

The tanker slipped its moorings at 9:12 and headed into Prince William Sound, under the direction of Pilot William Murphy. Everything seemed normal, and Captain Hazelwood left the bridge (even though company policy requires two officers to be on the bridge during the passage through the narrows). After the ship cleared the narrows, Pilot Murphy disembarked, leaving only one officer on the bridge. There apparently were no lookouts posted on the bow as required.

For reasons of convenience Captain Hazelwood requested permission to travel in the inbound lane as opposed to the outbound lane and was given permission to do so since there was no inbound traffic. The *Exxon Valdez* started to swing farther to the left, eventually crossing the inbound lane and continuing south toward Bligh Reef. The speed and heading were not reported to the Valdez traffic center as required.

Putting the ship on autopilot, Captain Hazelwood increased the engine speed to normal ocean travel and left the bridge, leaving Third Mate Gregory Cousins in charge. Cousins was supposed to have been relieved at midnight, but he decided to let his friend, Second Mate Lloyd LeCain, who had been busy with the loading during the day, sleep some more, and stayed on the bridge, again violating company policy. Later testimony indicated that Cousins had been awake for over 18 hours prior to the accident, and there is evidence that his fatigue contributed directly to the accident.

At about midnight, Cousins plotted the course for turning the *Exxon Valdez* into the inbound lane. About the same time, a lookout reported seeing the lights on Bligh reef off the *starboard* bow (starboard is right, port is left), when the light should have been seen off the *port* bow. The position of the light indicated that the *Exxon Valdez* had crossed the inbound lane and was heading directly on to the reef. Cousins gave the right rudder command and took the ship off autopilot, calling Captain Hazelwood to inform him of the change. When the ship did not turn rapidly

(continued)

(continued)

enough, Cousins ordered a further turn to starboard. But it was too late. At 4 minutes past midnight, the ship floundered on the rocks.

Captain Hazelwood tried vainly to power the ship off the rocks but to no avail. Stress and stability analysis indicated that the ship had lost enough oil and ballast to make it unstable if it had gotten itself off the rocks, but the captain nevertheless kept up efforts to free the tanker. Fortunately these were unsuccessful, and the crew settled in to await assistance, which was slow in coming. Although Exxon eventually sent massive quantities of machinery and materials to help in the cleanup, there was little done during the first crucial days. The helpless ship stuck on the rocks is shown below.

The 11 million gallons of oil fouled 1000 miles of beach, and eight years later, much of the shoreline in Prince William Sound was still contaminated. Although oil is biodegradable, the low temperature and lack of aquatic biomass in the fragile ecosystem prevented rapid biodegradation from taking place. Fourteen years after the spill, about 50% of the oil is thought to be have been either biodegraded or photolysed. Recovery operations accounted for about 14% of the oil, and about 13% sunk to the sea floor. Fourteen years after the accident over 2% of the oil is still on the beaches, identifiable as either oily slime or oil balls. Estimates of how long the remaining oil will affect wildlife and fishing are unknown. Some species such as the sockeye salmon are recovering because the fish were out to sea when the accident occurred, and the first few years after the accident the salmon were not harvested. Other species such as the common loon, however, are still struggling for survival in the fragile ecosystem.

when it was discovered that human milk fed to infants often contained 4 to 5 times the DDT content allowable for the interstate shipment of *cows'* milk!

Probably the greatest difference between people and all other living and dead parts of the global ecosystem is that people are unpredictable. All other creatures play according to well-established rules. Ants build ant colonies, for example. They will always do that. The thought of an ant suddenly deciding that it wants to go to the moon, or write a novel, is ludicrous. And yet people, creatures who are just as much part of the global ecosystem as ants, can make unpredictable decisions such as develop chlorinated pesticides, or drop nuclear bombs, or even change the global climate. From a purely ecological perspective, humans are indeed different, and potentially dangerous.

3.3.2 Lake Ecosystems

Lakes represent a second example of an ecosystem affected by people. A model of a lake ecosystem is shown in Figure 3.7. Note that the producers (algae) receive energy from the sun, and through the process of photosynthesis produce biomass and oxygen. Because the producers (the algae) receive energy from the sun, they are limited to the surface waters in the lake. Fish and other animals also exist mostly in the surface water since much of the food is there, but some scavengers are on the bottom as well. The decomposers mostly inhabit the bottom waters because this is the source of their food supply (the detritus).

Through the process of photosynthesis, the algae use nutrients and carbon dioxide to produce high-energy molecules and oxygen. The consumers, including fish, plankton, and many other organisms all use oxygen, produce CO_2, and transfer nutrients to the decomposers in the form of dead organic matter. The decomposers, including scavengers such as worms and various forms of microorganisms reduce the energy level further by the process of respiration, using oxygen and producing carbon dioxide. The nutrients, nitrogen and phosphorus (as well as other, nutrients often called micronutrients), are then again used by the producers. Some types of algae are able to fix nitrogen from the atmosphere, while some decomposers produce ammonia nitrogen, which bubbles out.

The only element of major importance that does not enter the system from the atmosphere is phosphorus. A given quantity of phosphorus is recycled from the decomposers back to the producers. The fact that only a certain amount of

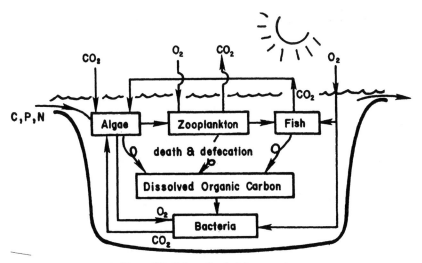

Figure 3.7. Schematic of a lake ecosystem.

phosphorus is available to the ecosystem limits the rate of metabolic activity. Were it not for the limited quantity of phosphorus, the ecosystem metabolic activity could accelerate and eventually self-destruct, since all other chemicals (and energy) are in plentiful supply. In order for the system to remain at homeostasis (steady state), some key component, in this case phosphorus, must limit the rate of metabolic-activity by acting as a brake in the process.

Note that in Figure 3.7 the algae must live on or near the surface because this is where the sunlight is, while the decomposition takes place on the bottom. Oxygen, from both the algae and the atmosphere, must travel through the lake to the bottom to supply the aerobic decomposers. If the water is well mixed, this would not provide much difficulty, but unfortunately most lakes are thermally stratified and not mixed throughout the full depth.

The reason for such stratification is shown in Figure 3.8. In the winter, with ice on the surface, the temperature of northern lakes is about 4°C, the temperature at which water is the densest. In warmer climates, where ice does not form, the deepest waters are still normally at this temperature. In the spring, the surface water begins to warm, and for a time the water in the entire lake is about the same temperature. For a brief time, wind can produce mixing that can extend to the bottom. As summer comes, however, the deeper, denser water sits on the bottom as the surface water continues to warm, producing a steep gradient and three distinct sections; *epilimnion, metalimnion,* and *hypolimnion.* The inflection point is called the *thermocline.* In this condition the lake is thermally stratified, and there is no mixing between the three strata. During this season, when the metabolic activity can be expected to be the greatest, very little oxygen gets to the bottom. As winter approaches and the surface water cools, it is possible for the water on top to be cooler and denser than the lower water, and a *fall turnover* occurs, a thorough mixing. During the winter, when the water on the top is again lighter, the lake is again stratified.

Consider now what would occur if an external source of phosphorus, such as from farm runoff, is introduced. The brake on the system would be released, and the algae would begin to reproduce at a higher rate, resulting in a greater production of food for the consumers, which in turn would grow at a higher rate. All of this activity would produce ever-increasing quantities of dead organic matter for the decomposers, which would greatly multiply.

Unfortunately, the dead organic matter is distributed throughout the body of water while the algae, which produce the necessary oxygen for the decomposition to take place, live only near the surface of the lake, where there is sunlight. The other supply of oxygen for the decomposers is from the atmosphere, also at the water surface. Because both sources of oxygen are at the top of the lake, and because the lake is thermally stratified during much of the year, not enough O_2 can be transported quickly enough to the lake bottom. With an increased food supply, oxygen demand by the decomposers increases, and finally outstrips the supply. The aerobic decomposers die out, and are replaced

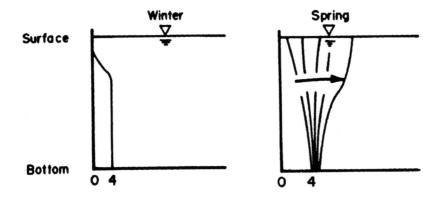

Temperature (°C)

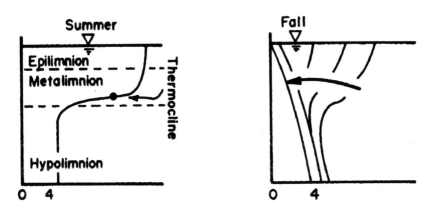

Figure 3.8. Lake stratification through the seasons.

by anaerobic forms which produce large quantities of biomass and result in incomplete decomposition. Eventually, with an ever-increasing supply of phosphorus, the entire lake becomes anaerobic, most fish life dies out, and the algae is concentrated on the very surface of the water, forming green slimy algae mats, called *algae blooms*. In time, the lake fills up with dead and decaying organic matter, and a peat bog results.

This process is called *eutrophication*. It is actually a naturally occurring phenomenon, since all lakes receive some additional nutrients from the air and overland flow. Natural eutrophication is however, usually very slow, measured

Rescue of "Lake Stinko"

The City of Seattle began discharging raw sewage into Lake Washington in 1900, and it wasn't until 1926 that the city began to treat its wastewater and divert the effluent to the Duwamish River that discharges into Puget Sound. By then, development had circled Lake Washington and ten treatment plants were constructed around the lake to handle the wastewater discharges. None of these plants removed nutrients, however, and the level of phosphorus in the lakewater was constantly climbing. In 1955, researchers at the University of Washington discovered a new kind of algae in the lake, the classic tell-tale blue-green algae that accompanies accelerated eutrophication and water quality deterioration. Using knowledge gained from accelerated eutrophication in lakes in Switzerland, the researchers pinpointed the problem as excessive phosphorus. As the studies continued, the water quality in the lake deteriorated. The Seattle Times in a 1963 article referred to the lake as "Lake Stinko."

But relief was on the way. A trunk sewer was constructed around the lake to intercept all of the wastewater and carry it out of the lake watershed, thus significantly reducing the level of phosphorus entering the lake. By 1971 the lake water quality was beginning to show distinct improvement, and the clarity was better than it had been in 1955. Today the lake is a model of what can be done with concerted public commitment.

The Long Island Duck Farms

In the 1940s the duckling industry on the southern shore of Long Island was booming, with new farms being established every year. During these years the residents began to notice a significant change in the water quality of the Long Island sounds. The oyster and hard-shell crabs disappeared and huge green algal blooms began to appear along the shores. The clams survived, but the levels of $E.$ $coli$ were so high that the clams were not commercially usable. Studies showed that the algae were not growing in the water but were washed from the bays and tributaries—and that the major source of pollution was the duck farms. Subsequent experiments showed that the limiting nutrient in the bays and tributaries was not phosphorus, but nitrogen! Ducks produce large quantities of uric acid which contains nitrogen, and this was what was responsible for the production of excessive algae.

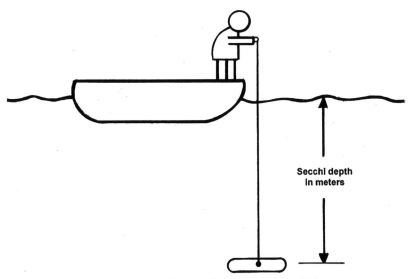

Figure 3.9. Secchi disc used for measuring water clarity.

in thousands of years before major changes occur. What people have accomplished is to speed up this process with the introduction of large quantities of nutrients, resulting in *accelerated eutrophication*. Thus, an aquatic ecosystem that might have been in essentially steady state (homeostasis) has been disturbed and an undesirable condition results.

The process of eutrophication can be reversed by reducing nutrient flow into a lake and flushing or dredging as much of the phosphorus out as possible. This is an expensive proposition.

Because excessive algal growth causes turbidity (interference with the passage of light), a quick and useful way of measuring the health of a lake is to determine changes in its turbidity. This is commonly done using a device called a Secchi Disc. As illustrated in Figure 3.9, the Secchi Disc is a white "garbage can lid" about two feet in diameter with a rope attached to the middle. The disc is lowered from a boat until it disappears from view. This depth is known as "Secchi Disk depth" and is an indication of the algal concentration in the lake. Trends in the health of a lake can be monitored if the test is performed over a number of years.

3.3.3 Stream and River Ecosystems

The primary difference between a stream and a lake is that the former is continually flushed out. Streams, unless they are exceptionally lethargic, are

PIETRO ANGELOS SECCHI

In 1865, the Pope decided that he wanted to test the quality of the water in the Mediterranean Sea and sent the commander of the Papal navy to investigate. At that time, Peitro Angelos Secchi (1818–1878) was the astronomer at the Vatican and he was asked to come up with a way to accomplish the task of measuring the water quality. Secchi devised a white iron disc that is lowered over the side of the boat and the depth at which the disc is no longer visible noted. The deeper that depth, the better is the light penetration, and the clearer is the water. On April 20, 1865 what became known as the Secchi Disc was first lowered over the side of the papal steam yacht, *l'Immaculata Concepzione*. The idea was so simple and worked so well that the Secchi Disc was soon adopted by water quality scientists all over the world.

Pietro Secchi was almost famous for another reason as well. He was the first to use photography to study solar bodies, and his pictures of Mars revealed lines which might have looked like canals. He did not suggest that these were artificial canals, but the imagination of science fiction writers took over and the myth of canals and civilization on Mars was born. The fascination with the possibility of life on Mars continues to this day.

therefore seldom highly atrophied. The stream can be pictured as in Figure 3.10, where a plug of water moves downstream at the velocity of the flow. Think of throwing an orange into the stream and following its progress downstream. In the same way, we can follow a plug of water as it flows down the stream.

In stream ecology, the greatest concern is the availability of oxygen to the microorganisms and other aquatic life. If a source of organic matter is introduced into the stream (such as a discharge from a wastewater treatment plant), the decomposers begin to use the organic material, and as long as it is available, use the oxygen that is dissolved in the water. At the same time, oxygen is replenished in the stream from the atmosphere, so there is oxygen being dissolved from the atmosphere as well as oxygen being depleted by the decomposers. The process by which oxygen is replenished from the atmosphere is called *reoxygenation*, and the process by which oxygen is being used by the decomposer is called *deoxygenation*.

Water can hold only a limited amount of a gas such as oxygen. The amount of oxygen that can be dissolved in water depends on the water temperature, atmospheric pressure, and the concentration of dissolved solids. The saturation

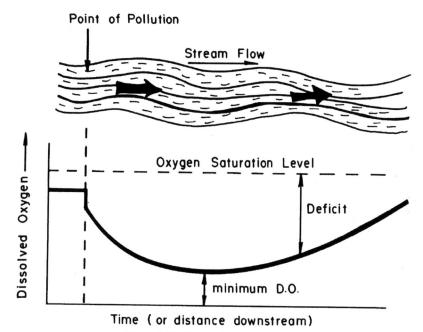

Figure 3.10. Dissolved oxygen sag curve.

levels of oxygen in deionized water at one atmosphere and at various temperatures is shown in Table 3.4.

In a stream loaded with organic material, the simultaneous action of deoxygenation and reoxygenation creates what is called the *dissolved oxygen sag curve*. The shape of the oxygen sag curve, as shown in Figure 3.10, is the result of algebraically adding the rate of deoxygenation, or oxygen use or consumption, and the rate of oxygen supply or reoxygenation. If the rate of use is great, as in the stretch of stream immediately after the introduction of organic pollution, the dissolved oxygen level drops because the supply rate cannot keep up with the use of oxygen, creating a deficit. The deficit (*D*) is defined as the difference between the oxygen concentration in the stream water (*C*) and the highest concentration possible, or saturation (*S*). That is,

$$D = S - C$$

where

D = oxygen deficit, mg/L
S = saturation level of oxygen in the water (the most it can ever hold), mg/L (see Table 3.4)
C = concentration of dissolved oxygen in the water, mg/L

TABLE 3.4. Oxygen Solubility in Fresh Water.

Temperature °C	Saturation of Dissolved Oxygen, mg/L	Temperature °C	Saturation of Dissolved Oxygen, mg/L
0	14.60	23	8.56
1	14.19	24	8.40
2	13.81	25	8.24
3	13.44	26	8.09
4	13.09	27	7.95
5	12.75	28	7.81
6	12.43	29	7.67
7	12.12	30	7.54
8	11.83	31	7.41
9	11.55	32	7.28
10	11.27	33	7.16
11	11.01	34	7.05
12	10.76	35	6.93
13	10.52	36	6.82
14	10.29	37	6.71
15	10.07	38	6.61
16	9.85	39	6.51
17	9.65	40	6.41
18	9.45	41	6.31
19	9.26	42	6.22
20	9.07	43	6.13
21	8.90	44	6.04
22	8.72	45	5.95

After the initial high rate of decomposition when the readily degraded material is used by the microorganisms, the rate of oxygen use decreases because only the less readily decomposable materials remain. Since so much oxygen has been used, the deficit, or the difference between oxygen saturation level and actual dissolved oxygen, is great, the supply of oxygen from the atmosphere begins to keep up with the use, and the deficit begins to level off. Eventually, the dissolved oxygen once again reaches saturation levels, creating the dissolved oxygen sag.

If this is not clear, consider this explanation. Suppose a stream receives a waste that can be assimilated by the microorganisms living in the stream. The rate of this assimilation can be measured by the rate of oxygen use since

$$\text{organic material} + O_2 \rightarrow CO_2 + H_2O + \text{inorganic material}$$

The easier-to-decompose organics are used first, thus using more oxygen, and the less available organics are used later, lowering the rate of oxygen use. If a stream receiving such a load of organic material did not have a means of

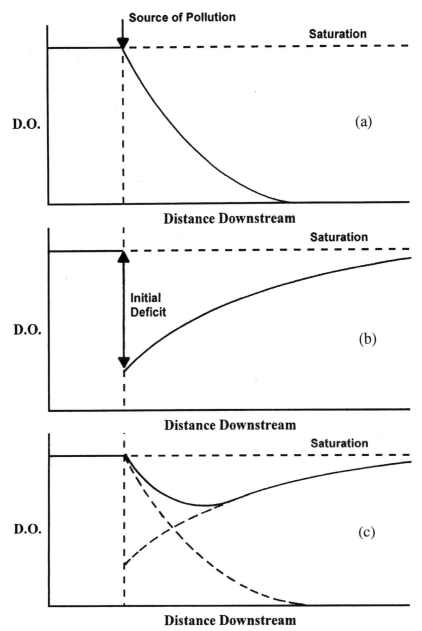

Figure 3.11. Algebraic construction of the dissolved oxygen sag curve.

replenishing the oxygen (such as from air through the stream surface), then the dissolved oxygen level would drop as shown in Figure 3.11(a) due to deoxygenation. The drop would be rapid at first and then slow down downstream.

Now consider another stream that contains no microorganisms that would use oxygen. Any organic material in the stream would simply remain there

The Destruction of the Rhine River

In 1986, a fire at the Sandoz Ltd. chemical plant in Switzerland resulted in the dumping of 30 tons of highly toxic organophophate pesticides called disulfoton and parathion into the Rhine River. The river flows north through Germany and the Netherlands and is both an important waterway and a source of much of the drinking water for many German and Dutch cities. The concentrated plug of pesticide flowed down the river, killing fish, wildlife, and aquatic plant for hundreds of miles and making the water unusable for domestic or industrial supply. For hundreds of kilometers the banks of the river were clogged with dead fish and even dead aquatic plants. The river reeked of decay and seemed totally dead. It took a few years for the effects of the poisons to wash through the river into the sea, and eventually life returned to the Rhine.

This was the most notorious destruction of water quality in Europe, and presented interesting political problems since the spill was in Switzerland, but the devastation was in Germany and the Netherlands. Sandoz Ltd. "cleaned up" their act by moving its pesticide chemical production to Brazil.

The Rhine has historically been the recipient of both human and industrial waste, and its water quality has been very poor. Our own Samuel Taylor Coleridge, as he watched the river flowing through Cologne (Köln) penned these verses:

> In Köln, a town of monks and bones,
> And pavements fanged with murderous stones,
> And rags, and bags, and hideous wenches;
> I counted two and seventy stenches,
> All well defined, and several stinks.

> Ye Nymphs that reign o'er sewers and sinks,
> The river Rhine, it is well known,
> Doth wash your city of Köln;
> But tell me Nymphs: What power divine
> Shall henceforth wash the River Rhine?

The Nashua River

In the 1960s, the North Branch of the Nashua River was an open sewer. It served the cities of Fitchburg and Leominster, Massachusetts and no fewer than five pulp and paper mills. The river was classified by the state as "U" or unfit for any designated uses and was considered one of the ten worst rivers in the entire nation. Regular releases of untreated sewage and industrial wastes into the river and its tributaries killed all the fish and other aerobic aquatic life. The only living aquatic organisms were sludge worms in the muddy bottom and the river stank to high heaven. The water was so full of paper industry effluents that it was solid enough for small mammals to cross over without getting wet. The river water also regularly changed color, taking on the color of the paper mill dyes that were released. Property values of riverfront property plummeted, and towns that used the Nashua as a source of drinking water had to look elsewhere. The high organic loading on the river caused anaerobic conditions, as shown in the dissolved sag curve below.

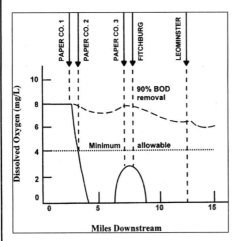

An engineering firm was called in to conduct a study and to make recommendations for improving the water quality. They concluded that 90% removal of effluent oxygen demand was necessary if the river was to attain the required dissolved oxygen level of 4 mg/L. This seemed like a huge financial burden, especially to the paper mills.

What the engineers did not know was that the paper mills had already planned to shut down these old an unprofitable mills, thus eliminating the worst effluents. When the cities of Fitchburg and Leominster built their wastewater treatment plants and removed the last significant point sources of oxygen demand, the Nashua River again became a source of pride for the people in Massachusetts.

since there are no microorganisms to use it as a food supply. If for some reason the dissolved oxygen in the stream at some point is low, there would be a steady increase of dissolved oxygen downstream as the stream is reaerated. The rate of this replenishment or reoxygenation would be most rapid at first since the difference between the saturated value of oxygen and the dissolved oxygen at that point would be greatest. So the curve for reoxygenation would look like Figure 3.11(b).

Now put these two processes together again as in Figure 3.11(c). When the organic material is introduced to the stream, there is a rapid decrease in dissolved oxygen as the microorganisms attack the organics and use oxygen. The rate of reoxygenation at that point is low, however, since the difference between saturation and the oxygen concentration is not high. The dissolved oxygen is being used up, but there is little coming into the water, and the oxygen levels drop. But as the rate of microbial use of oxygen (deoxygenation) decreases, the rate of reoxygenation actually increases because the difference between saturation and the dissolved oxygen concentration is great. There is a battle between the deoxygenation and reoxygenation, and the eventual winner will always be reoxygenation. The battle between oxygen use by the microorganisms and the supply of oxygen from the atmosphere create the oxygen sag curve.

The impact of organics on a stream is not limited to the effect on dissolved oxygen. As the environment changes, the competition for food and survival results in a change in various species of microorganisms in a stream, and the chemical make-up changes as well. Figure 3.12 illustrates the effect of an organic pollutant load on several measurable characteristics of a stream. Note especially the shift in nitrogen species through organic nitrogen to ammonia to nitrite to nitrate. Compare this to the previous discussion of nitrogen in the nutrient cycle (Figure 3.5). The changes in stream quality as the decomposers reduce the oxygen-demanding material, leading to a clean stream, is known as *self-purification*. This process is no different from what occurs in a wastewater treatment plant. In both cases energy is intentionally wasted. The organics contain too much energy, and they must be oxidized to more inert materials.

The River Cam

The River Cam is today one of the most picturesque of English rivers. A bridge across the river gave the name to the first and most famous university in England—Cambridge. Today students at Cambridge enjoy the river, especially for punting, or floating about in small boats on a Sunday afternoon.

But in the 19th century the River Cam, like the Thames, was not the best of places for a pleasure outing. During those days, wastewater from the town and the university flowed directly into the river, which acted as a large open sewer.

There is a tale of Queen Victoria being shown over Trinity by the Master of Cambridge, Dr. Whewell. As they crossed the bridge, the queen looked down into the water and asked: "What are all those pieces of paper floating down the river?" The Master, with great presence of mind, replied: "Those, ma'am, are notices that bathing is forbidden."

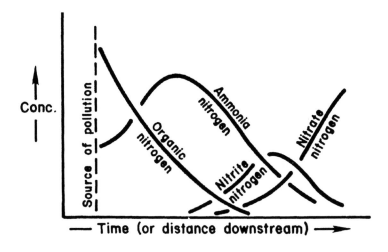

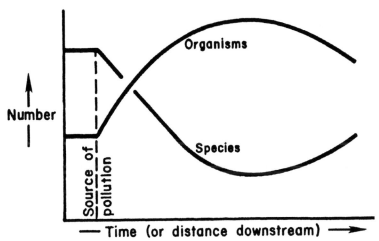

Figure 3.12. Varieties of species, chemicals, and numbers of organisms downstream of a point of pollution.

3.4 AVAILABILITY OF SURFACE WATER

While groundwater sources are fairly reliable for small water systems, they cannot provide the large quantities demanded by cities and large urban areas. The water supply for these systems has to be surface water. Communities, such as Chicago, have abundant supplies. It is unlikely that Chicagoans can use up Lake Michigan any time soon. But other cities such as New York and Los

Angeles must rely on surface water obtained many hundreds of miles away and impounded in reservoirs. The calculation of size for reservoirs is very important in the design of a dependable water system, and this calculation is done using the concept of *frequency analysis*, as introduced in the second chapter.

Flows in streams and rivers are measured with *stream gages* that record the depth of water in a stream, and these data are then used to estimate the flow in a stream. A record of a typical storm event is shown in the *hydrograph*, Figure 3.13. At first, the flow in the stream is essentially constant. Rain falling in the *watershed* or *catchment area* upstream of the stream gage eventually flows past the gage, and this is recorded on the plot as the storm hydrograph. Eventually the stormwater flows off the land, and the stream returns to its normal dry-weather depth.

The flow in streams and rivers fluctuates with time, as illustrated by the hydrograph, and therefore surface water supplies are not as reliable as groundwater sources. In addition to a wide fluctuation in flow rate during the course of a year or even a week, water quality is also affected by pollution sources. If a river has an average flow of 10 cubic meters per second, this does not mean that a community using the water supply can depend on having 10 m^3/s available at all times, and at any predictable level of water quality. The variation in flow may be so great that even a small demand cannot be met during dry periods, and storage facilities must be built to save water during wetter periods.

In order to save the excess water and make it available for use during dry

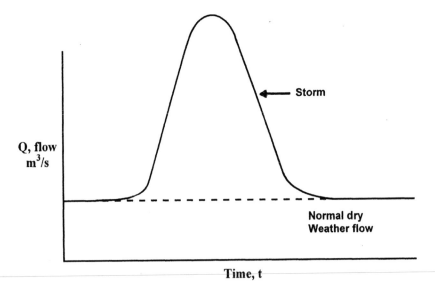

Figure 3.13. A stream hydrograph.

periods, reservoirs need to be constructed. The reservoirs must be large enough to provide dependable supplies, but since reservoirs are expensive, they should not be larger than what is needed by the community. The decision as to how large the reservoir should be hinges on how much dependability the community wants to buy. If the community is satisfied with running out of water once every 5 years (on average) then a small reservoir will suffice in most cases. If the community wants to spend its resources so that it will not run out of water more than say once ever 20 years, then a larger reservoir is necessary.

One method of estimating the proper reservoir size is use of a *mass curve* to calculate historical storage requirements and then to calculate risk and cost using statistics. Historical storage requirements are determined by summing the total flow in a stream at the location of the proposed reservoir, and plotting the change of total flow with time. The change of water demand with time is then plotted on the same curve. The difference between the total water flowing in and the water demanded is the quantity the reservoir must hold if the demand is to be met. The method is illustrated by Example 3.1.

Example 3.1

A reservoir is needed to provide a constant flow of 15 m^3/s. The monthly stream flow records, in total cubic meters, are

Month	J	F	M	A	M	J	J	A	S	O	N	D
Million m^3 of Water	50	60	70	40	32	20	50	80	10	50	60	80

The storage requirement is calculated by plotting the cumulative stream flow as in Figure 3.14. Note that the graph shows 50 million m^3 for January, $60 + 50 = 110$ million m3 for February, $70 + 110$ million m^3 for March, and so on. The demand for water is constant at 15 m^3/s, or

$$15 \text{ m}^3/\text{s} \times 60 \text{ s/min} \times 60 \text{ min/hr} \times 24 \text{ hr/day} \times 30 \text{ days/month}$$

$$= 38.8 \times 10^6 \text{ m}^3/\text{month}$$

This constant demand is represented in Figure 3.14 as a straight line with a slope of 38.8×10^6 $m^3/month$, and is plotted on the curved supply line. Note that the stream flow in May was lower than the demand, and this was the start of a drought. In July the supply increased until the reservoir could be filled up again, late in August. During this period the reservoir had to make up the difference between demand and supply, and the capacity needed for this time was 60×10^6 m^3. A second drought, from September to November required 35×10^6 m^3 of capacity. The municipality therefore need a reservoir with a capacity of at least 60×10^6 m^3 from which to draw water throughout the year.

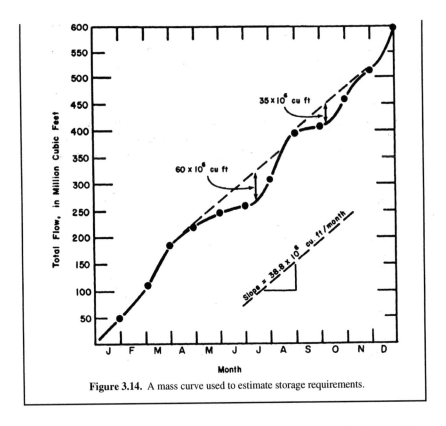

Figure 3.14. A mass curve used to estimate storage requirements.

A mass curve like Figure 3.14 is not very useful if only limited stream flow data are available. Data for one year yield very little information about long-term variations. The data in Example 3.1 do not indicate whether the 60 million m^3/s deficit was the worst drought in 20 years, or an average annual drought, or occurred during an unusually wet year.

Long-term variations may be estimated statistically when actual data are not available. Water supplies are often designed to meet demands of 20-year cycles, and about once in 20 years the reservoir capacity will not be adequate to offset the drought. The community may choose to build a larger reservoir that will prove inadequate only every 50 years, for example. A calculation comparing the additional capital investment to the added benefit of increased water supply will assist in making such a decision.

The *frequency analysis* technique, introduced in Chapter 2, can be used to provide more information on water availability. One calculation method requires first assembling required reservoir capacity data for a number of years, ranking these data according to the drought severity, and calculating the drought probability for each year. If the data are assembled for n years and the rank is designated by m, with $m = 1$ for the largest reservoir requirement during the most severe drought, the probability that the supply will be adequate for any year is given by *m/n*. For example, if storage capacity will be inadequate, on the average, one year out of every 20 years,

$$P = m/n = 1/20 = 0.05$$

If storage capacity will be inadequate, on the average, one year out of every 100 years,

$$P = m/n = 1/100 = 0.01$$

The probability of some event such as running out of water occurring is also expressed in terms of a the *return period*, or once in how many years is the event is expected to recur. If the annual probability of an even occurring is 5% ($m/n = 0.05$), then the event can be expected to recur once in 20 years, or have a return period of 20 years. Stated in equation form:

$$\text{Return period} = (1/P)$$

If the fractional probability of occurrence is 1 time in 100 years, or $1/100 = 0.01$, the return period is

$$1/0.01 = 100 \text{ years.}$$

That is, it is most likely that the event will "return" once every 100 years.

Example 3.2

A reservoir is needed to supply water demand for 9 out of 10 years. The required reservoir capacities, determined by the method of Example 3.1, are shown below.

Year	Required Reservoir Capacity, $m^3 \times 10^6$	Year	Required Reservoir Capacity, $m^3 \times 10^6$
1961	60	1971	53
1962	35	1972	62
1963	85	1973	73
1964	20	1974	80
1965	67	1975	50
1966	46	1976	38
1967	60	1977	34
1968	42	1978	28
1969	90	1979	30
1970	51	1980	45

These data must now be ranked, with the highest required capacity, or worst drought, getting rank 1, the next highest, 2, and so on. Data were collected for 20 years, so that $n = 20$. Next, m/n is calculated for each drought

Rank, m	Capacity $m^3 \times 10^6$	m/n	Rank, m	Capacity $m^3 \times 10^6$	m/n
1	90	0.05	11	50	0.55
2	85	0.1	12	46	0.60
3	80	0.15	13	45	0.65
4	73	0.20	14	42	0.70
5	67	0.25	15	38	0.75
6	62	0.30	16	35	0.80
7	60	0.35	17	34	0.85
8	60	0.40	18	30	0.90
9	53	0.45	19	28	0.95
10	51	0.50	20	20	1.0

These data are plotted in Figure 3.15 and a smooth curve is drawn through the data points. Note that the last point is not really accurate. Even though $m/n = 1.0$, this does not mean that the reservoir will never have to be less than 20 million cubic meters in size. This problem becomes negligible if the number of data points is large.

If the reservoir capacity is required to be adequate 9 years out of 10, it may be inadequate one year out of ten. Entering Figure 3.15 at $m/n = 1/10 = 0.1$, we find that the required capacity is 82 million cubic meters. There is a 10% probability that the reservoir of 82 million cubic meters will be inadequate for any given year. Had the community chosen to have enough water for 4 years out of 5, or $m/n = 0.2$ and, from Figure 3.15, a reservoir capacity of 73 million cubic meters would have sufficed.

The First American Public Water Works

The first American municipal waterworks was built in 1754 by Hans Christopher Christiansen for the Moravian settlement of Bethlehem, Pennsylvania. A wooden waterwheel, driven by the flow of Monocacy Creek coming down off South Mountain, powered wooden pumps that lifted spring water to a wooden reservoir on top of a hill where it was distributed by gravity. The pipes were hollowed out logs with iron straps on each end to prevent splitting. Every few hundred feet, the pipes had holes drilled into them and wooden plugs pounded into the holes. When there was a fire emergency the plug was removed, providing ready access for the bucket brigade. Even today we refer to our fire hydrants as "fire plugs."

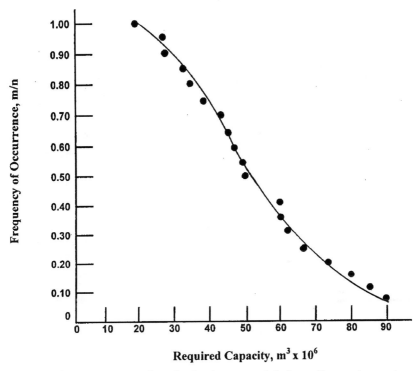

Figure 3.15. Frequency curve for estimating the return period of naturally occurring events.

Water Quality in Eastern Europe

Following the Second World War, the Eastern European countries, devastated by war, started to rebuild using Marxist ideology as a guide to economic development. This economic theory held that the production of energy and heavy goods such as steel was the most important priority, and that these plants should be constructed and operated by the state. Consumer goods were not a priority; nor was the preservation of environmental quality. The rush for production at any cost resulted in the worst environmental devastation known to date. What happened to Poland is typical of the conditions in Albania, Hungary, Rumania, Czechoslovakia, Bulgaria, and the Baltic States of Estonia, Latvia, and Lithuania.

About 90 percent of Poland's water supply comes from surface waters, and these are presently grossly polluted. The shores of Poland's main river, the Vistula, became the site for the construction of much heavy industry, and *none* of these plants had wastewater treatment. Likewise, the largest cities in Poland—Warsaw, Lódz, and Cracow—did not have wastewater treatment plants. When the Vistula arrives at Cracow it is totally devoid of biological life (even anaerobic!), mostly because of its extremely high salt content. The Vistula flows into the Baltic Sea, which is rapidly becoming the most highly polluted sea in the world. Over one fifth of the total BOD_5 discharged into the Baltic Sea comes from Poland, and the Vistula dumps 34 km^3 of untreated wastewater into the Baltic every year. [Pause for a moment and note the units of the last number. *Cubic kilometers!*]

The human health effects of this ignorant policy are equally devastating. By the time Polish children are ten years old, for example, most of them suffer from chronic respiratory or heart diseases. The rates of cancer deaths, birth defects, and miscarriages are also much higher in all of the former Communist countries than the European average.

How did things get to such a state? The answer is that one of the attributes of the Marxist economic system is that there is no such thing as "public opinion". The public was not allowed, under severe penalty, to express opinions. Such an organization as the Sierra Club simply could not have occurred in the Eastern Bloc countries because any such organizations would have been considered subversive and hence dangerous to the state. Without the force of public opinion, and with the mandate to increase production, the expendable cost for all industrial plants was environmental controls.

It will take billions of dollars to clean up the former Eastern Block countries to European standards, and it will take even more money (and a lot time) for the Baltic Sea to recover.

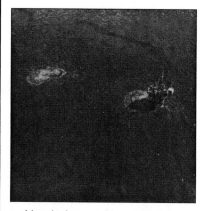

The Santa Barbara Oil Blowout

The geology of the California coast is complex, and frequent ground movements have created interlocking layers of highly permeable sand and less permeable clay. One of the largest of these fault-induced areas is off the coast of Santa Barbara. Oil deposits far below ground slowly percolate up to the surface, but if there is a layer of impervious clay they remain trapped in large lenses. In the Santa Barbara area, oil has been seeping to the surface since prehistoric times, and large oil slicks were reported even by 18th century explorers. The oil lenses trapped under the impervious clay, however, represent a usable if fragile resource. Given that oil can naturally percolate to the surface, it would not seem difficult to presume that any disruption of the geology could cause a massive rupture of the clay layers.

In 1969, the Union Oil company had a huge drilling rig in place five and a half miles off shore drilling its twenty-first well, when disaster struck. As the drilling was stopped and the pipe was to be removed, a massive amount of drilling mud came gushing out of the pipe, followed by natural gas. Had the gas exploded all would have been killed and the platform destroyed. But in about 10 minutes the crew was able to cap the well averting immediate disaster. A few minutes later they saw bubbles of gas breaking the surface of the water near the drilling rig, indicating that the lens that contained the trapped oil must have ruptured. Drilling into this fragile layer must have caused it to crack, and now it was releasing the trapped oil out of the fissures. This type of oil leakage is called a *blowout* and differs from other accidents in that the oil does not come from a pipe but instead leaks through the geological layers. There is no way to stop such a discharge of oil.

Union Oil Company at first tried to minimize the problem and put strict controls on information reaching water quality agencies, hampering remedial action. Eventually the full story got out, and the federal government became involved. They pumped massive amounts of concrete down the pipes into the oil lenses with the hope of stabilizing the geology. Whether this technique worked, or the leaking lens simply lost all of its oil, the blowout eventually dissipated. After the blowout, drilling in the Santa Barbara field was suspended (except for relief wells) and no new leases for prospecting were issued. This event, damaging as it was to the California coastline, provided useful information on how to clean up future oil spills.

3.5 SYMBOLS

C = concentration of oxygen in water, usually mg/L
D = oxygen deficit in water, mg/L
DO = dissolved oxygen
m = rank by size of a natural event
n = number of events
P = probability that an event will occur
S = saturation of oxygen in water, mg/L

3.6 PROBLEMS

3.1 Why does eutrophication generally not occur in streams and rivers? Explain.

3.2 The runoff in the Westfield River near Springfield MA for 15 months is as follows:

Month	Flow (million gallons per month)
1	100
2	100
3	50
4	5
5	8
6	2
7	2
8	10
9	6
10	20
11	50
12	120
13	40
14	25
15	30

The objective is to draw a constant draft of 23 million gallons per month from this river. How large would the impoundment have to be to provide such a water supply for Springfield during this drought?

3.3 The required reservoir sizes for guaranteeing a dependable supply for Springfield, NH are as follows:

Year	Required reservoir capacity, million gallons
1	5
2	23
3	12
4	0
5	0
6	10
7	12
8	28
9	15
10	20
11	8
12	6

a. If the town is not to run out of water more than 1 year in 20, what must the capacity of their reservoir be?

b. If the town decides that it cannot have a reservoir that big and agrees to a smaller reservoir, one that will be adequate for 9 out of 10 years, how big should this reservoir be?

c. Senator Blusterfuegel (R-NH), the chair of the Senate Porkbarrel Spending Committee, has declared he intends to build a reservoir for his constituents that is so big the town will *never* run out of water. How big would this reservoir be?

3.4 A car is rated as getting 20 miles to the gallon and has a 15-gallon gas tank. On a long trip through Outer Ebolia, the gas stations are spaced as follows:

Leg of trip	Miles between stations
1	50
2	230
3	187
4	320
5	100

If the car fills up at each gas station, will it be able to make the trip? Use a graphical mass curve to solve this problem.

3.5 Find out where the water supply for your home town comes from. You will need to call the city/town department of public works and talk to the water resources people.

3.6 Which well would be expected to have the higher yield if two wells are drilled, one in clay or one in sandstone. Why is this?

3.7 Assuming a sand aquifer with a groundwater table 30 meters below the ground surface. A well is drilled into this aquifer with the well point (the bottom of the well) being 40 meters below the ground surface.

    a. Draw this situation and note the groundwater table before the pump begins to pump water out of the well. Label this "A".

    b. Water is now being pumped out of the well and the yield of water is at a steady state. That is, the pump is able to pump water continuously for a long time. Draw the groundwater table as it would exist in this situation and label it "B".

    c. The rate or pumping is now increased, and after several weeks of pumping at this high rate, the well runs dry. That is, the pump is not able to produce the high water flow required and starts to suck air. Draw the groundwater table as it would exist in this situation and label it "C".

    d. The owner of the well decides to drop the well down another 10 meters and continues to pump at the high rate. Will this solve the problem? That is, will the new deep well be able to pump the high rate of flow or will it again run dry? State why for one alternative. Draw the situation that would exist and label the groundwater level as "D".

3.8 A farmer who has a well for his irrigation water hires you as a consultant, and asks you if he can withdraw more than he is presently withdrawing. You respond honestly that yes he can, but that the groundwater reserves will be depleted and that his neighbor's wells may go dry. He asks you if it would be possible for anyone to know that he is withdrawing more water than the maximum rate, and you again tell him honestly that it is unlikely than anyone will know. He then tells you that he plans to double the withdrawal rate, run as long as he can, and when the water gives out, abandon the farm and move to Florida. You appear to be the only person who knows of the farmer's plans.

    Driving home, you reach a decision on what you will do. What kind of decision did you make, and how did you make it?

3.9 Name three animals (other than humans) that are on top of their food chain. Identify some organisms in these food chains.

3.10 The dissolved oxygen sag curve in a stream receiving an effluent from a wastewater treatment plant drops the DO (dissolved oxygen) to about 1 mg/L. The DO upstream from the discharge is about 9 mg/L, which is also about saturation of oxygen in the stream.

    a. Draw this situation as a graphic, plotting DO on the *y* axis and distance downstream on the *x* axis. Label this curve "A".

    b. The wastewater treatment plant upgrades its treatment and discharges only about half of the organics it did before. What would the DO sag curve now look like? Label this curve "B".

    c. There is a terrible accident in the community and a huge amount of arsenic is released into the treatment plant. Arsenic does not biodegrade, and so it is released in the effluent from the wastewater treatment plant into the stream where it kills off all the stream microorganisms. What will the DO sag curve look like? Draw this curve and label it "C".

    d. [This is a thinking question.] Suppose the situation is again as in (b) above, or curve "B". But now it is summertime, and the water temperature increases. What would happen to the DO sag curve if the water temperature was higher? Label this curve "D". Why do you suppose most fish kills in streams occur in the summertime?

3.11 The Secchi disc measurements for a lake looked like this:

Year	Secchi depth (m)
2000	3.5
2001	3.6
2002	3.0
2003	2.3
2004	1.2
2005	0.9

What do you suppose is going on? What else would you expect to see when you go out on the lake?

CHAPTER 4

Water Treatment

W ATER is essential to life and an adequate supply of clean water is essential to human civilization. As human population increases, the availability of clean water becomes increasingly problematical in many regions of the world. One problem faced by environmental professionals is determining just what we mean by clean water. Because water has many uses, what is acceptable for one use may be totally unacceptable for another. Drinking water, for example, has to be both pleasing and safe (or *potable water*), whereas water used for irrigation has to have low salinity. In order to determine if a water is acceptable for specific use, we need quantitative measurements of water characteristics. Only then is it possible to determine if a water is sufficiently "clean" for an intended purpose.

In the water and wastewater profession, methods for measuring water quality have been standardized in a volume entitled *Standard Methods of the Examination of Water and Wastewater*. The title is commonly shortened to *Standard Methods*. This volume is continually revised and the tests are modified and improved as our skills in aquatic chemistry and biology develop. In the most recent volume, the 21st edition, the standard tests that have undergone rigorous review and parallel testing numbers over 500. In other words, there are at least that many standard, quantitative, and useful tests for expressing the quality of water. In this chapter we discuss only a few of these techniques for measuring water quality.

4.1 MEASURING WATER QUALITY

4.1.1 Turbidity

Water that is not clear and "dirty," in the sense that light transmission is inhibited, is known as *turbid* water. Many materials can cause turbidity,

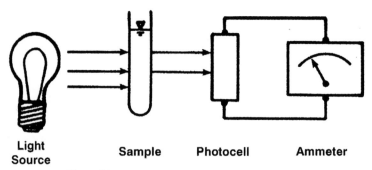

Light Source   **Sample**   **Photocell**   **Ammeter**

Figure 4.1. A simple photometer for measuring turbidity.

including clays and other tiny inorganic particles, algae, and organic matter. In the drinking water treatment process, turbidity is of great importance, partly because turbid water is aesthetically displeasing, and also because the presence of tiny colloidal particles makes it more difficult to remove or inactivate pathogenic organisms.

Turbidity in natural waters is measured with a Secchi Disc as previously discussed, but in a laboratory this obviously is impractical. In a laboratory, turbidity is measured using a *turbidimeter*, photometers that measure the intensity of scattered light. As shown in Figure 4.1, a simple photometer sends a beam of light through a sample of water and the light is measured with a photocell on the other side. The light hitting the photocell causes electricity to be generated, and this current is shown with a microammeter. Colloidal particles in the water scatter light and prevent the light from hitting the photocell. Any reduction in current is therefore an indication of turbidity. Newer turbidimeters actually measure the light that is scattered by the colloidal particles, but the idea is the same. Formazin polymer is currently used as the primary standard for calibrating turbidimeters, and the results are reported as *nephelometric turbidity units* (NTU).

4.1.2 Color, Taste and Odor

Color, taste, and odor are important measurements for determining drinking water quality. Color, taste, and odor are all important from the standpoint of aesthetics. If water looks colored, smells bad, or tastes swampy, people will instinctively avoid using it, even though it might be perfectly safe from the public health aspect. Color, taste, and odor problems in drinking water are often caused by organic substances such as algae or humic compounds, or by dissolved compounds such as iron.

Color can be measured visually by comparison with potassium chloroplatinate standards or by scanning at different spectrophotometric wavelengths. Turbidity interferes with color determinations, so the samples are filtered or centrifuged to remove suspended material.

Odor and taste are measured by successive dilutions of the sample with odor-free water until the odor or taste is no longer detectable. (Odor-free water is prepared by passing distilled, deionized water through an activated charcoal filter.) This test is obviously subjective and depends entirely on the olfactory senses of the tester. Panels of testers are used to compensate for variations in individual perceptions of odor.

4.1.3 Pathogenic Microorganisms

From the public health standpoint the bacteriological quality of water is as important as the chemical quality. A number of diseases can be transmitted by water, among them typhoid and cholera. However, it is one thing to declare that water must not be contaminated by pathogens (disease-causing organisms) and another to discover the existence of these organisms.

Seeking the presence of pathogens presents several problems. First, there are many kinds of pathogens and each pathogen has a specific detection procedure and must be screened individually. Second, the concentration of these organisms can be so small as to make their detection impossible. Looking for pathogens in most surface waters is a perfect example of the proverbial needle in a haystack. And yet only one or two organisms in the water might be sufficient to cause an infection if this water is consumed.

In the United States, the pathogens of importance today include *Salmonella*, *Shigella*, the hepatitis virus, *Entamoeba histolytica*, *Giardia lamblia* and *Cryptosporidium*, and a species of *Eschirichea coli* called *E.coli 157*.

Salmenosis is caused by various species of *Salmonella*. The symptoms of salmenosis include acute gastroenteritis, septicemia (blood poisoning), and fever. Gastroenteritis usually consists of severe stomach cramps, diarrhea, and fever, and although it makes for a horrible few days, is seldom fatal. Typhoid fever is caused by the *Salmonella typhi*, and this is a much more serious disease, lasting for weeks, and can be fatal if not treated properly. About 3% of the victims become carriers of *Salmonella typhi,* and although they exhibit no further symptoms, they pass on the bacteria to others, mainly through the contamination of water.

Shigellosis, also called *bacillary dysentery,* is another gastrointestinal disease, and has symptoms similar to sallmenosis. Infectious hepatitis, caused by the hepatitis virus, has been known to be transmitted through poorly treated water supplies. Symptoms include headache, back pains, fever, and eventually jaundiced skin color. While rarely fatal, it can cause severe debilitation. The hepatitis virus can escape dirty sand filters in water plants, and can survive for a long time outside the human body.

Amoebic dysentery, or *amebiasis*, is also a gastrointestinal disease, resulting in severe cramps and diarrhea. Although its normal habitat is in the large intestine, it can produce cysts, which pass to other persons through contaminated water and cause gastrointestinal infections. The cysts are resistant to disinfection and can survive for many days outside the intestine.

Originally known as "beaver disease" in the north country, *giardiasis* is caused by *Giardia lamblia*, a flagellated protozoan which usually resides in the small intestine. Out of the intestine, its cysts can inflict severe gastrointestinal problems, often lasting two to three months. Giardiasis was known as "beaver disease" because beavers can act as hosts, greatly magnifying the concentration of cysts in fresh water. Giardia cysts are not destroyed by usual chlorination levels, but are effectively removed by sand filtration. Backpackers should take care with drinking water on the trails without purifying with halogen tablets or small hand-held filters. Giardia has ruined more than one summer for unwary campers.

The latest public health problem has been the incidence of *cryptosporidiosis*, caused by *Cryptosporidium*, an enteric protozoan.

Outbreaks of *E.coli 157* attributed to contaminated water are rare, but this organism has been implicated in other incidents of infection due to poorly cooked meat. It is important to recognize that while this organism is a coliform, it differs from the many coliforms used as indicators for water purification in that it is toxic to humans. This is discussed further below.

Table 4.1 is a summary of the disease outbreaks in the United States over a 15-year period.

TABLE 4.1. Outbreaks of Disease in the U.S. from 1971–1985.

Illness	Number of Outbreaks	Cases of Illness
Gastroenteritis, undefined	251	61,478
Giardiasis	92	24,365
Chemical Poisoning	50	3,774
Shigellosis	33	5,783
Hepatitis A	23	737
Gastroenteritis, viral	20	6,524
Campylobacterosis	11	4,983
Salmonellosis	10	2,300
Typhoid	5	282
Yersiniosis	2	103
Toxigenic E. coli	1	1,000
Cryptosporidiosis	1	117
Cholera	1	17
Dermatitis	1	31
Amebiasis	1	4
Total	502	111,228

The Thames River Outing

Our knowledge of the presence of pathogenic microorganisms in water is actually fairly recent. Diseases such as cholera, typhoid, and dysentery were highly prevalent even in the mid- 19th century. At that time microscopes were able to show living organisms in water, but it was not at all obvious that the tiny critters caused diseases. In retrospect, the connection between contaminated water and disease should have been obvious, based on empirical evidence. For example, during the mid-1800s, the Thames River below London was grossly contaminated with human waste, and one Sunday afternoon a large pleasure craft capsized, throwing everyone into the drink. Although nobody drowned, most of the passengers died of cholera a few weeks later!

There are, of course, many pathogenic organisms that can be carried by water. How then is it possible to measure for bacteriological quality? The answer lies in the concept of indicator organisms. The indicator most often used is a group of microbes called *coliforms*. These microorganisms have five important attributes. They are:

- normal inhabitants of the digestive tracts of warm-blooded animals.
- plentiful, hence not difficult to find
- easily detected with a simple test
- generally harmless except in unusual circumstances (such as *E.coli 157*)
- hardy, surviving longer than most known pathogens.

Because of these five attributes, coliforms have become universal indicator organisms. But the presence of coliforms does not *prove* the presence of pathogens! If a large number of coliforms are present, there is a good chance of recent pollution by wastes from warm-blooded animals, and therefore the water may contain pathogenic organisms. Per se this is not proof of the presence of such pathogens.

The last point should be emphasized. The presence of coliforms does not mean that there are pathogens in the water. It simply means that there *might* be. A high coliform count in water is suspicious. Although the water may in fact be perfectly safe to drink, it should not be.

There are two principal ways of measuring for coliforms. The simplest is to filter a sample through a sterile filter thus capturing any coliforms. The filter is then placed in a petri dish containing a sterile agar that soaks into the filter and promotes the growth of coliforms while inhibiting other organisms. After 24 or 48 hours of incubation the number of shiny dark blue/green dots, indicating coliform colonies, is counted. If it is known how many milliliters of water were

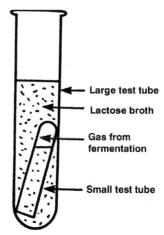

Figure 4.2. Test for measuring the presence of coliform organisms.

poured through the filter, the concentration of coliforms can be expressed as coliforms/100 mL.

The second method of measuring for coliforms is called the *most probable number* (MPN), a test based on the fact that in a lactose broth coliforms produce gas and make the broth cloudy. The production of gas is detected by placing a small tube upside down inside a larger tube (Figure 4.2), so as not to have air bubbles in the smaller tube. After incubation, if gas is produced, some of it will become trapped in the smaller tube, and this along with a cloudy broth will indicate that the tube had been inoculated with at least one coliform.

And here is the trouble. Theoretically, one coliform can cause a positive tube just as easily as a million coliforms can. Hence, it is not possible to ascertain the concentration of coliforms in the water sample from just one tube. The solution is to inoculate a series of tubes with various quantities of the sample, the reasoning being that a 10-mL sample would be 10 times more likely to contain a coliform than a 1-mL sample.

For example, using three different inoculation amounts, 10, 1, and 0.1 mL of sample, three tubes are inoculated with each amount, and after incubation an array might result as shown below. The plus signs indicate a positive test (cloudy broth with gas formation), and the minus signs represent tubes where no coliforms were found.

Amount of sample,	Tube Number		
mL put in test tube	1	2	3
10	+	+	+
1	−	+	−
0.1	−	−	+

The Walkerton, Ontario, *E.Coli* Outbreak

On May 19, 2000, the public health office at Walkerton, Ontario, received a call from the local hospital reporting two cases of severe diarrhea. The public health unit called the local water treatment plant and asked if the water supply was the cause, and was assured by the operators that all was well with the water. On the next day, the number of illnesses continued to rise, and the water treatment plant operators were again asked if the water quality was safe, and they assured the public health people that it was. On the third day, the first case of *E.coli 157* infection was confirmed, and the public health personnel determined that the only reasonable source had to be the drinking water.

It turned out that the chlorinator at the water treatment plant had been down for some time, and that the plant had stopped testing the water in their own labs. The independent laboratory that had responsibility for doing the bacteriological testing had informed the plant operators that excessive coliforms were in the water, a fact they would have known since the chlorinator was malfunctioning. Neither the laboratory nor the water treatment plant operators, however, contacted the public health agency to report the problem. As a result, there were 195 cases of *E.coli 157* infection and 118 cases of infection by other intestinal pathogens. Of those who were infected, 26 individuals developed Hemolytic Uremic Syndrome, a serious complication of the *E.coli 157* infection, resulting in swelling of the face and extremities, and abnormal kidney function, with some people requiring long-term dialysis.

The direct cause of the outbreak was the malfunctioning chlorinator, but the indirect cause was shown to be the 50% reduction in funding from the provincial government for water testing. As a result of the outbreak the water lines in Walkerton were flushed with high chlorine solutions, and some that could not be adequately cleaned had to be replaced—resulting in a cost far exceeding the estimated savings dictated by tax-cutting politicians.

Based on these data one would suspect that there is at least 1 coliform per 10 mL, but there still is no firm number. The solution to this dilemma lies in statistics. It can be proven statistically that such an array of positive and negative results will occur most probably if the coliform concentration is 75 coli/100 mL. A higher concentration would most probably result in more positive tubes, while a lower concentration would most probably result in more negative tubes. Thus 75 coli/100 mL is the most probable number (MPN).

4.2 WATER QUALITY STANDARDS

But what use is the quantitative analysis of water quality unless there exist some *standards* that describe the desired quality for various beneficial use of the water? There are, in fact, three main types of water quality standards.

The Minemata Mercury Disaster

After the Second World War Japan was rapidly industrializing and rebuilding its economic base. On the shores of Minemata Bay was the Shin Nitchitsu factory owned by Chisso Corporation. In the hospital owned and operated by the company, the doctors were intrigued by a strange new disease. They at first thought that the disease was contagious since often more than one person from a single household would have the disease, and at first placed the sick in isolation. After a while they had to conclude that this mystery disease was not contagious, but still had no idea what was causing the people to lose muscle control, to exhibit highly nervous symptoms, and eventually to become cripples. Epidemiological studies finally concluded that the disease was most likely to occur in people whose diet consisted mostly of the fish caught in the bay, and most likely it was some type of heavy metal contamination. One of the doctors at the hospital, Dr. Hosokawa Hajime began to suspect that the causative agent was mercury, and started to conduct tests using various animals to see if he could duplicate the symptoms. By 1962 he concluded that without doubt this strange new disease was mercury poisoning.

Dr. Hajime went to Shin Nitchutsu's chief engineer, Dr. Jun Ui, in charge of waste management and asked if mercury was being discharged into the bay. Dr. Ui discovered to his dismay that there were indeed direct discharges of mercury-laden wastes directly into the bay, and took this news to the plant management. He was promptly fired, and Dr. Hajime was pressured to retire. But Dr. Ui did not let the matter die and took the news of the disaster to the press.

The final toll of dead and permanently crippled people was staggering. Eventually the federal government built a special hospital to treat the victims. In the Japanese tradition, the head of the Shin Nitchitu plant took full responsibility for the disaster, but how do you say "I'm sorry" for having caused the disaster and then not having done anything about it for years after knowing what damage was being done?

Today, thousands of people in Japan are still suffering from the effects of mercury poisoning, now known as Minemata Disease. Many of the sufferers have been shunned under the false impression that the disease is contagious. Of the over 2000 people who lived around the bay and who were acknowledged victims, over half have died. Estimates are that over 20,000 others have suffered harm and need assistance. There is no cure for Minemata Disease.

The most important standards for protecting the American public from harm due to ingesting drinking water are the *U. S. EPA Drinking Water Standards,* as mandated by the Safe Drinking Water Act of 1974. Based on public health and epidemiological evidence, and tempered by a healthy dose of expediency, the EPA has established national drinking water standards for many physical, chemical and bacteriological contaminants.

One example of a physical standard is turbidity, or the interference with the passage of light. A water that has high turbidity is "cloudy," a condition caused by the presence of colloidal solids. Turbidity does not in itself cause a health problem, but the colloidal solids may prove to be convenient vehicles for pathogenic organisms. Additionally, water that is not clear is looked on with suspicion by the public (for good reason) and may force citizens to seek less safe or more expensive water sources.

The list of chemical standards is quite long and includes the usual inorganics (lead, arsenic, chromium, etc.) as well as some organics (e.g., DDT). Bacteriological standards for drinking water are written in terms of the coliform indicators.

Table 4.2 is a summary of some of the most important drinking water standards.

In addition to the *maximum contaminant level* (MCL) standards (often called *primary standards* since they reflect a concern with human health), there are other standards that are used to assure that the water is pleasing. For example, the level of chloride (salt) is set at 250 mg/L, not because water with higher chloride concentration is harmful, but because water with more than 250 mg/L tastes bad. Limits for iron and manganese similarly have no direct health benefit, but high iron and manganese concentrations stain white ceramics and clothing washed in that water.

The U. S. EPA drinking water standards have evolved in three ways over time: (1) they have become almost always stricter as new evidence linking drinking water and disease is discovered, (2) standards for more chemicals and other constituents have been added, again as the knowledge of disease transmission improves, and (3) the standards have become applicable to an increasing number of water supplies. An example of the first evolution has been the coliform standard, which in 1914 (the first Public Health Service drinking water standards) was set at 2 coli/100 mL. Most cities were producing water that had about 20 coli/100 mL, and therefore could not meet this standard. But others, by using filters and chlorine disinfection, were able to fall below the 2 coli/100 mL limit. By setting such a strict standard, the PHS was forcing technology on the cities. In 1925 the coliform standard was lowered to 1 coli/100 mL simply because most of the cities were able to meet this limit. The standard today is an impressive 0.05 coli/100 mL, calculated as not seeing any at all in a given set of samples.

The second change in the standards was in adding more constituents. An

TABLE 4.2. Highlights of the Safe Drinking Water Act.

Contaminant	Effects	MCL (mg/L)	SMCL (mg/L)	Source
PHYSICAL				
Turbidity	Interferes with Disinfection	0.5 NTU	<0.1 TU	discharge, erosion
Color	aesthetic		15 color units	dissolved organic matter
pH	corrosiveness		6.5–8.5	
Taste and Odor	aesthetics		3 threshold #	algae, H_2S
	0.25–1.0			
CHEMICAL				
Inorganic				
Total Dissolved Solids—Hardness	taste, limits soap effectiveness		500	groundwater
Lead	central nervous system, very toxic to pregnant women and infants	0.005	zero	leaches from lead pipes and solder from pipe joints
Copper	taste, staining porcelain	1.0	1.0	water pipes
Nitrate (as N)	methemoglobinemia = blue baby syndrome	10		fertilizer, feedlots, sewage
Fluoride	skeletal damage, dental fluorosis - brown discoloration	4.0	2.0	geological - groundwaterl
Iron	taste and staining of laundry, discolored water		0.3	groundwater
Manganese	aste and staining of laundry, discolored water		0.05	groundwater
Mercury	central nervous system, kidney	0.002		industrial, paint, paper, fungicides
Arsenic	central nervous system toxicity	0.05		pesticide residue, industrial waste

(continued)

134

TABLE 4.2 (continued). Highlights of the Safe Drinking Water Act.

Contaminant	MCL (mg/L)	SMCL (mg/L)	Effects	Source
CHEMICAL (*continued*)				
Organic				
Trihalomethanes (THMs)		500	carcinogens	chlorine reactions with natural organic matter
Benzene	0.005	zero	carcinogen	fuel, leaking underground storage tanks
2,4 D	0.1	1.0	Liver/kidney effects	herbicide used in agriculture, forests, and aqua.
BIOLOGICAL—pathogens				
Total Coliforms (indicator organism)	1 per 100 mL	2.0	Not necessarily disease producing themselves—used as indicator of org. that cause diseases such as dysentery, hepatitis, typhoid fever, cholera	human and animal contamination, untreated wastewater
Giardia lamblia	Surface Water Treatment Rule	0 cysts	extreme gastro—intestinal problems (days to months)—most common U.S. waterborne disease	human and animal contamination, beavers, muskrat
Cryptosporidium	Surface Water Treatment Rule		severe gastro-intestinal problems, can result in death of immuno-comprimised people	
Legionella	Surface Water Treatment Rule		pneumonia-like disease	human and animal contamination

MCL = maximum contaminant level (primary standard)
SMCL = secondary maximum contaminant level (secondary standard)

135

The Origin of the National Drinking Water Quality Standards

In the early 1900s, the death rate from typhoid, cholera, and other waterborne diseases was high in many American cities, partly from poor sanitation and contaminated milk and other food supplies, but almost certainly from contaminated drinking water. The city of Pittsburgh, for example, took water from the Allegheny River at locations that were immediately downstream from wastewater discharges. More wealthy people used bottled water and avoided this hazard, but the poor had little choice but to drink contaminated water

At this time, the Constitution of the United States was interpreted very narrowly, and any prospect of the federal government getting involved in setting and enforcing national drinking water standards was highly unlikely. Public health leaders within the U. S. Public Health Service (PHS), at that time an agency of the U. S. Department of Treasury, decided to try to improve the drinking water quality by implementing the interstate commerce clause which gives the federal government the right to facilitate interstate commerce. At that time the major interstate transportation was trains, and having trains stop at a community was a major economic advantage to the community. The scientists, engineers, and physicians at the PHS decided to develop drinking water quality standards that would be applicable only to communities where the trains stopped to take on water. The argument was that such water should be safe for interstate travelers. The public position of the PHS was that these standards, which for the first time used coliform organisms as a bacteriological indicator, were not intended to be national standards. The only club the PHS had was that if a community water supply did not meet these standards, the trains would not stop. But this was a very big club. As one chemist from the Iowa State Board of Health noted, "the effect of the posting of notices in (railroad) stations and the condemnation of local waters has been to force the standard upon the local plants through the action of governmental prestige and public opinion." And indeed, even communities where the railways did not stop had an incentive to clean up their drinking water supplies. The effect nationally was to reduce the death rate from typhoid and cholera to almost zero by 1940.

The history of drinking water standards in the United is summarized below:

Act	Agency	Year	Description
U.S. Treasury Stds	PHS	1914	Bacteria Plate count < 100 colonies per mL, MPN < 2
	PHS	1925	B. coli MPN < 1; copper, lead, zinc limits; color, taste and odor recommendations
	PHS	1942	Total coliform < 1; aesthetics requirements added
		1962	color and turbidity requirements added
Safe Drinking Water Act	EPA	1974	see previous page
SDWA	EPA	1996	Updated

136

example is hexavalent chromium, which was not known to be toxic when the first PHS standards were written. The distinction between trivalent chromium, CrIII, which is fairly benign and can even be thought of as a health supplement, and hexavalent chromium, CrIV, a strong toxin and carcinogen, was simply not known. And finally, as the states implemented the U. S. EPA drinking water standards, they applied them to smaller and smaller water supplies. In North Carolina, for example, any well serving more than ten households is considered a public water supply and subject to the U. S. EPA drinking water standards, resulting in expensive testing and monitoring.

ELLEN SWALLOW RICHARDS

Ellen Swallow was born into an age when the opportunities for women in higher learning were limited, but this did not prevent her from graduating from Vassar College in 1868 where she studied chemistry and biology. Because many believed that women could not get a college education without injuring their health, pioneers like Ellen Swallow had to do more than the men to prove herself, and she did. At one point she wrote home to her parents: "The only trouble here is that they won't let us study enough."

Upon graduation, she got herself admitted to MIT as an "experiment," and was allowed to work under the watchful eye of the only other woman on the staff. Her even demeanor and gentle helpfulness soon won over the faculty and she was awarded the first bachelor's degree from MIT given to a woman. She had continued to study at Vassar, and in the same year as her BS degree from MIT she received a Master of Science from Vassar. A few years later she married an MIT colleague, Robert Richards.

Ellen Swallow Richards was never paid a salary by MIT, but earned her own income through teaching and consulting projects. The students who went though her laboratory and learned analytical techniques of water chemistry reads like a who's who in early public health engineering. She herself became an expert in the chemical and biological quality of water and in the late 1880s conducted the first thorough stream survey for the State of Massachusetts, a survey that became a template for all of the stream quality projects for the other states.

She believed in including environmental questions in decision-making and found the public's apathy toward environmental destruction intolerable. Her world view and personal ethical outlook influenced a generation of public health professionals. At one point, she wrote:

"One of the most difficult lessons to learn is that our tolerance of evil conditions is not proof that the conditions are not evil."

4.3 WATER TREATMENT

While most groundwater sources for human use are useful and safe for human consumption without treatment, this is seldom the case with surface water sources. In almost all cases, water found in nature is unfit for direct consumption and has to be treated in a water treatment plant. A typical water treatment plant is diagrammed in Figure 4.3. Such plants are made up of a series of reactors or unit operations, with the water flowing from one to the next, and when stacked in series, achieve a desired end-product. Each operation is designed to perform a specific function, and the order of these operations is important. Described below are a number of the most important of these processes.

4.3.1 Coagulation and Flocculation

Raw (untreated) surface water entering a water treatment plant usually has significant turbidity caused by tiny colloidal clay and silt particles. These particles have a natural electrostatic negative charge which keeps them continually in motion and prevents them from colliding and sticking together. Chemicals such as alum (aluminum sulfate) are added to the water, first to neutralize the charge on the particles and then to aid in making the tiny particles "sticky," so they can coalesce and form large particles. The purpose is to clear the water of suspended colloidal solids by building larger particles from the stable colloidal solids, so that these larger and heavier particles could be readily settled out of the water.

Coagulation is the chemical alteration of the colloidal particles to make them stick together forming larger particles called *flocs*. When aluminum sulfate, $Al_2(SO_4)_3$, is added to water containing colloidal material, the alum initially dissolves to form an aluminum ion, Al^{+++}, and a sulfate ion, SO_4^-. But the aluminum ion is unstable and forms various types of charged species of aluminum oxides and hydroxides. The specific forms of these compounds is dependent on the pH of the water, the temperature, and the method of mixing.

Two mechanisms are thought to be important in the process of coagulation. The first is termed *charge neutralization* where the aluminum ions are used to counter the charges on the colloidal particles, pictured in Figure 4.4. The colloidal particles in natural waters are commonly negatively charged, and when suspended in water, repel each other due to their like charges. This causes the suspension to be stable and prevents the particles from settling out. If aluminum ions are added (regardless of what forms they may assume), some of them have a trivalent positive charge. As these are drawn to the negatively charged colloidal particles, they compress the net negative charge on the particles, making them less stable in terms of their charges. Such reduction in

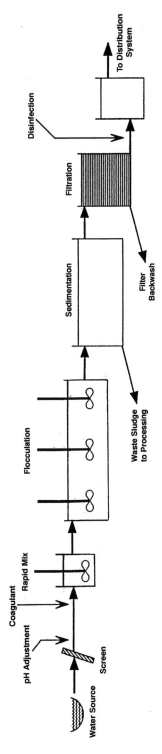

Figure 4.3. Typical water treatment plant.

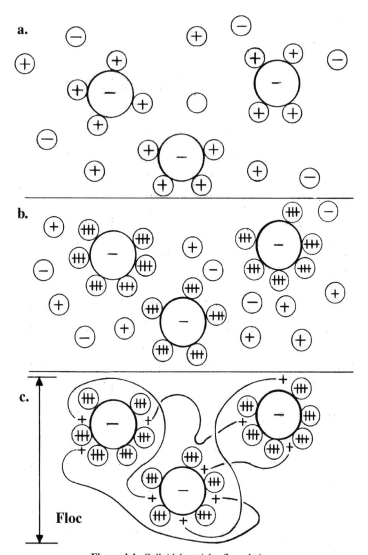

Figure 4.4. Colloidal particles flocculating.

negative charge makes the particles more likely to stick to each other if they should collide.

A second coagulation mechanism is termed *bridging* and involves the sticking together of the colloidal particles by virtue of macromolecules (polymers) formed by the aluminum hydroxides, as illustrated in Figure 4.4(c). The molecules have positive charge sites, by which the polymers attach themselves to various colloids, and then bridge the gap between adjacent particles, thereby creating larger particles. Because many of the desirable forms of aluminum hydroxides dissolve at low pH, lime [$Ca(OH)_2$] is often added to raise the pH. Some of the calcium precipitates as calcium carbonate [$CaCO_3$], assisting in the settling.

Both of these mechanisms are important in coagulation with alum. The net effect is that the colloidal particles are destabilized and have the propensity to grow into larger particles. The assistance in the growth of these larger particles is a physical process known as *flocculation.*

In order for particles to come together and stick to each other, either through charge neutralization or bridging, they have to move at different velocities. Consider for a moment the movement of cars on a highway. If all of the cars moved at exactly the same velocity, it would be impossible to have car-to-car collisions. Only if the cars have different velocities (speed and direction), with some cars catching up to others, would accidents be possible. The intent of the process of flocculation is to produce differential velocities within the water, so that the particles can come into contact. Commonly, this is accomplished in a water treatment plant by simply using a large slow-speed paddle that gently stirs the chemically treated water and produces the large particles of destabilized colloidal solids and aluminum hydroxides. Figure 4.5 shows a typical flocculator with slowly rotating paddles.

Once the flocculation is complete and the larger particles or flocs have been formed, the next step is to remove them using the process of settling.

4.3.2 Settling

When the flocs have been formed they must be separated from the water. This is invariably done in *gravity settling tanks* that simply allow the heavier-than-water particles to settle to the bottom. Settling tanks are designed

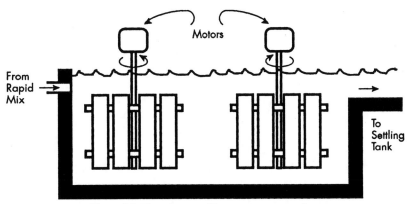

Figure 4.5. Flocculator used in a water treatment plant.

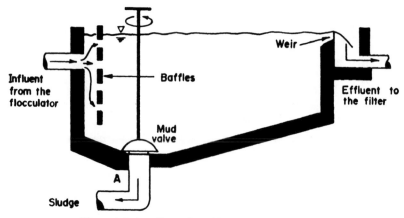

Figure 4.6. A settling tank used in a water treatment plant.

so as to approximate a plug flow reactor. That is, the intent is to minimize all turbulence. Figure 4.6 shows a typical settling tank used in water treatment. The water from the flocculation tank enters through baffles that are designed to slow down the flow, dissipate the energy, and spread out the flow so that the water moves through the settling tank as it would through a plug flow reactor.

Settling tanks work because the density of the solids exceeds that of the liquid. The movement of a solid particle through a fluid under the pull of gravity is governed by a number of variables, including

- particle size (volume)
- particle shape
- particle density
- fluid density
- fluid viscosity

The latter term may be unfamiliar, but it refers simply to the ability of the fluid to flow. Pancake syrup, for example, has a high viscosity, while water has a relatively low viscosity.

In settling tanks, it is advantageous to get particles to settle at the highest velocity. This requires large particle volumes, compact shapes, high particle and low fluid densities, and low viscosities. In practical terms, it is not feasible to control the last three variables, but coagulation and flocculation certainly results in the growth of particles and changes their density and shape.

The reason why coagulation/flocculation is so important in preparing particles for the settling tank is shown by some typical settling rates in Table 4.3. Although the settling velocity of particles in a fluid is also dependent on the particle shape and density, these numbers illustrate that even small changes in particle size for typical flocculated solids in water treatment can dramatically affect the efficiency of removal by settling.

TABLE 4.3. Typical Settling Rates.

Particle Diameter, mm	Typical Particle	Settling Velocity, m/s
1.0	Sand	0.2 m/s
0.1	Fine Sand	0.01 m/s
0.01	Silt	0.0001 m/s
0.001	Clay	0.000001 m/s

Settling tanks can be analyzed by assuming an *ideal tank*. Such ideal settling tanks can be thought of as plug flow reactors with uniform flow (the flow velocity does not vary with location), so that as the water enters the tank it has a velocity that is maintained all the way through the tank. This can be visualized as an imaginary column (a transparent cylinder) moving through the tank at some velocity v, with no intermixing (Figure 4.7). If a solid particle enters the tank at the top of the column and settles at a velocity of v_o, it should have settled to the bottom of the imaginary column right before the water exits the tank.

Several assumptions are required in the analysis of ideal settling tanks:

- Uniform flow occurs within the settling tank. All of the water flows horizontally through the tank at the same velocity.
- All particles settling to the bottom of the tank are removed.
- Particles are evenly distributed in the flow as they enter the settling tank.
- All particles still suspended in the water when the imaginary column of water reaches the far side of the tank are not removed and escape the tank.

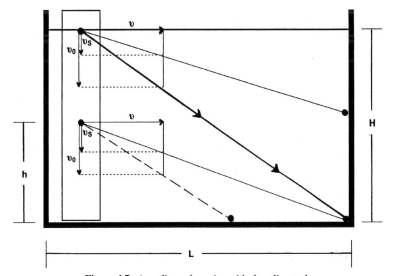

Figure 4.7. A settling column in an ideal settling tank.

Consider now a particle entering the settling tank at the water surface. This particle has a settling velocity of v_o and a horizontal velocity v such that its trajectory is as shown in Figure 4.7. In other words, the particle is just barely removed (it hits the bottom at the last instant). The velocity of this particle, v_o, is known as *critical settling velocity*. All particles entering this tank that have velocities of v_o or greater will be 100% removed in the settling tank. Some particles having settling velocities less than v_o will also be removed, depending on where they entered the tank.

With reference to Figure 4.7, if a particle enters the settling tank at some height h and has a settling velocity of v_s, its trajectory carries it to the bottom. If this particle enters the tank at a location higher than h, it will not be removed since this trajectory will carry it to the end of the tank without hitting the bottom. Since the particles entering the settling tank are assumed to be equally distributed, the proportion of those particles with a velocity of v_s removed is equal to h/H, where H is the height of the settling tank.

An analysis of settling within this ideal tank requires the understanding of two simple ideas. The first, the hydraulic retention time, defined previously in Chapter 2 as

$$\bar{t} = \frac{V}{Q} \tag{4.1}$$

or the volume V divided by the flow rate Q. The second is the *continuity equation*, also introduced in Chapter 2. The continuity equation expresses the fact that if a flow rate Q flows through any opening of area A the water will have a velocity v so that $Q = Av$.

The volume of rectangular tanks is calculated as $V = HLW$, where $W =$ the width of tank. Thinking of the tank as a short, fat pipe, the area through which the flow Q occurs is height times width, or $H \times W$, and v is the horizontal velocity through the tank. Substituting into Equation (4.1),

$$\bar{t} = \frac{V}{Q} = \frac{H \times W \times L}{H \times W \times v} = \frac{L}{v}$$

This equation says that the hydraulic retention time is also the time necessary for a particle in the influent to travel horizontally to the far end of the tank, which makes sense.

From Figure 4.7, the velocity of the critical particle is v_o and it takes this particle $\bar{t}$ time to reach the bottom of the tank, which is H high. In equation form,

$$v_o = \frac{H}{\bar{t}} \tag{4.2}$$

And substituting the definition of hydraulic retention time,

$$v_o = \frac{H}{t} = \frac{H}{V/Q} = \frac{H}{(HWL)/Q} = \frac{Q}{WL} = \frac{\text{flow rate}}{\text{surface area}} \tag{4.3}$$

The equation represents an important design parameter for settling tanks called the *overflow rate*. Note the units:

$$Q/(WL) = (m^3/s)/(m)(m) = m/s$$

which is velocity. So the overflow rate has the same units as velocity, but commonly the overflow rate is expressed as $m^3/day\text{-}m^2$. It is, however, a velocity term, and is in fact equal to the critical velocity for that tank. When the design of a clarifier is specified by overflow rate, what is really defined is the critical settling velocity.

Example 4.1

A water treatment plant settling tank has an overflow rate of 60 $m^3/day\text{-}m^2$ and a depth of 3 m. What is its retention time?

$$v_o = 60 \ m^3/d\text{-}m^2 = 60 \ m/d$$

$$t = \frac{H}{v_o} = \frac{3}{60} = 0.05 \text{ days} = 1.2 \text{ hours}$$

The overflow rate equation, $v_o = Q/(WL)$, is interesting in that a better understanding of settling can be obtained by looking at individual variables. For example, increasing the flow rate, Q, in a given tank increases the v_o; i.e., the critical velocity increases and thus fewer particles are totally removed since fewer particles have a settling velocity greater than v_o. It would, of course, be advantageous to decrease v_o, so more particles can be removed. This is done by either reducing the flow rate Q, or increasing the surface area WL. The latter term may be increased by changing the dimensions of the tank so that the depth is shallow, and the length and width are very large. But the shallowness of tanks is limited by problems of even distribution of flow, and the great expense in concrete and steel. Secondly, as particles settle they can flocculate, or bump into other slower moving particles and stick together, creating higher settling velocities and enhancing solids removal. As noted above, as the particles grow in size, they settle faster. Settling tank depth then is an important practical consideration, and typically tanks are built 3 to 4 m deep in order to take advantage of the natural flocculation that occurs during settling.

Example 4.2

A small water plant has a raw water inflow rate of 0.6 m^3/s. Laboratory studies have shown that the flocculated slurry can be expected to have a uniform particle size (only one size) and it has been found, through experimentation, that all of the particles settle at a rate of $v_s = 0.004$ m/s. (This is a pretty dumb assumption, of course.) A proposed rectangular settling tank has an effective settling zone of $L = 20$ m, $H = 3$ m and $W = 6$ m. Could 100% removal be expected? Remember that the overflow rate is actually the critical settling velocity. The critical settling velocity can be calculated as

$$v_o = Q/A = 0.6/(20 \times 6) = 0.005 \text{ m/s}$$

The critical settling velocity is greater than the settling velocity of the particle to be settled out, hence not all the incoming particles will be removed.

If the settling tank works well, the water leaving a settling tank is essentially clear. It is not, however, acceptable for domestic consumption and one more polishing step is necessary, usually using a rapid sand filter.

4.3.3 Sand Filtration

In the discussion of groundwater quality, we note that the movement of water through soil removes many of the contaminants in water. The soil particles help filter the groundwater, and through the years environmental engineers have learned to apply this natural process to water treatment and supply systems and developed what is now known as the *rapid sand filter*. The operation of a rapid sand filter involves two phases: filtration and backwashing.

A slightly simplified version of the rapid sand filter is illustrated in a cut-away drawing in Figure 4.8. Water from the settling basins enters the filter and seeps through the sand and gravel bed, through a false floor and out into a clear well that stores the finished water. During filtration valves *A* and *C* are open.

Rapid sand filters are designed on the basis of filter loading, expressed as the flow of water moving downward through the sand divided by the surface area of the sand, or

$$F = \frac{Q}{A} \tag{4.4}$$

where

F = filter loading, m^3 water/hour per filter area in m^2
Q = flow to the filter, m^3/hour
A = surface area of the filter, m^2

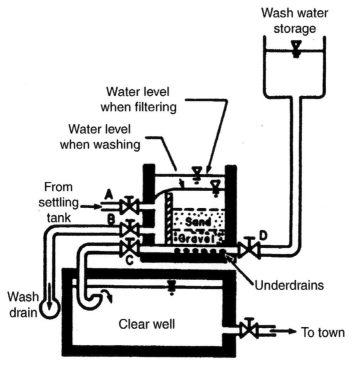

Figure 4.8. Rapid sand filter.

A typical filter loading used in municipal water treatment plants is about 10 m³/hour per m² of filter area.

The suspended solids that escape the flocculation and settling steps are caught on the filter sand particles and eventually the rapid sand filter becomes clogged and must be cleaned. This cleaning is performed hydraulically using a process called *backwashing*. Referring to Figure 4.8, the operator first shuts off the flow of water to the filter (closing valves A and C), then opens valves D and B which allow wash water (clean water stored in an elevated tank or pumped from the clear well) to enter below the filter bed. This rush of water forces the sand and gravel bed to expand and jolts individual sand particles into motion, rubbing them against their neighbors. The suspended solids trapped within the filter are released and escape with the wash water. After a few minutes, the wash water is shut off and filtration is resumed.

4.3.4 Membrane Filtration

Another type of filtration system becoming more common in drinking water

The Hamburg and Altona Cholera Epidemic

The most convincing proof of the effectiveness of water filtration was provided in 1892 by an experience gained in two neighboring cities in northern Germany, Hamburg and Altona, both of which drew their drinking water from the River Elbe. Hamburg was upstream from Altona and discharged all of its untreated wastewater into the Elbe.

Altona had decided to build sand filters to treat the drinking water, while Hamburg simply drew the water untreated from the river. When the river become infected from a camp of immigrants upstream of Hamburg, the city suffered from a cholera epidemic that infected one in thirty of its population and caused more then 7500 deaths. Altona, which ought to have had an even greater outbreak of cholera since it was using the water contaminated by the people living in Hamburg, escaped the epidemic almost unscathed. The filters proved to be effective in removing the infecting agents.

The Milwaukee *Cryptosporidium* Outbreak

In the spring of 1993, an estimated 40,000 people in Milwaukee, Wisconsin, become ill with diarrhea, abdominal cramping, nausea, vomiting, and fever. Of those, 50 people who were already ill and infirm and whose immune systems were not strong, died from the outbreak. The problem was clearly the water supply, but testing showed that the chlorine levels in the drinking water were at or above the minimums necessary to prevent disease transmission. The intestinal pathogen was soon identified as *Cryptosporidium*, a parasite that is particularly resistant to destruction by chlorine. If the water was tested for coliform organisms, it would have passed, and the presence of *Cryptosporidium* would never have been detected since it is more resistant to chlorine than coliforms.

Public health officials knew that the people who got ill from *Cryptosporidium* must have got the disease from drinking water, but by the time the officials figured this out the contaminated water had long passed through the distribution system. But they were lucky. An ice making house in town had used city water to make large blocks of ice which they were planning to sell as crushed ice in bags and these blocks were dated and stored in the ice house. The officials were able to chip away large chunks of these ice blocks and melt them down. And sure enough, many of the samples from the days when the people should have become infected contained *Cryptosporidium*. Mystery solved!

But they still did not know how these organisms got through the water treatment system. The disinfection was working properly, so what happened? The answer came when it was discovered that for some days the plant had had difficulty with its sand filters and some of them ran for many days without being washed. The *Cryptosporidum* had gotten through the dirty filters, and being more resistant to disinfectants than coliforms, were not killed with chlorination.

The Milwaukee outbreak sent a strong message to the water industry to make sure the rapid sand filters are properly operated. Moreover, we cannot rely solely on chlorination to prevent public health problems from occurring.

treatment is *membrane filtration*. Membranes have been used for many years for specialty applications, but their relatively high cost has limited their use in large-scale drinking water treatment systems. Recent advances in membrane production technology have reduced costs, making membrane treatment a more economical choice. Membrane filtration uses a permeable membrane to remove the contaminants of interest by a sieving mechanism. Water is forced through the membrane under pressure, and depending on the pore size of the membrane, different-sized particles and even ions and chemicals can be removed. A schematic of a membrane process is shown in Figure 4.9. Many membrane systems use a hollow-fiber membrane, similar to small pieces of tubing shown in Figure 4.9, that are encased in a hollow cylinder to create a filter module. In this system, water is forced into the filter module under pressure and a portion of it passes through the membrane wall into the hollow part of the membrane tube. The water that passes through the membrane is often called the *permeate* or *filtrate*, and this is the clean water. All the chemicals and particles that could not fit through the pores of the membrane are retained, and this flow is called the *retentate*. The retentate flow contains a concentrated solution of the incoming pollutants such as salts and particles. As a result, it is often called "brine". Brine can present a disposal issue because of the concentration of different compounds found in it. The flow pattern depicted in Figure 4.9 shows an "inside-out" flow pattern with the influent entering the inside of the membrane and flowing to the outside. Many systems have the opposite flow pattern, or an "outside-in" flow where flow is forced from the outside of the membrane tube or fiber into the hollow center. The filtrate then flows through the tube to the exit point.

One advantage of membrane filtration is the relatively well-defined pore sizes for different membrane types, thereby assuring removal of particles or compounds of interest. In addition, plants using membrane filtration are generally easier to operate. The disadvantages of membrane filtration systems are their relatively high cost and their propensity for clogging. In addition,

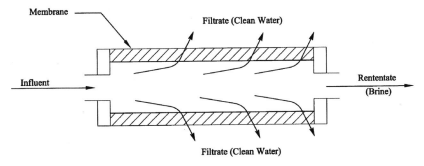

Figure 4.9. Membrane filter.

during the production of filtered water, a brine or waste flow is generated that has concentrated chemicals and contaminants. Disposal of this flow is therefore necessary.

Typically, membrane filtration systems are characterized based on the size of the particles that they are removing. The main categories of membranes in order of smaller particle removal are microfiltration, ultrafiltration, nanofiltration, and reverse osmosis.

Microfiltration membranes typically remove particles that are approximately 0.1 micron and larger (1 micron or μm is 10^{-6} m). Therefore, microfiltration is good for removing most bacteria as well as *Cryptosporidium* and *Giardia* as well as for clarifying the water. However, microfiltration will not remove viruses or dissolved chemicals in the water.

Ultrafiltration membranes has a typical particle size removal around 0.01 μm. As a result, ultrafiltration can remove most viruses as well as bacteria and protozoa. Also, ultrafiltration is able to remove larger macromolecules such as large proteins and polysaccharides, but is generally unable to remove chemicals such as pesticides and other compounds of similar size.

Nanofiltration is also referred to as low-pressure reverse osmosis. These membranes are not quite as dense as reverse osmosis membranes, and therefore can operate at lower pressures. Nanofiltration membranes can remove some larger molecules, and certainly viruses, bacteria, and protozoa. They can also capture divalent ions such as calcium (Ca^{2+}), but cannot capture monovalent ions such as sodium (Na^{+}). As a result, nanofiltration can be used for water softening (removal of multivalent cations such as Ca^{2+} and Mg^{2+}).

Reverse osmosis membranes are a dense membrane that retains almost all ions and low molecular weight compounds and larger. This treatment is the best filtration system for removal of the most compounds and pathogens from water. Reverse osmosis can be used for the desalinization of salt water, and has been used around the world for this application.

As the size of the particles being removed decreases, the pressure required to force the water through the membrane increase. Reverse osmosis requires high pressure to pump the water through the membrane compared to the other, larger pore sized membranes and therefore pretreatment by conventional drinking water treatment processes such as coagulation, flocculation, sedimentation, and filtration is required in order to prevent clogging.

4.3.5 Disinfection

Following filtration, and before storage in the clear well, the water produced in a water treatment plant is disinfected in order to destroy whatever pathogenic organisms might remain. Commonly, disinfection is accomplished using chlorine, which is purchased as a liquid under pressure, and released into the water as chlorine gas using a chlorine feeder system. The dissolved chlorine

---

Desalinization at Tampa Bay, Florida

In an earlier box the problems of the Tampa Bay Water district are discussed with regard to their "stealing" groundwater. Given the circumstances, and the continued demand for water by the growing population, Tampa Bay Water had only one choice left for providing water to the citizens of the bay area—desalinization. They chose membrane filters to do the job.

The Tampa Bay plant produces about 25 million gallons per day of drinking water and at the same time it also discharges 19 million gallons per day of concentrated seawater into the ocean, after mixing with cooling water from a power plant. The cost of water from this facility is listed at $2.08 per 1000 gallons, which is cheaper than water produced at most conventional drinking water plants and is reportedly the lowest cost of desalinization water in the world. It seems likely that increasing pressures on fresh water supplies will result in a steady increase in reverse osmosis systems such that eventually the huge supply of sea water can be used for drinking water purposes.

---

oxidizes organic material, including pathogenic organisms. The presence of a residual of active chlorine in the water is an indication that no further organics remain to be oxidized and that the water can be assumed to be free of disease-causing organisms. Water pumped into distribution systems usually contains a residual of chlorine in order to guard against any contamination in the distribution system. It is for this reason that water from drinking fountains or faucets has a slight taste of chlorine.

Chlorine has been used as a disinfectant in drinking water for almost a hundred years, and there is no evidence that it is harmful to the people who drink the disinfected water. There is, however, plenty of evidence of public health problems when water is not adequately disinfected.

But chlorine is not without problems. When the chlorine in drinking water comes into contact with organic molecules in the water, some of the chlorine forms methylated chlorine compounds such as trichloromethane, $CHCl_3$, which has been shown to be a human carcinogen. There again is no evidence that these very low concentrations of trichloromethane have been a public health problem, but the very presence of a carcinogen in drinking water, especially if it is made during the treatment of the water, is disconcerting to the public. The drinking water industry has developed many ways of reducing the amount of $CHCl_3$ produced during water treatment and the quantities consumed by the public are far below the accepted threshold.

And yet, with all that assurance, one would ask if there might not be a better way of disinfecting drinking water. The alternative is to use UV radiation, a very effective means of disinfection that leaves no taste or odor residual and requires no additional chemicals. The problem with UV is that it does not leave a residual in the waterpipes and thus if there is a break in the line that allows contaminated water to enter (a cross connection), there is no chemical in the water to take care of the pathogens that might be in the water. The decision as to method of disinfection is a public decision, and there is no single right answer that would be true for all water treatment plants.

4.3.6 Fluoridation

In the 1950s doctors in Texas noticed that children in what appeared to be similar communities had widely different rates of dental decay. The investigation showed that the difference could be attributed to the amount of fluoride that naturally was found in the public drinking water supplies. This was quite a surprise since fluoride is generally considered to be a poison. Stannous fluoride, for example, is a common rat poison. But in these communities, the kids drinking water that had less than about 0.5 mg/L fluoride, the rate of decay was high, while kids who drank water in the range of

The Piura, Peru, Cholera Epidemic

Threats to public health have to be balanced, and public policies promulgated to promote the greatest benefit. A classical example of where this balance was inappropriately managed is illustrated by the Piura, Peru, cholera epidemic of 1991.

Piura, Peru, is a medium-sized city (approx. 300,000 inhabitants) in the Andes that boasts a public water supply and a water treatment plant. Operators and managers of the system (it is unclear exactly how this decision was reached) decided that there was a grave threat to the public from the toxic byproducts of chlorination, particularly halogenated methanes (such as trichloromethane) which are known to be carcinogenic. This threat caused the municipality to turn off its chlorinators and pump undisinfected water into the municipal water system. The pipes in the city were old and poorly maintained, with numerous illegal connections and pipelines that resulted in cross connections. As a result, the drinking water supplied to the community became contaminated, and since there was no residual chlorine in the water, pathogens were widely distributed.

In late January of 1991, a cholera epidemic occurred, with over 8000 cases and 17 deaths. The route of transmission was traced to the water supply, including the drinking of beverages made with contaminated ice. The endemic nature of cholera in the city and the lack of effective wastewater collection had also caused the aquifer under the city to be contaminated, and the water from private wells was also found to contain *Vibrio cholerae*, the causative microorganism.

Placentina Bay Fluoride Disaster

In 1969, an industrial plant manufacturing elemental phosphorus for metal finishing accidentally discharged 22,800 pounds of fluoride in the form of hydrofluosilicic acid into the Placentina Bay in Newfoundland, Canada. The spill devastated the ecosystem in the bay, causing what some observers called a "biological desert." This form of fluoride is precisely that used in the fluoridation of drinking water.

1.0 to 2.0 mg/L had almost no tooth decay. But if the fluoride level was very high, such as 5 mg/L, the teeth were mottled—soft and discolored. These findings prompted public health officials to recommend the addition of fluoride to public drinking water in order to reduce the rate of dental decay in children and teenagers. This widely accepted public health measure is not without its detractors, and there are still ethical questions relating to forcing people to drink water with a known poison in it just to reduce dental cavities in kids.

4.4 DISTRIBUTION OF WATER

From the clear well in the water treatment plant the finished water, which is now both safe and pleasing to drink, is pumped into the distribution system.

Is Fluoridation of Drinking Water a Good Thing?

Fluorides are effective in dramatically reducing tooth decay, particularly for teenagers and young adults. But are there negative effects of fluoridation? What else do the fluorides added to drinking water do? Here are some disquieting facts:

1. The presence of fluoride reduces the synthesis of collagen resulting in destruction to various organs such as lungs and kidneys.
2. Fluoride disrupts the ability of white blood cells to fight off infection.
3. Fluoride increases the growth rate of cancerous tumors, especially bone cancer.
4. The formation of antibodies in blood is inhibited by the presence of fluoride.
5. Fluoride inhibits thyroid activity.
6. Fluoride causes premature aging.
7. Fluoride diminishes intelligence and the development of the human brain.
8. Fluorides (e.g. stannous fluoride) are effective rat poisons.

Given the negative effects, is the reduction in dental cavities in children worth the risk? This is a public policy question.

Such systems are under pressure, so that any tap into a pipe, whether it be a fire hydrant or domestic service, will yield water.

Because the demand for finished water varies with day of the week and hour of the day, storage facilities must be used in the distribution system. Most communities have an elevated storage tank (water tower) which is filled during the periods of low water demand. These tanks then supply water to the distribution system during periods of high demand. Distribution systems also include fire hydrants that are used for emergencies.

The essential elements of a *water distribution system* are shown in Figure 4.10. The water from the clear well at the water treatment plant is pumped to the pipes that distribute the water to the individual users. These pipes are always under pressure and thus any break in a pipe will result in a loss of water (and potential contamination). Generally at the far end of the distribution system is a water tower. The purpose of the water tower is severalfold:

- provide emergency storage in case of a fire or break in the pipes
- even out the supply to the individual water users
- maintain a constant pressure within the pipe system.

During the low water use periods such as the early morning hours in most communities, the pumps at the water pump are filling up the water tower. When the community requires more water than the pumps at the water treatment plant can provide, the water flows from the tower to the community and thus provides a sufficient flow at a required pressure.

Urban Water Supplies

In ancient times the very survival of cities depended on having a safe and secure water supply. As cities grew, the demand for water became critical, and canals and aqueducts were constructed to convey water over long distances. The most famous series of aqueducts was constructed by the Romans, as detailed in *De Aquis Urbis Romae*, written in 97 AD by the engineer Frontinus. At that time, with over a hundred years of development, Rome boasted several aqueducts with a total length exceeding 360 km. The canals were built of stone and lined with cement, and the water was distributed in lead pipes. (Historians have speculated that this may have been one cause of the fall of the Roman civilization—lead poisoning—but this is unlikely since water pipe develops a coating of calcium carbonate on the inside, and the water no longer comes into direct contact with the lead.)

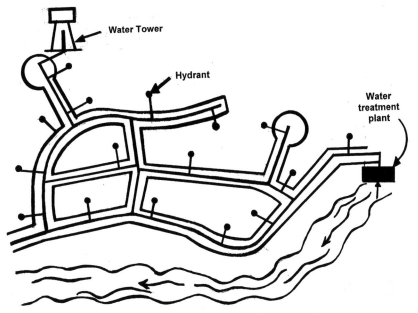

Figure 4.10. Water distribution system for a community.

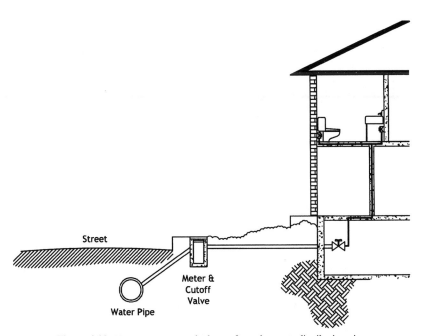

Figure 4.11. How water gets to the house from the water distribution pipes.

155

Communities go to great lengths to make sure drinking water distribution systems keep the water safe to drink, but sometimes an amazing coincidence of situations can occur that will result in contaminated water causing illness. One such example is the Holy Cross College football incident.

At the household level, the distribution pipes are tapped and the water enters the home through a water meter that measures the quantity of water used by the customer. After passing the meter, the water is distributed throughout the house as shown in Figure 4.11. The main shutoff valve right after the waterpipe enters the house is useful in case of an emergency. From there the water is distributed to the sinks, showers, and toilets.

The Holy Cross College Hepatitis Outbreak

Following the Dartmouth game, the first members of the 1969 Holy Cross College football team got sick. They had high fever, anorexia, nausea, abdominal pain, and were becoming jaundiced—all characteristics of infectious hepatitis. Over the next few days over 87 members of the football program, players, coaches, trainers and other personnel, became ill. The college cancelled the remainder of the football season, and became the focus of an epidemiological mystery. How could an entire football team have contracted infectious hepatitis?

The disease is thought to be transmitted mostly from person to person, but epidemics can also occur, often due to contaminated seafood or water supplies. There are several types of hepatitis virus, with widely ranging effects on humans. The least deadly is Hepatitis A virus which results in several weeks of feeling poorly and seldom has lasting effects, while Hepatitis B and C can result in severe problems, especially liver damage, and can last for many years. At the time of the Holy Cross epidemic the hepatitis virus had not been isolated and little was known of its etiology or effects.

When the college became aware of the seriousness of the epidemic, it asked for and received help from state and federal agencies, who sent epidemiologists to Worcester. The epidemiologists' first job was to amass as much information as possible about the members of the football team—where they were, what they ate, what they drank, and who they were with. The objective was then to deduce from the clues how the epidemic had occurred. Some of the things they knew of or found out were:

- Since the incubation period of hepatitis is about 25 days, the infection had to have occurred sometime before 29 August or thereabouts.
- Football players who left the team before 29 August were not infected.
- Varsity players who arrived late, after 29 August, similarly were not infected.
- Freshman football players arrived on 3 September, and none of them got the disease.
- Both the freshmen and varsity players used the same dining facilities, and since none of the freshmen became infected, it was unlikely that the dining facilities were to blame.

(continued)

(continued)

- There was no common thread of the players having eaten at restaurants where contaminated shellfish might have been the source of the virus.
- A trainer who developed hepatitis did not eat in the dining room.
- The kitchen prepared a concoction of sugar, honey, ice and water for the team (the Holy Cross version of Gator Ade), but since the kitchen staff sampled this drink before and after going to the practice field, and subsequently none of the kitchen staff developed hepatitis, it could not have been the drink.

The absence of alternatives forced the epidemiologists to focus on the water supply. The college receives its water from the city of Worcester, and buried lines provide water from Dutton Street, a dead-end street, to the football practice field where a drinking fountain is used during practices. Samples of water taken from that fountain however showed no contamination. The absence of contamination during the sampling did not, however, rule out the possibility of diseasetransmission through this water line, which ran to the practice field through a meter pit and series of sunken sprinkler boxes used for watering the field.

Two other bits of information turned out to be crucial. One of the houses on Dutton Street was found to have kids who had hepatitis. The kids also played near the sprinkler boxes during the summer evenings and often opened them, splashing around in the small ponds created in the pits. But how did the water in the play ponds, if it had been contaminated by the children, get into the water line, since the line was always under positive pressure?

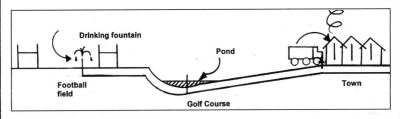

The final piece of the puzzle fell into place when the epidemiologists found that a large fire had occurred in Worcester during the evening of 28 August lasting well into the early hours of the next day. The demand for water for this fire was so great that the residences on Dutton Street found themselves without any water pressure at all. That is, the pumpers putting out the fire were pumping at such a high rate that the pressure in the water line became negative. If then the children had left some of the valves in the sprinkler boxes open, and if they had contaminated the water around the box, the hepatitis virus must have entered the drinking water line. The next morning, as pressure was resumed in the water lines, the contaminated water was pushed to the far end of the line—the drinking fountain on the football field—and all those players, coaches, and others who drank from the drinking fountain were infected with hepatitis.

This case illustrates a classical *cross connection*, or the physical contact between treated drinking water and contaminated water. One of the objectives of environmental engineering is to design systems in such a way as to avoid even the possibility of cross connections being created, although as the Holy Cross College incident shows, it is unlikely that all possibilities can be anticipated.

Tap Water vs. Bottled Water

A growing concern of the public is the safety of tap water. This concern has led to an explosion in bottled water sales, as well as the sale of home filtration systems. The sale of bottled water is a $4 billion dollar industry with over 10 billion bottles sold in 2003 and the bottled water industry continues to grow by about 10–20% each year even though the cost of bottled water is very expensive compared to tap water. A typical one liter bottle of water can cost about $1.25. This would buy about 230 gallons or 900 liters of tap water (assuming a cost of $5.27 per 1000 gallons—what the authors pay in 2004 in central Pennsylvania). A typical home filtration unit that uses reverse osmosis produces water at a cost of about $0.25 to $0.50 per gallon, which is cheaper than bottled water, but still significantly greater than tap water. Is it worth it? There are several issues to consider; the first is health and safety, and the second is aesthetics or taste and odor. Let's look at both of these issues, safety first.

Tap water is regulated by the EPA through the Safe Drinking Water Act. As discussed previously, a number of physical, chemical, and biological parameters are regulated to maintain a safe and generally pleasing water through maximum contaminant levels and treatment technologies. However, the Safe Drinking Water Act does not cover certain emerging pollutants, such as pharmaceuticals and other contaminants. With its structure, the act cannot keep up with the proliferation of new compounds being found in water. In contrast, the Food and Drug Administration (FDA) is responsible for regulating the bottled water industry and setting the maximum contaminant levels in bottled water. However, if the bottled water is made and sold only in one state (no interstate commerce), then the bottled water is regulated by state agencies.

The EPA and FDA limits are similar for most contaminants, but they do differ, with the EPA generally having more stringent requirements for testing. The presumption is that if the tap and bottled water are meeting the regulations, then the water is most likely safe.

One way to compare the safety of tap water is to look at the number of waterborne disease outbreaks. The Center for Disease Control maintains records for outbreaks of waterborne disease in the U.S. and according to their data there have been 730 outbreaks of waterborne disease associated with public drinking water from 1971 to 2000. The typical agents for disease are *E.coli 157*, *Salmonella*, *Cryptosporidium*, and *Giardia*, although many outbreaks occur where the agent is not identified. While these numbers seem high, this is a relatively small percentage of the overall population. Many of the outbreaks occur in the summer and are associated with small, recreational facilities such as campgrounds. They are a result of inadequate treatment and cross-connections. Based on these data, the chance for developing a waterborne disease from tap water appears relatively slim. Bottled water has a similar track record in that very few outbreaks of disease have been associated with bottled water.

A more difficult problem to quantify is the health impacts due to long-term exposure to the low-levels of contaminants in water such as chlorinated by-products and pharmaceuticals. Recent studies have reported long-term

(continued)

(continued)

consumption (40 or more years) of chlorinated drinking water is associated with bladder cancer. A "moderate" relationship exists between disinfection by-products and miscarriage in pregnant women as well as neural tube defects.

The comparison is clouded by the problems of defining the source of bottled water. One of the safest or cleanest water types is one that has gone through reverse osmosis. Several types of bottled water are simply tap water that is then treated with reverse osmosis technology and often in combination with carbon filtration and UV disinfection. However, government studies have found that about 25% of bottled water is simply tap water with no further treatment. Your bottled water may be nothing more than your tap water, but at a much higher price!

We can conclude that even though tap water and bottled water are meeting the required regulations, they may still pose some health risk, and a comparison of the quality of tap water and bottled water based on disease outbreaks or health risk is not useful.

A second way of comparing the two is by analyzing the water quality. A study by the Natural Resources Defense Council (NRDC) examined over 1000 samples and 103 brands of bottled water and concludes that "most . . . were found to be of high quality . . ." and that ". . . contamination posing immediate risks to healthy people is rare." However, about 22 percent of the samples had concentrations of contaminants greater than the California State limits (considered the most stringent), and 17 percent had bacterial heterotrophic plate counts that exceed recommended guidelines.

One reason people buy bottled water is they believe it tastes better. Tap water can have a somewhat unpleasant taste due to the residual chlorine that is required to be present to prevent contamination in the distribution system. This leads customers to buy bottled water because they perceive that it tastes better.

Another way of improving tap water's taste is by using in-home filters. This industry has exploded in the last 10 years and most grocery stores and department stores sell some type of water filtration unit. People hook up an activated carbon filter or reverse osmosis system or similar type of filtration unit to further treat their tap water (similar to some bottled waters), and these systems generally provide water that tastes better due to removal of residual chlorine and other contaminants.

But is this water any safer that untreated tap water? A recent study by the Public Health Department at the University of California, Berkeley, compared the number of gastrointestinal episodes reported by people in their study group, where some individuals had in-home filters installed and others did not. To be sure the effects were not imagined, the researchers installed filtration units in all the homes, but half the units were shams. In other words, they did not treat the water at all. Halfway through the study, the researchers switched the populations, so that the group which previously did not have actual filtration systems now had the units. The participants did not know whether they had a working filtration unit or not.

The researchers reported no difference in gastrointestinal disease between the two groups; in fact, the groups receiving untreated tap water both had a slightly lower number of illnesses, although this was not statistically significant. The researchers concluded that home-filtration units did not provide extra protection

(continued)

(continued)

from gastrointestinal illness, but with an important caveat, and that is when the tap water was produced from a well run drinking water plant meeting all required regulations.

The other health issue as discussed with bottled water is exposure to low-levels of contaminants that are legally present in tap water. The advantage of home filtration units, especially those with activated carbon and/or reverse osmosis is that these systems will remove the majority of these chemicals which reduced the risks associated with long-term exposure to some of the chemical contaminants in drinking water.

The general conclusion seems to be that tap water is safe to drink from the perspective of microbial disease as well as acute illness from chemical contaminants. However, so little is known about the health impacts of low-level exposure to many chemicals found in water that a prudent person would likely want to minimize their exposure to these compounds. For this reason, tap water that has gone through extra treatment (activated carbon and reverse osmosis) to remove these compounds seems to be a safe and prudent choice.

4.5 SYMBOLS

A = filter surface area
F = filter loading, for example, m³/hour per m² filter area
h = height at which a particle enters the settling tank
H = height of the settling tank
L = length of the settling tank
Q = flow rate
$\bar{t}$ = retention time
V = volume
v_o = critical settling velocity
v = horizontal velocity in an ideal settling tank
v_s = settling velocity of any particle
W = width of the settling tank

4.6 PROBLEMS

4.1 Suppose you ran a multiple tube coliform test and got the following results: 10 mL samples, all 5 positive; 1 mL samples, all 5 positive; 0.1 samples, all 5 negative. Use the table in *Standard Methods* to estimate the concentration of coliforms.

4.2 If coliform bacteria are to be used as an indicator of viral pollution as well as an indicator of bacterial pollution, what attributes must the coliform organisms have (relative to viruses)?

4.3 During the years following the Civil War, the New Orleans Water Company installed sand filters to filter the water out of the Mississippi River and sell it to the folks in New Orleans. The filters resembled the rapid sand filters in use today, and water from the river was pumped directly into the filters. Unfortunately, after the plant was built, the facility failed to produce the expected quantity of water and the company went bankrupt. Why do you think this happened? What would you have done as company engineer to save the operation?

4.4 A typical colloidal clay particle suspended in water has a diameter of 1.0 μm. If coagulation and flocculation with other particles manages to increase its size 100 times its initial diameter (at the same shape and density), how much shorter will be the settling time in 10 feet of water, such as a settling tank?

4.5 A settling tank in a water treatment plant has an inflow of 2 m^3/min and a solids concentration of 2100 mg/L. The effluent from this settling tank goes to sand filters. The concentration of sludge coming out of the bottom (the underflow) is 18,000 mg/L, and the flow to the filters is 1.8 m^3/min.

a. What is the underflow flow rate?
b. What is the solids concentration in the effluent?
c. How large must the sand filters be (in m^2)?

4.6 A settling tank is 20 m long, 10 m deep and 10 m wide. The flow rate to the tank is 10 m^3/minute. The particles to be removed all have a settling velocity of 0.1 m/minute.

a. What is the hydraulic retention time?
b. Will all the particles be removed?

4.7 A water treatment plant is being designed for a flow of 1.6 m^3/s. How many rapid sand filters, using only sand as the medium, are needed for this plant if each filter is 5 m × 5 m?

4.8 A water treatment plant has 6 settling tanks that operate in parallel (the flow gets split into six equal flow streams), and each tank has a volume of 40 m^3. If the flow to the plant is 10 million gallons per day, what is the retention time in each of the settling tanks? If instead, the tanks operated in series (the entire flow goes first through one tank, then the second, and so on), what would be the retention time in each tank?

4.9 A settling tank in a wastewater treatment plant has a volume of 50,000 L, and it is designed for a retention time of 2 hours. How much flow can it receive without exceeding its design retention time?

4.10 A settling tank in a water treatment plant receives a flow of 12 m³/hour. If the dimensions of the tank are Length = 20 m, Width = 3 m, and Height = 2 m,

    a. What is the hydraulic retention time?

    b. What is the critical velocity for this tank?

    c. Suppose the dimensions of the tank were changed so that Length = 10 m, Width = 6 m, and Height = 2 m, would this tank then be able to settle out more solids, less solids, or the same amount of solids? Explain.

    d. Suppose the tank in (c) above now receives double the flow rate. Is the critical velocity now is higher, lower, or the same as before? Explain.

    e. With double the flow rate, will the tank be able to settle out more solids, less solids, or the same amount of solids as before? Explain.

4.11 "Whew! This drinking water sure stinks today."

"It's chlorine, Grandma."

"Chlorine, schmerine! It stinks. They are doing something wrong at the water plant. Probably poisoning the whole town."

"Maybe. I guess in a way they are. But the chlorine is necessary."

"Why do they put chlorine in the water anyway? And what do you mean it might be bad for us?"

How to respond to Grandma?

CHAPTER 5

Wastewater Treatment

ONE of the beneficial uses of water, in addition to drinking, bathing, and cooking, is carrying away wastes. There seems something intuitively wrong with treating water to drinking water standards and then using it to flush a toilet. But alternatives, such as having two water systems, are much too expensive, and until we find some other way of getting rid of wastes, we will use water as the carrier. When water is thusly contaminated it is known as wastewater, and it must be treated before it is discharged to a watercourse. This chapter covers the collection of wastewater, the techniques we use to measure its strength, and finally the means available to treat it so that the water can be safely disposed of into the environment.

5.1 COLLECTION OF WASTEWATER

The term *sewage* is used here to mean only domestic wastewater, which includes everything that goes "down the drain," including waste from the kitchen sink, the shower, and the toilet. The piping used to collect these wastewaters is shown in Figure 5.1.

From the individual homes, sewage flows into *sewers* that collect the sewage from the customers. In addition to sewage, sewers also must carry industrial wastes, as well as be able to handle water that might flow into the sewers as the result of storms, or seep into the sewers due to cracks in the pipes. Because sewers carry more than just domestic sewage, the water flowing in sewers is more generally referred to as *wastewater.*

Drainage ditches in older cities were constructed for the sole purpose of moving stormwater out of the cities. In fact, discarding human excrement into these ditches was illegal. Eventually, the drainage ditches were covered over and became what we now know as *storm sewers.* As water supplies developed and the use of the indoor water closet increased, the need for transporting

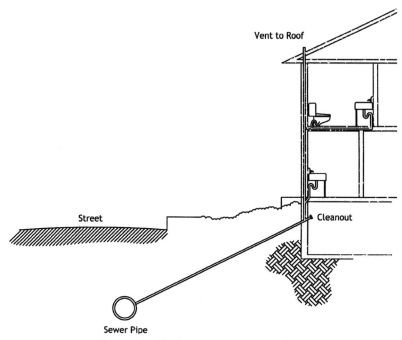

Figure 5.1. Collection of wastewater from a home.

domestic wastewater became necessary. The obvious solution was to use the storm sewers. These pipes then carried both sewage and stormwater and were known as *combined sewers*. Eventually a new system of underground pipes, known as *sanitary sewers*, was constructed for removing the sewage. Cities and parts of cities built in the twentieth century almost all have separate sewers for sewage and stormwater. Older cities still have combined sewers, which discharge into a wastewater treatment plant. Peak flows during heavy rainfall can result in overloading and poor treatment, or even flooding of the sewage treatment plant. In the United States the separation of existing combined sewers was mandated by federal water quality legislation in 1972.

Infiltration is the flow of groundwater into sanitary sewers. Sewers are often placed below the groundwater table, and any cracks in the pipes allows water to seep in. Infiltration is minimal in new, well-constructed sewers, but can be as high as 500 m^3/km-day. For older systems, 700 m^3/km-day is the commonly estimated infiltration. Infiltration flow is detrimental since the extra volume of water must go through the sewers and the wastewater treatment plant. It should be reduced as much as possible by maintaining and repairing sewers and keeping sewerage easements clear of large trees whose roots can severely damage the sewers.

The flow in sewers during dry weather (logically enough) is called *dry weather flow*, and this flow includes domestic and industrial wastewater as well as infiltration. The dry weather flow has a typical pattern over a 24-hour period. The flow is least during the late night hours when most people are asleep, and peaks during the morning after the high water use as people get up, take showers, and make breakfast. A second peak occurs when most people return home from work.

The flow in sewers changes during wet weather, however. *Inflow* is storm water collected unintentionally by the sanitary sewers. A common source of inflow is a perforated manhole cover placed in a depression so that stormwater flows into the manhole. Sewers laid next to creeks and drainageways that rise up higher than the manhole elevation, or where the manhole is broken, are also a major source of inflow. Illegal connections to sanitary sewers, such as roof drains, can also substantially increase the flow. The flow in sewers during wet weather is called *wet weather flow*, and the ratio of dry weather flow to wet weather flow is usually between 1:1.2 and 1:4.

Sewers that collect wastewater from residences and industrial establishments almost always operate as open channels or gravity flow conduits. Pressure sewers are used in a few places, but these are expensive to maintain and are useful only when there are severe restrictions on water use or when the terrain is such that gravity flow conduits cannot be efficiently maintained.

A typical sewerage system for a residential area is shown in Figure 5.2. In the United States, building connections are usually made with plastic pipe, 4 inches

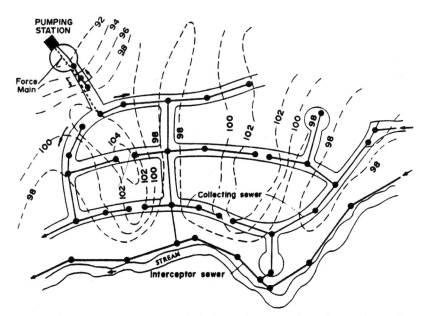

Figure 5.2. A typical sewerage system, including manholes, pumping station, and force main.

in diameter, to the *collecting sewers* that run under the street. Collecting sewers are sized to carry the maximum anticipated peak flows without surcharging (filling up) and are ordinarily made of PVC or cast iron. They discharge in turn into *trunk sewers, intercepting sewers,* or *interceptors,* that collect from large areas and discharge finally into the wastewater treatment plant. Collecting and intercepting sewers must be placed on a sufficient slope to allow for adequate flow velocity during periods of low flow, but not so steep as to promote excessively high velocities when flows are at their maximum.

A sewerage system must have access for repair and cleaning. Access is by manholes, usually placed every 120 to 180 m in the street, is necessary whenever the sewer changes grade (slope), size, or direction. The placement of manholes in a sewerage system is shown in Figure 5.2. Note on this figure how the sewers are straight between manholes.

Gravity flow may be impossible in some locations, or it may be uneconomical to use only gravity drainage. In these instances the wastewater must be pumped. The location of pumping stations in a sewerage network is shown in Figure 5.2. The gravity sewers flow down to the lowest point, from which the sewage has to be pumped. The line leading from the pumping station is called a *force main,* and it invariably leads to a manhole where the wastewater is carried once again by gravity. A force main, shown in Figure 5.2, is a closed pipe where the wastewater does not flow by gravity. Hence, a force main does not have to be straight and can follow the contours of the streets.

EDWIN CHADWICK

Edwin Chadwick (1800–1890) launched in the 1840s the "great sanitary awakening," arguing that filth was detrimental and that a healthy populace would be of higher value to England than a sick one. He had many schemes for cleaning up the city, one of which was to construct small-diameter sanitary sewers to carry away wastewater, a suggestion that did not endear himself to the engineers. A damaging confrontation between Chadwick, a lawyer, and the engineers ensued, with the engineers insisting that their hydraulic calculations were correct and that Chadwick's sewers would get plugged up, collapse, or otherwise be inadequate. The engineers wanted to build large-diameter egg-shaped brick sewers for both wastewater as well as stormwater that would be large enough to allow human access for cleaning and repair. These sewers were, however, anywhere from three to six times more expensive than Chadwick's vitrified clay conduits. Eventually the answer was a compromise, with pipes used for the collecting (small-diameter) sewers and the interceptors constructed of brick.

GEORGE E. WARING, JR.

George E. Waring Jr. (1833–1898) was educated at College Hill, Poughkeepsie, and then studied agriculture with a private tutor. In 1857 he was appointed agricultural and drainage engineer of Central Park, New York City, and later received a commission in the U. S. Army, eventually rising to the rank of colonel in the cavalry. Waring was involved in numerous Civil War battles. After the war he settled in Newport, Rhode Island, where he became the manager of a large farm, but his knowledge of drainage led him to become a full-time engineering consultant.

After the destruction and deprivation of the Civil War, many southern cities were poor and unsanitary. In Memphis, Tennessee, the death rates from communicable diseases were so high that this problem caught the attention of the nation and a commission was formed to study the health problems in the city. One of the recommendations was to construct a sewerage system that would be limited only to household wastewater. Instead of a combined system that would carry both storm water and human waste, Waring proposed a system that would allow the stormwater to run off in channels while the household sewage was collected in small-diameter sewers with flush tanks. His plan, which he based on the small-diameter sewers first promoted by Edwin Chadwick in London, was to be only one tenth the cost of constructing a combined system. After much political fighting, the city decided to adopt Waring's plan and a system consisting of 6-inch vitrified clay pipes with flush tanks leading into increasingly larger collecting sewers was constructed. The main argument advanced by Waring was that these sewers were necessary in order to enhance public health. Waring believed in the miasma theory of disease, that people became ill because they came into contact with sewer gas, and thus the totally buried and tight system was supposed to reduce the incidence of cholera, typhoid, and other such diseases by not allowing the miasma to waft into the community.

After the sewers had been constructed the incidence of communicable diseases dropped markedly and Waring claimed the system to be a success. But most people considered the system a failure because of operational problems. The small lines from the households often clogged and had to be cleaned with snakes, and when the collecting lines clogged, streets had to be dug up to unblock the sewers. Eventually manholes were constructed which, if factored into the original cost, would have significantly increased the price of the system.

Nevertheless, the controversy as to whether a city should build separate sewers instead of combined sewers raged for decades, with most engineers favoring combined systems since they were less expensive to build than two separate systems. At that time, there were no wastewater treatment plants, and all water—storm and wastewater—went to the same convenient place, such as the Mississippi River in the case of Memphis. Only when the polluted waterways became a national concern did the cities recognize that Edwin Chadwick and George Waring had been right all along.

The Great Louisville Sewer Explosion

Most wastewater sewers function silently and efficiently, with little maintenance or concern. But if volatile material is dumped into them, they can become spectacular bombs. One of the most memorable sewer explosions occurred in Louisville, Kentucky, in 1981. In those days there were limited controls over what industries could discharge into the sewerage system, and large corporations were able to get away with dumping what today would be illegal. In Louisville the southern part of the sewerage system collected waste from several industries, the largest one being a Ralston-Purina Company plant that used hexane to extract oil from soybeans. Waste and spillages had occurred in the past, but a containment system was used to capture the hexane and return it to the plant. On one cold night, however, the containment system was not functioning, and a large amount of hexane escaped into the sewers (estimates ranged from a few hundred to several thousand gallons). In the sewers the hexane volatilized and started to seep out of manholes. The combination of hexane and oxygen from the air formed an explosive mixture and all it took was one hot muffler or spark from a tailpipe to set off the explosion.

The entire trunk sewer exploded, with manhole covers being flung two stories into the air. Huge craters developed in the middle of the street as if the street had been bombed. In all, two miles of trunk sewer were completely destroyed and had to be replaced. Water pipes were severed and many residents had to do without water for weeks. Fearing an explosion at the wastewater treatment plant, the city bypassed its raw wastewater into the Ohio River for 12 hours, causing great concern from downstream communities such as Evansville, Indiana, which draws its drinking water supply from the Ohio. Fortunately, there were few injuries and no deaths, but many residents were evacuated, and several schools, including the University of Louisville, were closed until the situation was deemed under control.

Death in Manholes

Hydrogen sulfide, H_2S, one of the end-products of anaerobic decomposition, at low concentrations has a rotten odor, but at higher levels the olfactory senses are overwhelmed and H_2S becomes odorless. This gas attacks the nervous system and causes almost instant death. Before venting of manholes became mandatory by blowing air in from the top before someone climbs into the manhole, numerous sewer workers would be overcome by H_2S in the manholes and die. Typically, the accidents occurred when a slug of wastewater that had been stagnant started to move. The hydrogen sulfide would have been supersaturated in the wastewater because it is produced by anaerobic microorganisms, and with the release of pressure the slug of gas would come out of solution and "burp" out of a manhole, killing anyone who was in the manhole at the time. Often fellow workers, seeing their friends stagger and fall, would climb in to help them, only to be themselves overcome by the gas.

5.2 MEASUREMENT OF WASTEWATER QUALITY

Quantitative measurements of pollutants are obviously necessary before water pollution can be controlled. Measurement of these pollutants is, however, fraught with difficulties. Sometimes specific materials responsible for the pollution are not known. Moreover, many pollutants are generally present at low concentrations, and very accurate methods of detection are required.

Many water pollutants are measured in terms of milligrams of the substance per liter of water (mg/L). In older publications pollutant concentrations are often expressed as parts per million (ppm), a weight/weight parameter. If the only liquid involved is water, ppm is identical to mg/L, since one liter (L) of water weighs 1,000 grams (g). For many aquatic pollutants, ppm is approximately equal to mg/L; however, because of the possibility that some wastes have a specific gravity different from water, mg/L is preferred to ppm.

5.2.1 Sampling

In some situations the measurments must be conducted at the site because the process of obtaining a sample may change the measurement. For example, to measure the dissolved oxygen in a stream or lake, the measurement should either be conducted at the site or the sample must be extracted with great care to ensure that there has been no loss or addition of oxygen as the sample is exposed to air. Similarly, it is better to measure pH at the site if the water is poorly buffered.

Most tests may be performed on a water sample taken from effluet channel or the stream. The process by which the sample is obtained may greatly influence the result. The three basic types of samples are: grab samples, composite samples, and flow-weighted composite samples.

The *grab sample*, as the name implies, measures water quality at only one sampling point and one moment in time. Grab samples accurately represent the water quality at the moment of sampling, but say nothing about the quality before or after the sampling. A *composite sample* is obtained by taking a series of grab samples and mixing them together. The *flow-weighted composite* is obtained by taking each grab sample so that the volume of the sample is proportional to the flow at that time. The last method is especially useful when daily loadings to wastewater treatment plants are calculated. Whatever the technique or method, however, the analysis can only be as accurate as the sample, and often the sampling methods are far more sloppy than the analytical determinations.

5.2.2 Dissolved Oxygen

One of the most important measures of water quality is dissolved oxygen. Oxygen, although poorly soluble in water, is fundamental to aquatic life. Without free dissolved oxygen, streams and lakes become uninhabitable to aerobic organisms, including fish and most invertebrates. The importance of dissolved oxygen in streams and rives is discussed in Chapter 3. Recall that dissolved oxygen is inversely proportional to temperature and that the maximum amount oxygen that can be dissolved in water at 0°C is 14.6 mg/L. The saturation value decreases rapidly with increasing water temperature.

The simplest (and historically the first) type of oxygen measurement device is shown in Figure 5.3. The principle of operation is that of a galvanic cell. If lead and silver electrodes are placed in an electrolyte solution and connected to a microammeter, the reaction at the lead electrode is:

$$Pb + 2OH^- \rightarrow PbO + H_2O + 2e^-$$

At the lead electrode, electrons are liberated that travel through the microammeter to the silver electrode where the following reaction takes place:

$$2e^- + 1/2O_2 + H_2O \rightarrow 2OH^-$$

The reaction does not occur, and the microammeter does not register any current, unless free dissolved oxygen is available. The meter must be constructed and calibrated so that the electricity recorded is proportional to the concentration of oxygen in the electrolyte solution.

In commercial microammeter, the electrodes are insulated from each other with nonconducting plastic and are covered by a permeable membrane with a few drops of electrolyte between the membrane and electrodes. The amount of oxygen that travels through the membrane is proportional to the dissolved

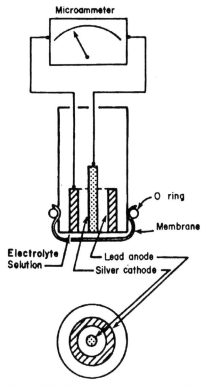

Figure 5.3. Original dissolved oxygen probe.

oxygen concentration. Dissolved oxygen probes are convenient for fieldwork, but need careful maintenance and calibration. Most oxygen probes are sensitive to changes in temperature and have thermisters attached to the probe so that temperature adjustments can be made in the field.

5.2.3 Biochemical Oxygen Demand

The rate of oxygen use is commonly referred to as biochemical oxygen demand (BOD). Biochemical oxygen demand is not a specific pollutant, but rather a measure of the amount of oxygen required by bacteria and other microorganisms engaged in stabilizing decomposable organic matter over a specified period of time.

The BOD test is often used to estimate the impacts of effluents that contain large amounts of biodegradable organics, such as those from food processing plants and feedlots, municipal wastewater treatment facilities, and pulp mills. A high oxygen demand indicates the potential for developing a dissolved oxygen sag (see previous chapter) as the microbiota oxidize the organic matter in the effluent. A very low oxygen demand indicates either clean water, or the presence of a toxic or non-biodegradable pollutant.

The BOD test was first used in the late 1800s by the Royal Commission on Sewage Disposal as a measure of the amount of organic pollution in British rivers. At that time, the test was standardized to run for 5 days at 18.3°C. These numbers were chosen because none of the British rivers had travel times (headwater to sea) greater than 5 days, and the average summer temperature for the rivers was 18.3°C. Accordingly, this should reveal the "worst case" oxygen demand in any British river. The BOD incubation temperature was later rounded to 20°C, but the 5-day test period remains the current, if somewhat arbitrary, standard.

In its simplest version, the *five-day BOD test* (BOD$_5$) begins by placing water or effluent samples into a standard 300 mL BOD bottle (Figure 5.4). The initial dissolved oxygen concentration is measured and the second BOD bottle is sealed and stored at 20°C in the dark. (The samples are stored in the dark to avoid photosynthetic oxygen generation.) After five days the amount of dissolved oxygen remaining in the sample is measured. The difference between the initial and ending oxygen concentrations is the BOD$_5$. In equation form,

$$BOD_5 = IDO - FDO \qquad (5.1)$$

where

BOD_5 = five-day biochemical oxygen demand, mg/L
IDO = initial dissolved oxygen, mg/L
FDO = final dissolved oxygen (at five days), mg/L

If the dissolved oxygen concentrations were measured daily, the results might produce curves like those shown in Figure 5.5. In this example, Sample A had an initial dissolved oxygen concentration of 8 mg/L, which dropped to 2 mg/L in 5 days. The BOD$_5$ therefore is 8 − 2 = 6 mg/L. Sample B also had an

Figure 5.4. BOD bottle.

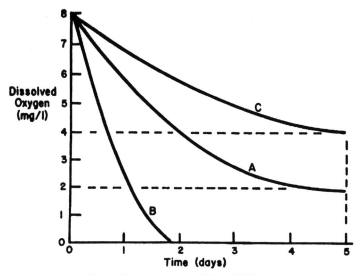

Figure 5.5. Example of a DO in a BOD test.

initial dissolved oxygen concentration of 8 mg/L, but the oxygen was used up so fast that it dropped to zero by the second day. Since there is no measurable dissolved oxygen left after 5 days, the BOD_5 of sample B must be more than 8 − 0 = 8 mg/L, but we don't know how much more because the organisms in the sample might have used more dissolved oxygen if it had been available.

For samples like Sample B, dilution is required. The assumption made is that the oxygen used is in direct proportion to the strength of the sample. The dilution water used in the BOD test is basically deionized water to which minerals have been added to facilitate microbial growth. It has no BOD itself, since there is nothing in the water to degrade. If a sample is diluted 1:10, for example, the strength of the mixture is 0.1 of the original sample, and the oxygen used in the bottle is assumed to be one tenth of the oxygen that would have been used in an undiluted sample. With dilution, the BOD is calculated as

$$BOD_5 = (IDO - FDO)D \qquad (5.2)$$

where

BOD$_5$ = five day biochemical oxygen demand, mg/L
IDO = initial dissolved oxygen, mg/L
FDO = final dissolved oxygen, mg/L
D = dilution

The dilution, D, is calculated as the number of times the sample has been diluted by its own volume. For example, if we put 10 mL of sample into a 100 mL bottle and fill the bottle with dilution water, the dilution is 1:10, or $D = 10$. We have diluted the sample 10 times.

Suppose sample C in Figure 5.5 is sample B diluted by 1:10. The BOD_5 for sample B would be

$$BOD_5 = (IDO - FDO)D = (8 - 4)10 = 40 \text{ mg/L}$$

It is possible to measure the BOD of any organic material (e.g., sugar) and thus estimate its influence on a stream, even though the material in its original state might not contain the microorganisms necessary to break down organic matter. Seeding is a process in which the microorganisms that oxidize organic matter are added to the BOD bottle. Seeding also facilitates measurement of very low BOD concentrations. The seed source can be obtained from unchlorinated domestic wastewater or surface water that receives degradable wastewater effluents.

Suppose we use the water previously described in curve A as seeded dilution water, since it obviously contains microorganisms (it has a 5-day BOD of 6 mg/L). We now put 100 mL of an unknown solution (the sample) into a bottle and add 200 mL of seeded dilution water, thus filling the 300 mL bottle. We then place the bottle into the 20°C incubator for 5 days. At the end of the 5 days the final DO is measured. Let's say it is 1 mg/L, or FDO = 1 mg/L. The total oxygen consumed in the sample bottle is $(8 - 1) = 7$ mg/L, but some of this oxygen was used by the microorganisms decomposing the seeded dilution water, because it also has a BOD, and therefore only a portion of that 7 mg/L difference was due to the decomposition of the sample. Because 2/3 of the bottle was seeded dilution water, the oxygen consumed as a result of the seeded dilution water is

$$(6)(2/3) = 4 \text{ mg/L}$$

since the total oxygen use by the seeded dilution water was 6 mg/L. The remaining oxygen consumed $(7 - 4 = 3$ mg/L) must be due to the unknown material. Equation 5.3 shows how to calculate the BOD_5 for a diluted, seeded effluent sample.

$$BOD(mg/L) = \left[(IDO - FDO) - (IDO' - FDO')\frac{X}{Y} \right] D \qquad (5.3)$$

where

BOD = biochemical oxygen demand, mg/L
IDO = initial dissolved oxygen in the bottle containing both effluent sample and seeded dilution water, mg/L
FDO = final dissolved oxygen in the bottle containing the effluent and seeded dilution water, mg/L

IDO′ = initial dissolved oxygen of the seeded dilution water bottle, mg/L
FDO′ = final dissolved oxygen of the seeded dilution water bottle, mg/L
X = volume of seeded dilution water in sample bottle, mL
Y = total volume in the bottle, mL
D = dilution of the sample

Example 5.1

Calculate the BOD_5 of a water sample, given the following data:

Initial dissolved oxygen is saturation, at 9.2 mg/L
Dilution is 1:30, with seeded dilution water
Final dissolved oxygen of seeded dilution water bottle is 8 mg/L
Final dissolved oxygen bottle with sample and seeded dilution water bottle is 2 mg/L
Volume of BOD bottle is 300 mL

Since the BOD bottle has a volume of 300 mL, a 1:30 dilution with seeded water would mean 10 mL of sample and 290 mL of seeded dilution water was in the bottle, and, by Equation 5.3,

$$BOD_5(mg/L) = [(9.2 - 2) - (9.2 - 8)(290/300)]30 = 183 \text{ mg/L}$$

BOD is a measure of oxygen use, or potential oxygen use. An effluent with a high BOD may be harmful to a stream, if the oxygen consumption is great enough to cause anaerobic conditions. Obviously, a small trickle going into a great river will have negligible effect, regardless of the BOD concentration involved. Conversely, a large flow into a small stream may seriously affect the stream even though the BOD concentration might be low.

The discussion thus far is about a 5-day BOD test, but it should be clear that we can measure the final DO at any time, put the bottle back in the incubator, and measure it again some days later. If we measured the dissolved oxygen each day we might get a curve like that shown in Figure 5.6. Note that some time after 5 days the curve turns sharply upward. This discontinuity is due to the demand for oxygen by the microorganisms that decompose nitrogenous organic compounds to inorganic nitrogen. For example, when microorganisms decompose a simple organic nitrogen compound, urea (NH_2CONH_2), ammonia is released (NH_3; NH_4^+ in ionized form), which is further decomposed into nitrite (NO_2^-) and nitrate (NO_3^-). In summary, the equation is

$$NH_2CONH_2 + 4O_2 \rightarrow 2NO_3^- + H_2O + 2H+ + CO_2$$

The important point to remember is: as nitrogenous materials such as urea decompose to nitrate, oxygen is required. The microorganisms that mediate this

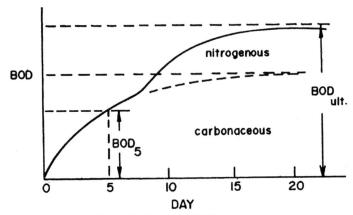

Figure 5.6. Long term BOD test results.

conversion are not as hearty as the carbonaceous microorganisms, and thus the reaction is delayed, producing the bump in the long-term BOD curve. The BOD curve can therefore be divided into *nitrogenous* and *carbonaceous* BOD areas. The *ultimate BOD*, as shown in Figure 5.6, includes both nitrogenous and carbonaceous BOD.

For streams and rivers with travel times greater than about 5 days, the ultimate demand for oxygen must include the nitrogenous demand. The ultimate BOD or BOD_{ult} (carbonaceous plus nitrogenous) is the BOD where both the carbonaceous and nitrogenous BOD curves level off, and this is often assumed to occur at 20 days. Often in wastewater treatment it is necessary to describe the biological loading on a treatment operation or a recipient—that is, how much of a demand for oxygen is the treatment operation, river, or lake required to provide in order to decompose the organic matter. For this, the BOD is expressed in terms of kg BOD/day, and calculated by multiplying BOD in mg/L by the flow rate:

$$mg\ BOD/s = (BOD\ in\ mg/L)(flow\ in\ L/s)$$

The L units cancel and we have mg/s, which is a mass flow rate. Usually this term is expressed as kg BOD/day, so a conversion factor is needed to convert seconds to days and mg to kg. Also, flow is usually expressed in m^3/s.

$$W = 86.4\ (BOD_5)\ Q \qquad (5.4)$$

where

W = kg BOD/day
Q = flow rate, m^3/s
BOD_5 = five day BOD, mg/L

American engineering continues to use the antiquated British units, and in these units, if the BOD is still in mg/L but the flow is in million gallons per day, the pounds of BOD/day is expressed as

$$W = 8.34 \, (\text{BOD}_5) \, Q \qquad (5.5)$$

where

W = pounds of BOD/day
Q = flow rate, million gallons per day

Note that the conversion factor is different in this equation.

Example 5.2

A wastewater treatment plant discharges an effluent of 20 mg/L BOD_5 with a flow rate of 2 m^3/s into a river. What is the BOD loading on the river in terms of kg BOD/day?

$$W = 86.4 \, (\text{BOD}_5) \, Q$$

$$= 86.4 \, (20 \text{ mg/L})(2 \text{ } m^3/s) = 3456 \text{ kg BOD/day}$$

5.2.4 Other Measures of Carbon in Water

One problem with the BOD test is that it takes 5 days to run. If the organic compounds are oxidized chemically instead of biologically, the test could be shortened considerably. Such oxidation can be accomplished with the *chemical oxygen demand* (COD) test. Because nearly all organic compounds are oxidized in the COD test, while only some are decomposed during the BOD test, COD results are always higher than BOD results. One example of this is wood pulping waste, in which compounds such as cellulose are easily oxidized chemically (high COD) but are very slow to decompose biologically (low BOD).

The standard COD test uses a mixture of potassium dichromate and sulfuric acid to oxidize the organic matter (HCOH), with silver (Ag^+) added as a catalyst. A simplified example of this reaction is illustrated below, using dichromate ($Cr_2O_7^{2-}$) and hydrogen ions (H^+):

$$2Cr_2O_7^{2-} + 3HCOH + 16H^+ \xrightarrow{\text{heat} + Ag^+} 3CO_2 + 11H_2O + 4Cr^{3+}$$

A known amount of a solution of $K_2Cr_2O_7$ in moderately concentrated sulfuric acid is added to a measured amount of sample, and the mixture is boiled in air. In this reaction, the oxidizing agent, hexavalent chromium (CrVI), is reduced to trivalent chromium (CrIII). After boiling, the remaining CrVI is

---

Problems in BOD Measurement

Before the time of the DO probe, dissolved oxygen was measured by a chemical technique called the Winkler test, which required the addition of four different chemicals and a titration—not a simple undertaking.

In the 1960s we were learning how estuaries flushed themselves, and one of the variables was how the BOD of the water in the estuary was progressively reduced by the flushing. The Delaware Estuary study was a large project which would show how the effects of pollutants (and their control) would affect water quality. As a part of this study, a group of young interns were detailed to run up and down the estuary taking samples at specified locations, bringing these samples back to the laboratory, and running the BODs.

As the first large batch (well over 100) bottles came in, the intern assigned to conduct the laboratory measurements ran the initial dissolved oxygen in all the bottles and placed these into an incubator. Five days later he took the bottles out and measured the final dissolved oxygen. Much to his surprise the final dissolved oxygen numbers were *higher* than the initial dissolved oxygen values. Suspecting chemical error he remixed all the chemicals very carefully, following exactly the procedures in *Standard Methods*. But the results did not change. The final DO values were still higher, and many bottles were actually supersaturated (higher than saturation).

Swallowing his pride, he finally asked for help. The director of the project decided to be a good mentor and to lend a hand. But after he had remixed the chemicals and run the tests, the results were the same.

After eliminating all possibilities that would explain the anomalous result, the intern looked into the incubator. When he opened the door, a light came on. But when he depressed the switch in the door that was supposed to turn off the light, the light remained illuminated. With the door closed, it was impossible to know that the light had stayed on during the five days of incubation and provided the energy for the algae in the bottles to produce oxygen. The light bulb was removed, and the project continued, with one very smug intern.

---

titrated against a reducing agent, usually ferrous ammonium sulfate. The difference between the initial amount of CrVI added to the sample and the CrVI remaining after the organic matter has been oxidized is proportional to the chemical oxygen demand.

Total organic carbon is measured by oxidizing the organic carbon to CO_2 and H_2O and measuring the CO_2 gas using an infrared carbon analyzer. The oxidation is done by direct injection of the sample into a high temperature (680–950°C) combustion chamber or by placing a sample into a vial containing an oxidizing agent such as potassium persulfate, sealing and heating the sample to complete the oxidation, then measuring the CO_2 using the carbon analyzer.

5.2.5 Solids

The separation of solids from the wastewater is one of the primary objectives of treatment, and the tests for solids are important in establishing the degree of treatment.

Total solids include any material left in a container after the water is removed by evaporation, usually at 103–105°C. Total solids can be separated into *total suspended solids* (solids that are retained on a 2.0 µm filter) and *total dissolved solids* (dissolved and colloidal material that passes through the filter). The difference between total suspended solids and total dissolved solids is illustrated by salt and pepper in water. If a teaspoonful of table salt is dissolved in a glass of water, a clear solution results. Although the water is clear, if the water evaporates, the salt will remain in the bottom of the class as a dry white powder. If a teaspoonful of pepper is put into the same glass, the particles of pepper will not dissolve and will remain as visible grains, giving the water a turbid appearance. If the water is evaporated, the pepper will also remain behind. In this example, the salt is a dissolved solid, whereas the pepper is a suspended solid.

Suspended solids are separated from dissolved solids using filtration through a membrane filter. The water sample is drawn through the filter with the aid of a vacuum and the suspended material is retained on the filter, while the dissolved fraction passes through. If the initial dry weight of the filter paper is known, the subtraction of this from the total weight of the filter and the dried solids caught in the filter yields the weight of suspended solids, expressed in milligrams per liter.

Solids may be classified in another way: those that are volatilized at a high temperature (550°C) and those that are not. The former are known as *volatile solids*, the latter as *fixed solids*. Volatile solids are usually organic compounds. At 550°C some inorganics are also decomposed and volatilized, but this is not considered a serious drawback. Example 5.3 illustrates the relationship between total solids and total volatile solids.

Example 5.3

The weight of a clean and dry filter paper is 8.6212 g; 100 mL of sample is placed in a dish and evaporated. The weight of the dish and dry solids = 8.6432g. The dish is placed in a 550°C furnace, then cooled. The weight = 8.6300g. Find the total, volatile, and fixed solids.

$$\text{Total solids} = \frac{(\text{dish} + \text{dry solids}) - (\text{dish})}{\text{sample volume}} = \frac{8.6432 - 8.6212}{100}$$

$$= (220)10^{-6}\,g\,/\,mL = (220)10^{-3}\,mg\,/\,mL = 220\;mg\,/\,L$$

$$\text{Fixed solids} = \frac{(\text{dish + unburned solids}) - (\text{dish})}{\text{sample volume}}$$

$$= \frac{486300 - 486212}{100} = 88 \text{ mg / L}$$

Total volatile solids = Total solids – Total fixed solids

$$= 220 - 88 = 132 \text{ mg / L}$$

5.2.6 Nitrogen and Phosphorus

Recall that nitrogen and phosphorus are important nutrients for biological growth. Nitrogen occurs in five major forms in aquatic environments: organic nitrogen, ammonia, nitrite, nitrate, and dissolved nitrogen gas. Phosphorus occurs almost entirely as organic phosphate and inorganic orthophosphate or polyphosphates.

Ammonia is one of the intermediate compounds formed during biological metabolism. Together with organic nitrogen, it is considered an indicator of recent pollution. Aerobic decomposition of organic nitrogen and ammonia eventually produces nitrite (NO_2^-) and finally nitrate (NO_3^-). High nitrate concentrations, therefore, may indicate that organic nitrogen pollution occurred far enough upstream that the nitrogenous organics have had time to oxidize completely. Similarly, nitrate may be high in groundwater after land application of organic fertilizers, if there is sufficient residence time (and available oxygen) in the soils to allow oxidation of the organic nitrogen in the fertilizer.

Because ammonia and organic nitrogen are pollution indicators, these two forms of nitrogen are often combined in one measure, called *Kjeldahl nitrogen*, after the scientist who first suggested the analytical procedure. A popular alternative to the technically difficult Kjeldahl test is to measure total nitrogen and nitrate+nitrite separately. The difference between the two concentrations equals organic nitrogen plus ammonia nitrogen.

Phosphorus is usually measured as total phosphorus (all forms combined) or dissolved phosphorus (portion that passes through a 0.45 μm membrane filter). Dissolved orthophosphate (PO_4^-) is an important indicator of water pollution because it is easily and rapidly taken up by biota, and therefore is almost never found in high concentrations in unpolluted waters.

The various forms of nitrogen and phosphorus can all be measured analytically by colorimetric techniques. In colorimetry, the ion in question combines with a reagent to form a colored compound; the color intensity is proportional to the original concentration of the ion. The color is measured photometrically, or occasionally by visual comparison to color standards.

A spectrophotometer, illustrated in Figure 5.7, consists of a light source, a filter, the sample, and a photocell. The filter allows only those wavelengths of light to pass through that the compounds being measured will absorb. Light passes through the sample to the photocell, which converts light energy into electric current. An intensely colored sample will absorb a considerable amount of light and allow only a limited amount of light to pass through and thus create little current. On the other hand, a sample containing very little of the chemical in question will be lighter in color and allow almost all of the light to pass through, and set up a substantial current.

Although most nitrogen and phosphorus analyses are done using a spectrophotometer, other techniques are growing in acceptance. Selective ion electrodes are available for measuring ammonia, nitrite, and nitrate. Ion chromatography (ICP) can be used to measure nitrite, nitrate, and phosphate, as well as total nitrogen and total phosphorus, if the samples are first digested (oxidized) to convert all forms of nitrogen or phosphorus to nitrate and phosphate. Ion chromatography involves passing a water sample through a series of ion exchange columns that separate the anions, so that they are released to a detector at different times. For simple (i.e., not particularly accurate) measurements, field kits are now available that provide pre-measured packets of chemicals for testing nitrogen and phosphorus in water and soil samples. These kits usually use colorimetric techniques similar to the more sophisticated versions used in analytical labs, but rely on color reference cards rather than a spectrophotometer for determining chemical concentrations.

5.2.7 Heavy Metals

Heavy metals such as arsenic, copper, and mercury can harm aquatic organisms, or bioaccumulate in the food chain, even when the metal concentration in water is relatively low. Consequently, the method of measuring metals in water must be very sensitive. There are a large variety of methods available to measure metals in water samples, and the choice of method often depends on the desired sensitivity as well as cost. Heavy metals

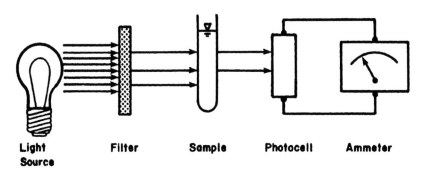

Light Source   **Filter**   **Sample**   **Photocell**   **Ammeter**

Figure 5.7. Spectrophotometer for measuring light adsorption in phosphate test.

Acute Mercury Poisoning

In September 1969 a poor farmer, Ernest Huckelby, picked up floor sweepings from a New Mexico granary. Unfortunately, he collected mainly millet seeds, which had been treated with methyl-mercury fungicide to prevent destruction of the seed by fungi when planted. Treatment of seed in this manner was common in agriculture in those years. The seed was dyed red as prescribed by law to warn against eating it. Huckelby, however, fed the seed to 17 of his hogs. When some of the hogs became ill he butchered one of them and fed the meat to his family. In a few weeks three Huckelby children fell ill from the meat, which contained as much as 27 parts per million of mercury. The three children suffered severe and partly irreversible neurological damage.

---

are usually measured using flame, electrothermal (graphite furnace), or cold-vapor atomic absorption (AA), inductively coupled plasma (ICP), and inductively coupled plasma/mass spectrometry (ICP/MS), and colorimetric techniques. Samples can be filtered and analyzed for dissolved metals or digested using strong acids to measure total metals.

In flame AA a solution of lanthanum chloride is added to the sample, and the treated sample is sprayed into a flame using an atomizer. Each metallic element in the sample imparts a characteristic color to the flame and the intensity of the color is measured spectrophotometrically. Graphite furnace AA methods use an electrically heated device to atomize metal elements, and can measure much lower concentrations of metals than flame AA, but often have "matrix" interference problems caused by salts and other compounds in the sample. Cold vapor AA is used primarily to measure arsenic and mercury. Inductively coupled plasma and ICP/MS are less sensitive to matrix problems and cover a wide range of concentrations.

5.2.8 Other Organic Compounds

One of the most diverse (and difficult) areas of pollution assessment is the measurement of toxic, carcinogenic, or other potentially harmful organic compounds in water. These organics encompass the disinfection byproducts mentioned earlier in conjunction with chlorination, as well as pesticides, detergents, industrial chemicals, petroleum hydrocarbons, and degradation products (e.g., DDT biodegrades to hazardous DDD and DDE).

Some of the methods described earlier in this chapter can be used to assess the overall content of organics in water. *Gas chromatography* (GC) and *high-performance liquid chromatography* (HPLC) are effective methods for measuring minute quantities of specific organics. Gas chromatography uses a mobile phase (carrier gas) and a stationary phase (column packed with an inert granular solid) to separate organic chemicals. The organics are vaporized, and then allowed to move through the column at different rates that are unique to each organic chemical. After separation in the column, the amount of each organic is measured using a detector sensitive to the type of organic that is being measured. Peak height and travel time are used to identify and quantify each organic. High-performance liquid chromatography is similar to gas chromotography except that the mobile phase is a high-pressure liquid solvent.

5.3 CENTRAL WASTEWATER TREATMENT

The most common contaminants found in domestic wastewater that can cause damage to natural watercourses or create human health problems are:

- organic materials, as measured by the demand for oxygen (BOD)
- nitrogen (N)
- phosphorus (P)
- suspended solids (SS)
- pathogenic organisms (as estimated by coliforms)

The demand for oxygen can be great enough to strip a river or lake of dissolved oxygen, creating undesirable anaerobic (without oxygen) conditions. The reduction of the demand for oxygen (or conversely, the destruction in BOD) is probably the most important aspect of centralized wastewater treatment. Nitrogen and phosphorus are important contaminants because of their role in accelerated eutrophication. Wastewaters that eventually discharge into lakes or estuaries are usually treated to remove nitrogen and phosphorus. Finally, pathogenic microorganisms in sewage can become a public health concern, and their destruction is a central part of wastewater treatment. Table 5.1 lists the typical values of some of the major constituents in wastewater. Which of these constitutes must be modified or removed depends on where the water is to be discharged. If, for example, a wastewater plant effluent from a small community is discharged into a large river that flows into the ocean, then the desired values might be as shown in the second column of Table 5.1. If, on the other hand, the discharge will be to a lake, the numbers might be more like the third column. Finally, if the effluent is discharged into a very small stream that has sensitive aquatic species, or one that runs dry for a significant part of the year, the fourth column numbers might be appropriate. The objective of a wastewater treatment plant is to achieve the effluent limits most appropriate for the environmental conditions of the receiving body into which the treated water is discharged.

The effectiveness of any plant in reducing the concentration of a specific

TABLE 5.1. Characteristics of a Typical Domestic Wastewater.

Parameter	Domestic untreated wastewater	To a large river or ocean	To a lake	To a sensitive or partially dry river or creek
Biochemical oxygen demand (BOD)	180 mg/L	20 mg/L	20 mg/L	5 mg/L
Suspended solids (SS)	160 mg/L	20 mg/L	20 mg/L	5 mg/L
Total solids (TS)	700 mg/L	200 mg/L	200 mg/L	200 mg/L
Phosphorus (P)	10 mg/L	10 mg/L	<1 mg/L	10 mg/L
pH	6.8	not important	6.5–8.0	6.5–8.0

parameter is known as *removal*, and this is expressed as a percent. Percent removal can be calculated as

$$R = \frac{C_o - C}{C_o} \times 100 \qquad (5.5)$$

where

R = percent removal of a pollutant
C_o = initial (influent) concentration, mg/L
C = final (effluent) concentration, mg/L

The calculation of percent removal is illustrated in Example 5.4.

Example 5.4

Using the values in Table 5.1, what percent removal of BOD would be required if the effluent were to be discharged into a large river.

The raw wastewater BOD is 180 mg/L. This is C_o in the equation. The effluent concentration is 20 mg/L. Hence the percent removal is

$$R = \frac{C_o - C}{C_o} \times 100 = \frac{180 - 20}{180} \times 100 = 89\%$$

Treatment plants used for the treatment of municipal wastewater must respond to these objectives and treat the wastewater to remove the objectionable characteristics. The typical wastewater treatment plant is divided into five main areas:

- Preliminary treatment—removal of large solids in order to prevent damage to the remainder of the unit operations.

The Lawrence Experiment Station

Massachusetts was the cradle of American environmental engineering and sciences. In 1869 the state created the Massachusetts State Board of Health, the first state agency to seriously investigate water pollution and wastewater treatment methods. Its first chairman, Henry I. Bowditch, made a study of sewage disposal practices in England and guided the agency's investigations in its formative years. The second annual report of the MSBH in 1871 explored the causes of typhoid fever and observed the causal relationships between pollution and disease. The agency's 1873 report concluded that water supplies were threatened by stream contamination and recommended that sewage be applied to land areas to abate water pollution. These initial studies were followed by investigations of stream self-purification and wastewater treatment processes.

The agency's most important sanitary engineering experiments were conducted at the Lawrence Experiment Station, which was founded in 1887. This group of scientists and engineers, initially headed by Hiram Mills, represented the first American efforts to scientifically investigate waste disposal practices. Its 1890 report, which documented the facility's engineering research, established a technical basis for the purification of water and the treatment of sewage and industrial wastes. Perhaps the most lasting contribution of the station was bringing together a brilliant group of young chemists, engineers, and biologists. Men such as Allen Hazen, George Whipple, Earle Phelps, George W. Fuller, and William T. Sedgwick acquired their early training at the station and become leaders in the water pollution control field.

- Primary treatment—removal of solids by settling. Primary treatment systems are usually physical processes, as opposed to biological or chemical.
- Secondary treatment—removal of the demand for oxygen. These processes are commonly biological.
- Tertiary treatment—a name applied to any number of polishing or cleanup processes, one of which is the removal of nutrients such as phosphorus. These processes can be physical (e.g., filters), biological (e.g., oxidation ponds) or chemical (e.g., precipitation of phosphorus).
- Solids treatment and disposal—the collection, stabilization, and subsequent disposal of the solids removed by other processes.

These steps in the wastewater treatment scheme are shown in Figure 5.8.

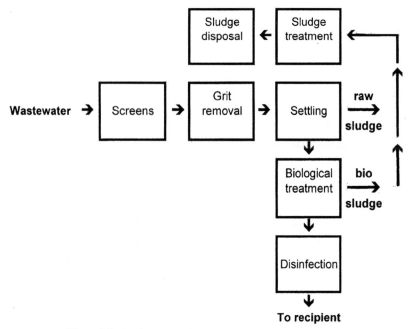

Figure 5.8. Configuration of a typical wastewater treatment plant.

5.3.1 Preliminary Treatment

The most objectionable aspect of discharging raw sewage into watercourses is the presence of floating material. It is only logical, therefore, that *screens* were the first form of wastewater treatment used by communities, and even today, screens are used as the first step in treatment plants. Typical screens, shown in Figure 5.9 consist of a series of steel bars, which might be about 2.5 cm apart. The purpose of a screen in modern treatment plants is the removal of materials that can damage equipment or hinder further treatment. In older treatment plants some screens are cleaned by hand, but mechanical cleaning equipment is used in almost all new plants. The cleaning rakes are automatically activated when the screens become sufficiently clogged to raise the water level in front of the bars.

The second treatment step involves the removal of grit or sand. This is necessary because grit can wear out and damage equipment such as pumps and flow meters. The most common *grit chamber* is simply a wide place in the channel where the flow is slowed down sufficiently to allow the heavy grit to settle out. Sand is about 2.5 times as heavy as most organic solids and thus settles much faster than light solids. The objective of a grit chamber is to remove sand and grit without removing the organic material. The latter must be

---

Making Water Run Uphill

At the turn of the 20th century, the city of Chicago drew its drinking water from Lake Michigan. As the connection between waste disposal and epidemics of cholera and typhoid began to be better understood, the practice of discharging wastewater into Lake Michigan seemed untenable. Chicago decided to keep the lake as its source of drinking water, and instead make the wastewater flow uphill into the Chicago River that flows southward into the Illinois River and eventually into the Mississippi above St. Louis. A 28-mile canal was constructed, and this permanently reversed the flow of the river from Lake Michigan to the Mississippi.

The people of the city of St. Louis were understandably perturbed by this development, and belatedly sued Chicago. Ironically, it was later discovered that the water in the Illinois River as it approached St. Louis was purer than the water in the Mississippi, and St. Louis decided to build the water treatment plant there. Some people have observed that this might be the reason why beer made in St. Louis has particularly good body.

---

further treated in the plant, but sand can be dumped as fill without undue odor or other problems. One way of assuring that the light biological solids do not settle out is to aerate the grit chamber, allowing the sand and other heavy particles to sink, but keeping everything else afloat. Aeration has the additional advantage of driving some oxygen into the sewage, which may have become devoid of oxygen in the sewerage system.

5.3.2 Primary Treatment

Following the grit chamber, most wastewater treatment plants have a *settling tank* to settle out as much of the solid matter as possible. These tanks in principle are no different from the settling tanks used in water treatment plants. The objective is to again operate the settling tanks as plug flow reactors, with uniform flow and a minimum of turbulence. The solids settle to the bottom and are removed through a pipe, while the clarified liquid escapes over a *V-notch*

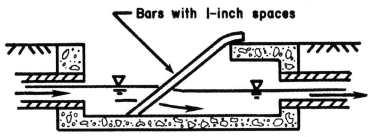

Figure 5.9. Typical screens used in a wastewater treatment plant.

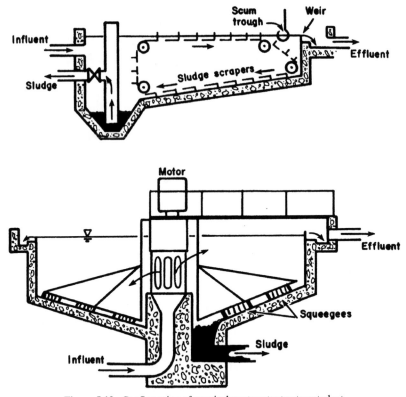

Figure 5.10. Configuration of a typical wastewater treatment plant.

weir, a notched steel plate over which the water flows, promoting equal distribution of the liquid discharge all the way around a tank. Settling tanks can be circular or rectangular (Figure 5.10).

Settling tanks are also known as *sedimentation tanks* and often as *clarifiers*. The settling tank that follows preliminary treatment such as screening and grit removal is known as the *primary clarifier*. The solids that drop to the bottom of a primary clarifier are removed as *raw sludge*, a name that doesn't do justice to the undesirable nature of this stuff.

Raw sludge is generally odoriferous, can contain pathogenic organisms, and is full of water, three characteristics that make its disposal difficult. It must be stabilized to reduce its possible public health impact and to retard further decomposition, and dewatered for ease of disposal. In addition to the solids from the primary clarifier, solids from other processes must similarly be treated and disposed of. The treatment and disposal of wastewater solids (sludge) is an important part of wastewater treatment and is discussed further later in this chapter.

Primary treatment, in addition to removing about 60% of the solids, also removes about 30% of the demand for oxygen and perhaps 20% of the phosphorus (both as a consequence of the removal of raw sludge). If this removal is adequate and the dilution factor in the watercourse is such that the adverse effects are acceptable, then a primary treatment plant is sufficient for wastewater treatment. Governmental regulations, however, have forced all primary plants to add secondary treatment, whether needed or not.

5.3.3 Secondary Treatment

The water leaving the primary clarifier has surrendered much of the suspended organic matter but still contains a high demand for oxygen due to the dissolved organics: i.e., it is composed of high energy molecules that will decompose by microbial action, creating a biochemical oxygen demand (BOD). This demand for oxygen must be reduced (energy wasted), if the discharge is not to create unacceptable conditions in the watercourse. The objective of secondary treatment is to remove BOD, while, by contrast, the objective of primary treatment is to remove solids. Except in rare circumstances, almost all secondary treatment methods use microbial action to reduce the energy level (BOD) of the waste. The basic differences among all these alternatives are how the waste is brought into contact with the microorganisms.

Although there are many ways the microorganisms can be put to work, the first really successful modern method of secondary treatment was the *trickling filter*. The trickling filter, shown in Figure 5.11, consists of a filter bed of fist-sized rocks over which the waste is trickled. An active biological growth forms on the rocks, and the organisms obtain their food from the waste stream dripping over the bed of rocks. Air is either forced through the rocks or, more commonly, air circulation is obtained automatically by a temperature difference between the air in the bed and ambient temperature. The wastewater

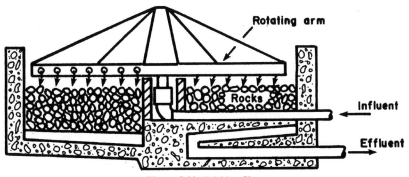

Figure 5.11. Tricking filter.

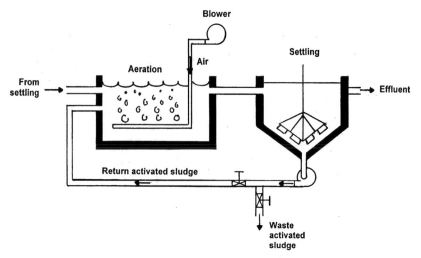

Figure 5.12. Activated sludge system.

is sprayed onto the rocks using a rotating arm that moves under its own power, like a lawn sprinkler, distributing the waste evenly over the entire bed. Often the flow is recirculated, obtaining a higher degree of treatment. The name *trickling filter* is a misnomer, since no filtration takes place.

An alternative biological treatment device is the *activated sludge system,* where the microorganisms, instead of being attached to surfaces like rocks, are allowed to float freely in the suspension. The key to the activated sludge system is the reuse of microorganisms. The system, illustrated in Figure 5.12, consists of a tank full of waste liquid (from the primary clarifier) and a mass of microorganisms. Air is bubbled into this tank (called the *aeration tank*) to provide the necessary oxygen for the survival of the aerobic organisms. The microorganisms come into contact with the dissolved organics and rapidly adsorb these organics on their surface. In time, the microorganisms use the energy and carbon by decomposing this material to CO_2 and H_2O, as well as a number of stable compounds. In the process more microorganisms are produced. The production of new organisms is relatively slow, and most of the aeration tank volume is used for this purpose. Once most of the food has been used up, the microorganisms are separated from the liquid in a settling tank, sometimes called a *secondary* or *final clarifier.* The liquid escapes over a V-notch weir and can be discharged into the receiving stream. The separated microorganisms exist on the bottom of the final clarifier without additional food and become hungry, waiting for more dissolved organic matter for food. These microorganisms are "activated"; hence this process is termed *activated sludge.* When these settled and hungry microorganisms are pumped to the head of the aeration tank, they find more food (organics in the effluent from the primary clarifier), and the process starts all over again. The sludge pumped from the bottom of the final clarifier to the aeration tank is known as *return activated sludge*.

The activated sludge process is a continuous operation, with continuous sludge pumping and clean water discharge. Unfortunately, one of the end products of this process is excess microorganisms. If none of the microorganisms are removed, their concentration eventually increases to the point where the system is clogged with solids. It is therefore necessary to waste some of the microorganisms, and this *waste activated sludge* must be processed and disposed of. Its disposal is one of the most difficult aspects of wastewater treatment.

Activated sludge systems are designed on the basis of loading, or the amount of organic matter (food) added relative to the microorganisms available. This ratio is known as the *food to microorganisms ratio* (F/M) and is a major design parameter. Unfortunately, it is difficult to measure either F or M accurately, and engineers have approximated these by BOD and the suspended solids in the aeration tank, respectively. The combination of the liquid and microorganisms undergoing aeration is known (for some unknown reason) as *mixed liquor*, and the suspended solids are called *mixed liquor suspended solids* (MLSS). The ratio of incoming BOD to MLSS, the F/M ratio, is also known as the *loading* on the system, and is calculated as kg of BOD/day per kg of MLSS in the aeration tank. A typical F/M ratio for an activated sludge system is 0.2 to 0.5 kg BOD/day per kg MLSS. Recall that kg of BOD can be calculated by multiplying BOD in mg/L by the flow rate, with appropriate conversion factors. kg BOD = (BOD in mg/L)(flow in m^3/s)(86.4) where 86.4 is the conversion factor.

Example 5.5

The BOD of the flow from the primary clarifier is 120 mg/L at a flow rate of 0.05 m^3/s. The aeration tank is $20 \times 10 \times 3$ m, and the MLSS = 2000 mg/L. Calculate the F/M ratio. The conversion factors are listed in the conversion table in the appendix.

$$kg\ BOD/day = 120\ mg/L \times 0.05\ m^3/s \times 86.4 = 518\ kg/day$$

$$kg\ MLSS = (20 \times 10 \times 3)m^3 \times 2000\ mg/L \times 1000\ L/m^3 \times 10^{-6}\ kg/mg$$
$$= 1200\ kg$$

$$F/M = 518/1200 = 0.43\ kg\ BOD/day/kg\ MLSS$$

If the F/M ratio is low (little food for lots of microorganisms) and the aeration period (retention time in the aeration tank) is long, the microorganisms

WILLIAM DIBDIN

In the 1880s the Thames River was a sorry sight. The pollution was so bad that Parliament had to stuff rags into the window crevices to keep out the foul odors, and anyone unlucky enough to fall in was more likely than not to come down with cholera. Finally the Metropolitan Board, responsible for the river, built two interceptor sewers on either side of the Thames to take the wastewater to Barking Creek on one side and Crossness on the other. This did some good, but the tidal nature of the Thames brought much of the wastewater back into the city. It was quite clear that some form of wastewater treatment was necessary.

The chief chemist working for the Board at that time was William Joseph Dibdin (1850–1925). Dibdin, a self-educated son of a portrait painter, began work with the Board in 1877, rising to chief chemist in 1882, but with the responsibilities of the chief engineer. In seeking a solution to the wastewater disposal problem at the Barking Creek and Crossness outfalls, he initiated a series of experiments using various flocculating chemicals such as alum, lime and ferric chloride to precipitate the solids before discharging them to the river. This was not new, of course, but Dibdin discovered that using only a little alum and lime was just as effective as using a lot, a conclusion that appealed to the stingy Metropolitan Board.

Dibdin recognized that the precipitation process did not remove the demand for oxygen, and he had apparently been convinced that it was necessary to maintain positive oxygen levels in the water in order to prevent odors. Dibdin decided to add permanganate of soda (sodium permanganate) to the water in order to replenish the oxygen levels. Because Dibdin's recommendations were considerably less expensive than the alternatives, the Board went along with his scheme.

Dibdin's plan was adopted, and in 1885 construction of the sewage treatment works at the Barking outfall commenced. Given the level of misunderstanding at the time, there were a great many who doubted that Dibdin's scheme would work, and he had to continually defend his project. He argued that the presence of the addition of the permanganate of soda was necessary in order to keep the odor down, and he began to explain this by suggesting that it was necessary to keep the aerobic microorganisms healthy. Christopher Hamlin, a historian who has written widely on Victorian sanitation, believes that this was a rationalization on Dibdin's part and that he did not yet have an insight into biological treatment. The more Dibdin was challenged by his detractors, however, the more he apparently became an advocate of beneficial aerobic microbiological activity in the water since this was his one truly unique contribution that could not be refuted.

At about this time the Lawrence Experiment Station in Lawrence, Massachusetts, was completing a fascinating series of experiments in which they

(continued)

(continued)

tested various soils for use in intermittent filters. In their experiments they used a number of Massachusetts soils, and apparently incidentally used gravel in one of the intermittent filters. To their surprise, this filter worked the best of all. They concluded that the gravel became coated with microorganisms and that this microbial slime was what produced the high degree of treatment. Dibdin was very much impressed with these experiments and recognized the importance of the presence of microorganisms on the gravel. Years later he would insist that his insights into biological treatment occurred before the Lawrence Experiment Station results were known, but this is problematical. Although the Lawrence Experiment Station report was not published until 1890, research information was readily shared between the American researchers and their British counterparts. The preliminary results from the Lawrence Experiment Station no doubt gave Dibdin additional ammunition to fight off his detractors.

But Dibdin was not totally successful in convincing the Board that his ideas were right. Many scientists argued that odor control could only be achieved by killing the microorganisms, still believing in the evils of the microbial world. These scientists managed in 1887 to wrest control of the treatment works from Dibdin and initiated a summer deodorization control suggested by a college professor that involved antiseptic treatment with sulfuric acid and chloride of lime. This process failed and Dibdin was vindicated. The use of beneficial microorganisms for wastewater treatment soon became the standard means of water pollution control.

make maximum use of available food, resulting in a high degree of treatment. Such systems are known as *extended aeration*, and are widely used for isolated sources (e.g., motels, small developments). An added advantage of extended aeration is that the ecology within the aeration tank is quite diverse, and little excess biomass is created, resulting in little or no waste activated sludge to be disposed of—a significant saving in operating costs and headaches. At the other extreme is the high rate system, where the aeration periods are very short (thus saving money by building smaller tanks) and the treatment efficiency is lower.

The two principal means of introducing sufficient oxygen into the aeration tank are by bubbling compressed air through *porous diffusers*, or by beating air in by *mechanical aeration* (Figure 5.13). In both cases, the intent is to transfer one gas (oxygen) from the air into the liquid, and simultaneously transfer another gas (carbon dioxide) out of the liquid.

The success or failure of an activated sludge system often depends on the performance of the final clarifier, the usual method of separating the solids grown in the aeration tank from the liquid, which is then discharged. If the final settling tank is not able to achieve the required return sludge solids, the solids concentration returned to the aeration tank will be low and the MLSS in the

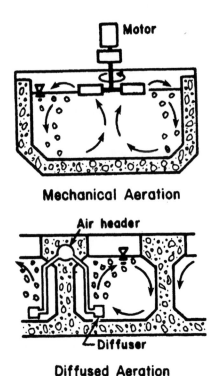

Mechanical Aeration

Diffused Aeration

Figure 5.13. Porous diffusers and mechanical aeration used in activated sludge.

aeration tank will drop, and the treatment efficiency will be reduced because there will be fewer microorganisms to do the work. It is useful to think of the microorganisms in the aeration tank as workers in an industrial plant. If the total number of workers is decreased, the production is cut. Similarly, if fewer microorganisms are available, less work is done.

The types of microorganisms grown in the aeration system can sometimes be very difficult to settle out, and the sludge is said to be a *bulking sludge*. Often this condition is characterized by a biomass comprised almost totally of filamentous organisms that form a kind of lattice structure with the filaments and refuse to settle. Treatment plant operators normally keep a close watch on settling characteristics because a trend toward poor settling can be the forerunner of a badly upset (and hence ineffective) plant. The settleability of activated sludge is most often described by the sludge volume index (SVI), which is determined by measuring the ml of volume occupied by a sludge after settling for 30 minutes in a 1liter cylinder, and calculated as

$$SVI = \frac{(1000)(\text{volume of sludge after 30 min, mL})}{\text{mg / L of SS}} \qquad (7.6)$$

The Birth of the Activated Sludge System

Probably no other single wastewater treatment process has had as much success as the activated sludge process. This system had an interesting beginning.

The first experiments with aeration were performed at the Lawrence Experiment Station in Massachusetts in 1912 using one gallon glass jugs. At the end of each run, when nitrification had been achieved (which was the primary measure of treatment in those days), researchers would empty the jugs, fill them up with new wastewater, and resume aerating. They were astonished to find that the time necessary to achieve nitrification decreased with every test, but did not understand that by not washing out the jugs at the conclusion of a run, they were leaving a film of microorganisms attached to the jug wall and that these microorganisms were the active culture that then promoted the next test. They had invented the activated sludge system but did not know it.

Dr. Gilbert Fowler from the University of Manchester in England happened to be visiting the experiment station and heard about the experiments. He apparently understood the significance of what had occurred, but did not reveal his ideas to the Americans. Instead, he went back to England and had two of his associates, Edward Arden and W. T. Lockett, set up a series of tests in which they would aerate a sample of waste for some days, settle out the sludge solids, decant the liquid off the top, and refill the container with new wastewater. Using this procedure, the time to nitrification was reduced from 5 weeks to 24 hours! It was a simple matter to go from such a fill- and draw-operation to a continuous system where the sludge was settled in a clarifier and then pumped back to the aeration system. Arden and Lockett are given credit for inventing the activated sludge system (and for giving it the name "activated sludge"), but the original work was done in Massachusetts. It just took a clever researcher like Fowler to recognize what was going on.

Treatment plant operators usually consider sludges with an SVI of less than 100 as well-settling sludges and those with an SVI of greater than 200 as potential problems, since these sludges will settle poorly in the final clarifier.

Example 5.6

A sample of mixed liquor was found to have suspended solids = 4000 mg/L and, after settling for 30 minutes in a 1liter cylinder, occupied 400 mL. Calculate the SVI.

$$SVI = (400 \times 1000)/4000 = 100$$

The causes of poor settling (high SVI) are not always obvious, and hence the solutions are elusive. Wrong or variable F/M ratios, fluctuations in temperature, high concentrations of heavy metals, and deficiencies in nutrients

The Deer Island Wastewater Treatment Plant

In the 1980s, Boston Harbor was one of the most polluted bodies of water in the United States. Beaches were closed to swimming and fishing banned. The entire aquatic ecosystem was on the verge of collapse. A wastewater treatment plant had been built in 1968 on Deer Island but the plant had only primary treatment and discharged its effluent directly into the shallow Boston Harbor. The sludge was digested but it also was discharged into the harbor. The combination of effluent and sludge represented a daily loading of over 120 tons into the harbor. The scum floated to the shore, and the sludge settled to the bottom where if formed a "black mayonnaise" that was devoid of all normal aquatic life.

The Clean Water Act of 1972 mandated that cities construct secondary wastewater treatment facilities, but coastal cities that had long outfalls and significant dilution were given exemptions. Boston, however, did not have such an outfall and was told by the U. S. EPA that it needed a secondary treatment plant.

The responsibility for cleaning up the harbor fell to the Metropolitan Distinct Commission, an agency with little enthusiasm for doing anything. Because its funds came from the State of Massachusetts, the governor at the time, Michael Dukakis, was responsible for initiating corrective action. During two terms in office, however, the water quality in the inner harbor remained an embarrassment to the people of Boston.

In 1988, Michael Dukakis was nominated by the Democratic Party to run for the presidency of the United States, and his opponent was George Bush, who did not have much of a record on environmental issues. Recognizing an opportunity, however, George Bush made a famous political ad in which he called the Boston Harbor "the filthiest harbor in America". He went on to suggest that Dukakis' environmental record was questionable since he could not even clean up the harbor in his own state. By criticizing his opponent's inaction on cleaning up the harbor, Bush was able to deflect attention from his own minimal commitment to the environment.

When George Bush was elected, the clean-up of the Boston Harbor became a first priority, and the U. S. EPA became very involved in the design and construction of the treatment facilities. Obviously, George Bush, who wished to run in four years, did not want his new opponent to stand at the same location and criticize him for not cleaning up the harbor.

The end result was that the Deer Island plant was greatly expanded to include secondary treatment (activated sludge). The effluent from the plant is now discharged through a 9.5 mile-long outfall into the deeper ocean, there to be diluted by the currents. There is some question as to the wisdom of having built both, the secondary treatment plant to treat 350 million gallons of wastewater a day and the 9.5 mile-long outfall. Probably either one would have had the same beneficial effect on the harbor (which is still receiving non-point source stormwater runoff). Building both the secondary plant and the outfall were overkill, but this was not an engineering decision, but a political one, at the highest level.

in the incoming wastewater have all been blamed for bulking. Cures include reducing the F/M ratio, changing the dissolved oxygen level in the aeration tank, and dosing with hydrogen peroxide (H_2O_2) to kill the filamentous microorganisms. When the sludge does not settle, the return activated sludge becomes thin (low suspended solids concentration) and the concentration of microorganisms in the aeration tank drops. This results in a higher F/M ratio (same food input, but fewer microorganisms) and a reduced BOD removal efficiency.

Effluent from secondary treatment often has a BOD of about 15 mg/L and a suspended solids concentration of about 20 mg/L. This is most often quite adequate for disposal into watercourses, since the BOD of natural stream water can vary considerably, from about 2 mg/L to far greater than 15 mg/L. Effluent from a well-run wastewater treatment plant can actually dilute the stream.

Following secondary treatment, the BOD and solids levels in the treated water are low, and especially with tertiary treatment such as filtration, represent no danger to the aquatic ecosystem into which it is discharged. But there is still the possibility that human pathogens survived the treatment, and modern wastewater treatment plants are required to disinfect the effluents in order to reduce the possibility of disease transmission. Often chlorine is used for disinfection since it is fairly inexpensive. The use of chlorine is not without problems, however. The effluent must be assimilated into the aquatic ecology and chlorine would be toxic to many of these organisms. Further, dosing wastewater treatment plant effluents with chlorine results in the production of chlorinated compounds such as chloroform, a carcinogen. And finally, there is no epidemiological evidence that unchlorinated treatment plant effluents cause any public health problems. As a result, newer methods of disinfection have been developed, the most widely used being ultraviolet (UV) radiation. This

---

Chlorination of Wastewater Effluents

The disinfection with chlorine of wastewater treatment plant effluents has been standard practice in the United States since the 1960s. Recent studies in Canada have shown that chlorinated effluents have caused lethality to fish and resulted in changes in aquatic ecosystems, such as reduction in diversity and shifts in species distribution. These effects, which persist for up to 0.5 km downstream, can be caused by chlorine levels as low as 0.02 mg/L. When residual chlorine levels exceed 0.1 mg/L, laboratory tests show that the effluents are lethal to fish and invertebrate species. There appears to be no evidence that the chlorination of effluents has had any effect, either positive or negative, on human health. The Minister of National Health in Canada has thus concluded that chlorinated effluents are to be considered "toxic" wastes, as defined under the Canadian Environmental Protection Act.

process leaves no residual and is effective in destroying the large bulk of pathogenic organisms.

5.3.4 Tertiary Treatment

There are situations when secondary treatment (even with chlorination) is inadequate to protect the watercourse from harm due to a wastewater discharge. One concern is that nutrients such as nitrogen and phosphorus may still cause a problem if the effluent is discharged into a still body of water. If the water downstream is used for recreational purposes, a high degree of treatment is necessary, especially in solids and pathogen removal. When such situations occur, the effluents from secondary treatment are treated further to achieve whatever quality is required, or the primary + secondary plant is upgraded to produce a higher-quality effluent. Such processes are collectively called tertiary treatment.

Rapid sand filters

Rapid sand filters similar to those in drinking water treatment plants are used to remove residual suspended solids and to polish the water.

Oxidation ponds

Oxidation ponds are commonly used for BOD removal. The oxidation or polishing pond is essentially a hole in the ground, a large pond used to confine the plant effluent before it is discharged into the natural watercourse. Such ponds are designed to be aerobic, and since light penetration for algal growth is important, a large surface area is needed. The reactions occurring within an oxidation pond are depicted in Figure 5.14. Oxidation ponds are sometimes used as the only treatment step if the waste flow is small and pond area is large.

Activated carbon adsorption

Another method of BOD removal is activated carbon adsorption, and this process has an added advantage in that inorganics as well as organics are removed. The mechanism of adsorption on activated carbon is both chemical and physical, with tiny crevices catching and holding colloidal and smaller particles. An activated carbon column is a completely enclosed tube with dirty water pumped up from the bottom and the clear water exiting at the top. As the carbon becomes saturated with various materials, these must be removed from the column. At that point the carbon is regenerated, or cleaned. Removal is often continuous, with clean carbon being added at the top of the column. The cleaning or regeneration is usually done by heating the carbon in the absence of

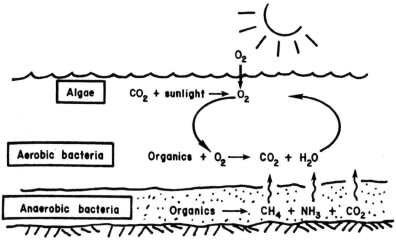

Figure 5.14. Oxidation pond.

oxygen, driving off the organic matter. A slight loss in efficiency is noted with regeneration, and some virgin carbon must always be added to ensure effective performance.

Nitrogen removal

The removal of nitrogen is accomplished by first treating the waste thoroughly enough in secondary treatment to oxidize all of the nitrogen to nitrate. This usually involves longer detention times in secondary treatment, during which bacteria such as *Nitrobacter* and *Nitrosomonas* convert ammonia nitrogen to NO_3, a process called *nitrification*. These reactions are

$$2NH_4^+ + 3O_2 \xrightarrow{\text{Nitrosomonas}} 2NO_2^- + 2H_2O + 4H^+$$

$$2NO_2^- + O_2 \xrightarrow{\text{Nitrobacter}} 2NO_3^-$$

Both of these reactions are slow and require sufficient oxygen and long retention times in the aeration tank. The rate of microorganism growth is also low, resulting in low net sludge production, making washout a constant danger.

Once the ammonia has been converted to nitrate, it can be reduced by a broad range of facultative and anaerobic bacteria, such as *Pseudomonas*. This reduction, called *denitrification*, requires a source of carbon, and methanol (CH_3OH) is often used for that purpose. The sludge containing NO_3^- is placed in an *anoxic* condition where the microorganisms, using methanol as the source of carbon, convert the nitrate to nitrogen gas, N_2, which then bubbles out of the sludge into the atmosphere.

Phosphorus removal

The removal of phosphorus is accomplished by either chemical or biological means. In wastewater, phosphorus exists as orthophosphate (PO_4^{3-}), polyphosphate (P_2O_7) and organically bound phosphorus. Polyphosphate and organic phosphate may be as much as 70% of the incoming phosphorus load. In the metabolic process, microorganisms use the poly- and organo-phosphates and produce the oxidized form of phosphorus, orthophosphate. Phosphate removal is either a chemical precipitation process or a biological process designed as part of the activated sludge system.

Chemical phosphorus removal requires that the phosphorus be fully oxidized to orthophosphate, and hence the most effective chemical removal occurs at the end of the secondary biological treatment system. The most popular chemicals used for phosphorus removal are lime $Ca(OH)_2$, and alum, $Al_2(SO_4)_3$. The calcium ion, in the presence of high pH, will combine with phosphate to form a white, insoluble precipitate called calcium hydroxyapatite which is settled out and removed. Insoluble calcium carbonate is also formed and removed and can be recycled by burning in a furnace.

The aluminum ion from alum precipitates out as poorly soluble aluminum phosphate,

$$Al^{+++} + PO_4^{---} \rightarrow Al(PO_4)$$

and also forms aluminum hydroxides

$$Al^{+++} + 3OH^- \rightarrow Al(OH)_3$$

which are sticky flocs and help to settle out the phosphates. The most common point for alum dosing is in the final clarifier.

Biological methods or phosphorus removal are widely used because they can become a part of the activated sludge system. The aeration tank is divided into sections that are anoxic (no dissolved oxygen) and those that are aerobic. Figure 5.15 shows one such configuration for an activated sludge system that includes biological phosphorus removal. Basically, the return activated sludge is returned to the anoxic chamber where the microorganisms are stressed, finding no oxygen to use in their metabolic activity. One of their survival mechanisms is to release phosphorus into the surrounding liquid as a means of maintaining life. At the end of the anoxic zone the majority of phosphorus is dissolved as phosphate ions. When the microorganisms then enter the aerobic zone they undergo rapid metabolic activity and need as much phosphorus for energy transfer processes as they can get, and assimilate the dissolved phosphorus into their cell mass. This "gorging" is sometimes known as *luxury uptake* of phosphorus, because the cells take much more than the actually need.

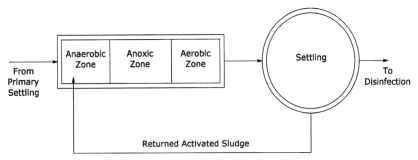

Figure 5.15. Activated sludge with biological phosphorus removal.

Then, once the phosphorus is contained in the microbes, the content of the aeration tank flows to the secondary clarifier where the excess sludge, now containing much phosphorus, is separated as waste activated sludge.

Finally, when discharge to a watercourse is not either practical or where the discharge would result in unacceptable environmental damage to the watercourse, alternative disposal schemes are used. Probably the most promising alternative disposal technique is irrigation. The amount of land area required is substantial, however, and disease transmission is possible since the waste carries pathogenic organisms. Commonly, from 1000 to 2000 hectares of land are required for every 1 m³/sec of wastewater flow, depending on the crop and soil. Nutrients such as N and P remaining in the secondary effluent are of course beneficial to the crops.

5.3.5 Sludge Treatment and Disposal

Two types of sludges are produced in common wastewater treatment plants—*raw primary sludge* and *biological* or *secondary sludge*. The raw primary sludge comes from the bottom of the primary clarifier, and the biological sludge is either solids that have grown on the fixed film reactor surfaces and sloughed off the rocks or other surfaces, or waste activated sludge grown in the activated sludge system.

Estimating the production of sludge in a wastewater treatment plant can be done using various rules of thumb. For example, a good first estimate of total sludge production is that wastewater treatment plants are expected to produce about 0.17 lb dry solids/capita/day. A better technique for estimating sludge production in wastewater treatment is to estimate production of various type of sludge using empirical data. The production of primary sludge, for example, can be estimated by assuming that a certain fraction of the suspended solids will settle out (the *settleable solids*). Thus,

$$P = Q \, (R) \, (SS) \times 10^{-5} \qquad (5.6)$$

The Santee Lakes

Closing the loop in water use has intrigued engineers for many years. Several projects have developed in the United States that demonstrated that using adequately treated wastewater for useful purposes was both economically attractive but also acceptable to the public. The most famous of these cases is the Santee Lakes.

Santee is a bedroom community outside of San Diego, with now about 100,000 residents. In 1958, with only a fraction of the present population, the community chose to not join the San Diego Metropolitan Sewer District and to build its own wastewater treatment plant. The major impetus for this was the donation of land for the plant by a golf course developer, who wanted to use the treated wastewater for irrigation. The scheme they developed was to construct a series of three lakes through which the water flowed. By the time the water reached the third lake, the developer believed, it would meet all the health requirements for irrigation water. This system was constructed in a valley so that the water flowed by gravity from lake to lake, and the geology was such that the clay soil prevented almost all percolation into the ground. The system met the most optimistic predictions. Not only did the golf course obtain all the water it needed, but there was considerable water left over for other secondary uses.

After the lakes had been constructed they were surrounded by chain-link fences to keep the public out. Eventually the health department was convinced that the water quality was acceptable for both water contact sports as well as fishing. The fences came down, and the people of Santee had a marvelous recreational facility within their community.

Because Santee received its drinking water from the Colorado River by means of canals it never considered the logical next step of tapping the lowest lake as a source of drinking water.

---

where

P = production of primary sludge, dry solids, kg/day
Q = influent to the treatment plant, m³/day
R = removal rate, as percent of suspended solids removed
SS = concentration of suspended solids, mg/L

The removal rate increases with decreasing retention time, averaging about 60%. Solids removal (and primary sludge production) can be enhanced by adding flocculating chemicals to the clarifier influent. Ferric chloride, lime, aluminum sulfate, and sometimes organic polyelectrolytes have been used for such "enhanced primary clarification" and up to 90% of the suspended solids can be removed.

If the wastewater treatment plant has secondary (biological) treatment, most of the suspended solids remaining after primary treatment are removed.

However, the biological processes often produce excess biological sludge. The type of system used dictates the amount of this sludge produced. Because excess biological sludge has properties quite different from primary sludge, its estimation is necessary for proper design of sludge stabilization and dewatering. The activated sludge system produces the most biological sludge, and the *yield* of excess activated sludge is calculated on the basis of the mass of solids produced per mass of BOD destroyed. As a rule of thumb, the yield from an activated sludge process is about 0.5 kg waste solids (dry basis) produced per kg BOD destroyed.

A great deal of money could be saved, and troubles averted, if sludge disposal could be done as the sludge is drawn off the main process train. Unfortunately, sludges have three characteristics that make such a simple solution unlikely: they are aesthetically displeasing, they are potentially harmful, and they have too much water.

The first two problems are often solved by *stabilization,* which may involve anaerobic or aerobic digestion. The third problem requires the removal of water, either by thickening or dewatering. The next three sections cover the topics of stabilization, thickening, and dewatering, followed by considerations of ultimate sludge disposal.

The objective of sludge stabilization is to reduce the problems associated with sludge odor and putrescence, as well as reduce the hazard presented by pathogenic organisms. Sludge may be stabilized using lime, aerobic digestion, or anaerobic digestion. The most common method of sludge stabilizaton is using anaerobic microorganisms. *Anaerobic digestion* is a staged process:

$$\text{Organic materials} \xrightarrow{\text{enzymes}} \text{solublilzed organics}$$

$$\text{Soluble organics} \xrightarrow{\text{acid formers}} \text{organic acids}$$

$$\text{Organic acids} \xrightarrow{\text{methane formers}} CH_4 + CO_2$$

The solubilization of organic compounds by extracellular enzymes is followed by the production of organic acids by a large and hearty group of anaerobic microorganisms known, appropriately enough, as the *acid formers.* The organic acids are in turn degraded further by a group of strict anaerobes called *methane formers.* These microorganisms are the prima donnas of wastewater treatment, getting upset at the least change in their environment. The success of anaerobic treatment depends on maintenance of suitable conditions for the methane formers. Since they are strict anaerobes, they are unable to function in the presence of oxygen and are very sensitive to environmental conditions like pH, temperature, and the presence of toxins. A digester goes "sour" when the methane formers have been inhibited in some way. The acid formers keep chugging away, making more organic acids, thus further lowering the pH and making conditions even worse for the methane formers. Curing a sick digester requires suspension of feeding and, often, massive doses of lime or other antacids.

Most treatment plants have both a primary and a secondary digester (Figure 5.16). The primary digester is covered, heated, and mixed to increase the reaction rate. The temperature of the sludge is usually about 35°C (95°F). Secondary digesters are not mixed or heated, and are used for storage of gas produced during the digestion process (e.g., methane) and for concentrating the sludge by settling. As the solids settle, the liquid supernatant is pumped back to the main plant for further treatment. The cover of the secondary digester often floats up and down, depending on the amount of gas that has accumulated. The

KARL IMHOFF

One of the most influential wastewater engineers ever was a German named Karl Imhoff (1876–1965), who was responsible for training many of the leading American sanitary engineers such as Gordon Maskew Fair at Harvard. Imhoff's most lasting contribution is an unassuming little treatment system now called the Imhoff tank.

Pictured below, the Imhoff tank incorporates primary settling and sludge digestion all in one tank. The sludge solids settle to the bottom into the cone-shaped hopper where they begin to digest. The gases they produce will bubble up, but the really clever part of the tank is how

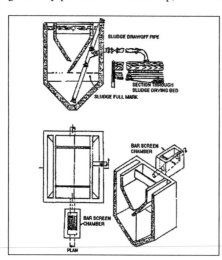

deflectors force the gas bubbles to the side, thus allowing the solids to continue to settle unhindered by the rising gas bubbles. The sludge is drawn out periodically and placed on a sand drying bed. The tank has no moving parts and is able to achieve at least 40% solids and BOD removal. Many Imhoff tanks are still in use, even in the United States. They just don't wear out, and in small communities where effluents from the tanks can be pumped to holding ponds prior to discharge, the system works beautifully— 100 years later!

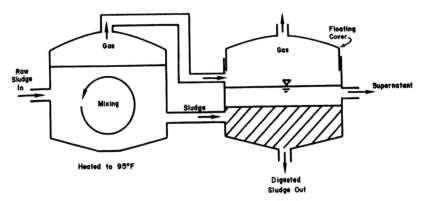

Figure 5.16. Primary and secondary anaerobic digesters.

gas produced during anaerobic digestion contains enough methane to be used as a fuel, and is frequently used to heat the primary digester.

Various means of sludge dewatering are used to get water out of sludge before ultimate disposal. In the United States, the usual dewatering techniques are sand beds, belt filters, and centrifuges. *Sand beds* have been used for a great many years and are still the most cost-effective means of dewatering when land is available. The beds consist of tile drains in gravel, covered by sand. The sludge is poured on the sand and the water is removed by seepage through the sand into the tile drains and by evaporation. Although seepage into the sand results in a substantial loss of water, it only lasts for a few days. The sand pores are quickly clogged, and drainage into the sand ceases. The mechanism of evaporation takes over, and this process is actually responsible for the conversion of liquid sludge to solid. As the surface of the sludge dries, deep cracks develop that facilitate evaporation from lower layers in the sludge mat.

If dewatering by sand beds is impractical, mechanical dewatering techniques must be used. Two general processes are used for mechanical dewatering: filtration and centrifugation. The belt filter, shown in Figure 5.17, has a moving fabric belt through which the water is both drained and then forced by pressure. Belt filters are quite effective in dewatering many different types of sludges and are widely used by small wastewater treatment plants. Larger plants generally use centrifuges for dewatering. A centrifuge is a horizontal bowl that rotates rapidly on its longitudinal axis, forcing the sludge solids to the inside wall. A large screw scrapes the solids out of the machine, allowing the clear liquid to flow out the opposite end. A typical sludge centrifuge is shown in Figure 5.18.

For the most part, sludge in the United States is either landfilled with municipal refuse, or applied to land on which crops may or may not be cultivated. The ability of land to absorb sludge and to assimilate it depends on

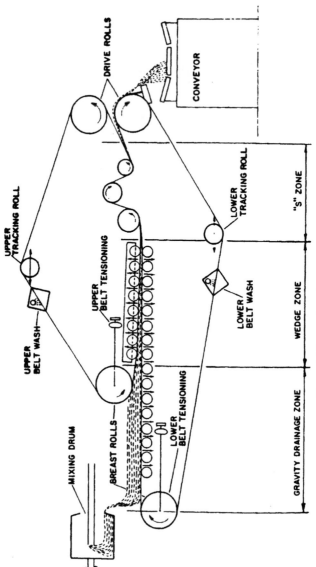

Figure 5.17. Belt filter for sludge dewatering.

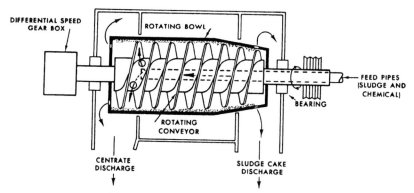

Figure 5.18. Centrifuge for sludge dewatering.

variables such as soil type, vegetation, rainfall, slope, etc. In addition, the important variable of the sludge itself will influence the capacity of a soil to assimilate it. Generally, sandy soils with lush vegetation, low rainfall, and gentle slopes have proven most successful. Mixed digested sludges have been spread from tank trucks, and activated sludges have been sprayed from both fixed and moving nozzles. The application rate has been variable, but 100 dry tons/acre-yr is not an unreasonable estimate. Most unsuccessful land application systems may be traced to overloading the soil. Given enough time, and the absence of toxic materials, soils will assimilate sprayed liquid sludge.

A Rose By Any Other Name

In recent years the sludge treatment and disposal field has started to call waste sludge from wastewater treatment plants *biosolids*. It seems like a silly thing to do on the surface. You can call it whatever you want, but it is still sludge.

Well it turns out that the word "sludge" has some very negative connotations. H. L. Mencken, famous literary critic, observed that words are made up of elements of other words, and many of these formations are used in particular combinations of consonants or vowels that convey a particular meaning, or picture. One of these sounds is the "sl" sound, particularly at the beginning of the word. Go through your dictionary sometime and check the words beginning with that combination; "slick, slimy, slither, slop, slippery, slough, slovenly, sluggish, slum, slut, slurp, sly, etc." Is the reason the public has such an adverse reaction to sludge simply its name? How can anything that has a name like—*sludge*—be beneficial?

Hence the move to use the word "biosolids" in place of sludge. But a rose by any other name smells just as sweet.

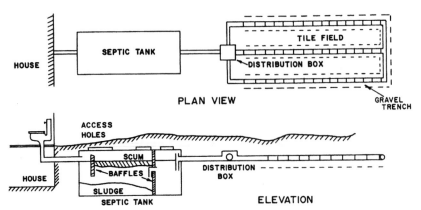

Figure 5.19. Septic tank and seepage field for an on-site disposal system .

5.4 ON-SITE WASTEWATER DISPOSAL

The original onsite system for human waste disposal is the *pit privy*, glorified in song and fable. The privy, still used in camps and other temporary residences, consist simply of a deep (perhaps 2 m or 6 ft) pit into which human excrement is deposited. When the pit fills up it is covered and a new pit is dug. The logical extension of the pit privy idea is a composting toilet that accepts not only human waste but food waste as well and produces a useful compost. With such a system, wastewaters from other sources such as washing and bathing (called *gray water*) are discharged separately. Composting toilets are odor free and quite effective in producing a useful product from waste. They do require maintenance, however, and a place to put the compost. Thus, they are not useful in an urban environment.

By far the most common onsite wastewater disposal method is a *septic tank* with a *drain field*. As shown in Figure 5.19, a septic tank consists of a concrete box that removes the solids and promotes partial decomposition. The solid particles settle out and eventually fill the tank and thus necessitate periodic cleaning. The water overflows into a drain field that allows the water to seep into the ground. The drain field consists of pipe laid in a trench. The pipes are plastic with holes on the bottom to allow the water to seep out. The system depends on the ability of the soil to accept the wastewater, or to percolate into the groundwater. many areas in the United States have soils that percolate poorly, and septic tank/drain field systems are inappropriate.

5.5 WATER POLLUTION LAW

Water law in the United States evolved from the *riparian doctrine* developed within the English common law. The basic doctrine is that the owner of land also owns the water on the land or the water that flows through the land. The legal responsibility of every *riparian owner* is therefore to not reduce the amount or not contaminate the water that flows into the land of the downstream riparian owner. In legal terms, riparian owners can expect that water flowing into their land is "undiminished in quantity and unimpaired in quality."

The riparian doctrine was useful and fair as long as the primary use of water was for watering livestock or other such agricultural uses. But when water began to be used by heavy industry such as coal mining and steel production it was clear that the downstream users were no longer receiving the water they had every legal right to expect.

Through the next 100 years the riparian doctrine continued to be violated more than respected and it became clear that common law was not adequate to meet the needs of society to enjoy clean water. States appeared to be powerless to pass significant legislation, partly because so many of the watersheds crossed state boundaries and it was unclear who was responsible for water quality.

Mrs. Sanderson vs. The Pennsylvania Coal Company

One of the most significant lawsuits to test the riparian doctrine was the suit brought by a Mrs. Sanderson against the Pennsylvania Coal Company in 1868.

Mrs. Sanderson was a riparian owner on Meadow Creek in Scranton, Pennsylvania, and used the water from the creek to fill a pond and to use as a domestic supply. She had settled on this land partly because of the purity of the water in Meadow Creek.

At about the same time, upstream from her the Pennsylvania Coal Company opened a new mine and used the water from the creek to wash the coal, discharging the untreated waste into the stream. The water became unusable both for her pond and as a source of domestic supply, and Mrs. Sanderson went to court to have her riparian rights asserted.

She seemed to have an open and shut case. The coal company was making the water unfit for use in her farm, and she demanded that the Pennsylvania Coal Company cease polluting the water.

To the surprise of many, she lost the case. The judge argued that mining coal was a legitimate use of one's land, and water was necessary to mine the coal. He observed that the coal company was using the land "naturally" in that it had not brought anything (such as chemicals) on to the land, and what the company was doing was lawful use of the land. There was no way to clean up the water before it left the mine site (so the lawyers argued) and therefore the company had the right to use the stream and not worry about polluting the stream. There is no record of what Ms. Sanderson did after the verdict came down. We now honor her as a courageous riparian owner who was willing to take on the powerful Pennsylvania Coal Company.

In the Western United States, water law was not based on English riparian doctrines, but instead on the *first-use principle*. While in the East, water belonged to the landowner only as it flowed across the land, in the West water was a commodity like any other resource. The purpose was to put water to use in its most productive form, and thus it was possible to use (literally consume) the water without worrying about downstream users. This can best be illustrated using a hydrograph, or a picture of the flow of water in river, such as Figure 5.20. The curve shows the flow in the river at any time over the year, and is sliced into layers. The first layer (a) is 1 million gallons per day (mgd), and the owner of this water is able to use that water all through the year. The right to use it belongs to the owner because he/she was there first to put the water to use and to claim the water right, much as gold mines were claimed during the gold rush. The owner of this water can sell the water rights to it to someone else, but the only way he/she would lose the rights is if they no longer used it for a beneficial purpose.

The second slice (b) is also owned by someone else, but note that at one time during the year there was not enough water in the river to allow this owner to draw water out of the river. This owner could only draw water as long as the 1 mgd remained in the river for the first owner to use. The third owner (c) only had water available during one time of the year. The placement of the owners along the river is not important, so owner (b) and (c) could be upstream but would not be allowed to use any water if the flow was less than 1 mgd.

The most important aspect of this relative to environmental pollution is that the *quality* of this water was of little, if any, importance. Water was water, and

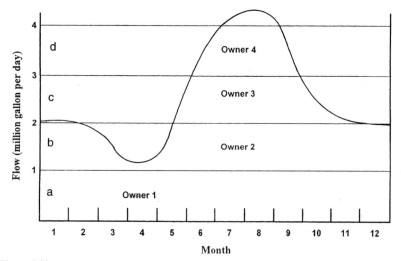

Figure 5.20. A hydrograph showing the ownership of water under the first-use doctrine of water law.

pollution simply did not matter. Only in later years in the development of this first-use doctrine, when water was being withdrawn for public supplies, did pollution by such uses as irrigation cause legal problems. The Colorado River, for example, is totally used in the United States, and when it finally reaches the Rio Grande, what water is left in the river is of such high salinity from leaching during irrigation that it is not usable for agriculture or most any other use. A large treatment plant has been constructed in the lower Colorado to remove some of this salinity and to restore the entire river (or what's left of it) to a useable condition.

The conclusion we have to come to is that the two prevailing water law doctrines, based on common law, are simply inadequate to protect water quality, and legislative law is required.

The first significant action on the federal level was the Clean Water Act of 1972, legislation championed by Senator Edwin Muskie of Maine. The present laws and regulations all derive their origins from this act. The landmark legislation set about to clean up the surface waters in the United States, but had little to do with drinking water quality and public health. In response to the need for stronger controls on water consumed by the public, the 1974 Congress passed the Safe Drinking Water Act that greatly expanded the role of the Environmental Protection Agency in protecting drinking water quality. These

---

EDMUND MUSKIE

Edmund Muskie (1914–1996) was a long-time senator from Maine. After getting his law degree from Cornell University, he served in the Navy during World War II and then entered politics, serving in the Maine Legislature before being elected governor in 1955. He was the first Democrat to be elected to the U. S. Senate from Maine.

During the 1960s he championed legislation for cleaner water and air, and was instrumental in the passage of both the Clean Air Act and the Clean Water Act. He became known as "Mr.Clean" both for his integrity and his stance on pollution issues. He was a front-runner for the Democratic presidential nomination but lost it when he broke down one snowy day defending his wife's integrity following a scurrilous editorial by the *Manchester Union Leader*. Much of the wisdom and effectiveness in the various environmental acts were the direct result of his work and energy. The main ideas for the control of pollution developed by Muskie and his staff have become the template for environmental legislation all over the world.

two pieces of legislation form the basic fabric of legislated water regulations in the United States.

Taking into account the various legislated laws relating to water, three primary types of water standards have evolved:

- drinking water standards
- effluent standards
- stream standards

Based on public health and epidemiological evidence, and tempered by a healthy dose of expediency, the EPA has established national drinking water standards for many physical, chemical, and bacteriological contaminants. Included in these standards are recommended standards for physical characteristics such as turbidity and taste, limits of chemicals such as inorganics (lead, arsenic, chromium, etc.) as well as some organics (e.g. DDT), and bacteriological standards for protecting people from ingesting pathogenic microorganisms.

Setting *drinking water standards* is exceedingly difficult since good data on the effect of the multiple constituents in water is seldom available. Consider the problem of arsenic in drinking water. We know that arsenic is a toxin and that drinking water containing 0.05 mg/L or arsenic (the 1942 U. S. Public Health Service standard) has been associated with causing cancer of the liver, bladder, lungs, skin, kidney, nasal passages, and prostate. Non-cancer effects on the ingestion of arsenic at low concentrations for long periods include cardiovascular, pulmonary, immunological, neurological, and endocrine effects. The European community has for many years used a standard of 0.01 mg/L, acknowledging that arsenic is toxic at any concentration. With the weight of new epidemiological information and the European experience, the EPA set the new drinking water standard for arsenic at 0.01 mg/L.

Why was this number chosen? Why not 0.005 mg/L, or even zero? We have the technology to achieve these levels, why not set the level so that we know that arsenic in our drinking water will not cause harm to the public? The answer lies in what is known as the *principle of expediency*. At the 0.01 mg/L level, about 95% of all public water supplies in the United States will be able to meet this standard, and only five percent of the public water treatment plants will have to adjust their treatment or change their raw water sources. Had the EPA set the new standard at 0.00 mg/L of arsenic, *all* public water supply plants would have been unable to meet it, and the cost to the public would have been enormous.

Another excellent example of the principle of expediency at work is the moving coliform standard in drinking water. Coliforms are microorganisms that live in our intestinal tract. While coliforms themselves are seldom harmful, their presence in drinking water suggests that the water may have been contaminated from a human source and that pathogenic organisms might be present. When the first coliform standard was set by the U. S. Public Health Service in 1914 at 2 coliforms/100 mL, many communities were not able to produce drinking water that met this strict standard. But technology was available, and using both filtration and chlorine disinfection, all cities were eventually able to meet this standard (resulting in a dramatic decrease in water-borne disease). In 1925 the PHS published a new drinking water standard and set the coliform limit at 1 coliform/100 mL. Why were they able to get away with setting such a strict standard? Because most cities were already producing water that met this standard, it was expedient to do so. As we learn more about the treatment of drinking water, and new chemicals and technologies become available at reasonable cost, water quality standards will continue to get increasingly strict. The coliform standard has continued to drop, and at the present time the coliform standard is at essentially 0.05 coliforms/100 mL, a dramatic drop from the 2 coliforms/100 mL of 90 years ago. The principle of expediency is at work.

EARLE PHELPS

Earle Phelps (1876–1953) graduated from MIT with a degree in chemistry. He was a student of William Sedgwick, who also mentored other notable early sanitary engineers and public health scientists. After graduation, Phelps worked for a while with the Massachusetts Board of Health at the Lawrence Experiment Station and was closely involved in the development of new treatment technology. He moved on first to be a faculty member at MIT and then to the U. S. Public Health Service, where he did his most influential work on stream pollution, authoring the classic text *Stream Pollution*, which for decades was considered the definitive text on the subject.

Phelps was immensely practical. He recognized that pollution would always exist, but the objective would be to reduce the effect to some reasonable level that can be economically attained using available technology. As he stated: "It is wasteful and therefore inexpedient to require a nearer approach to [the optimal] than is readily obtainable under current engineering practices and at justifiable costs." From this reasoned approach to pollution was born the "Principle of Expediency."

Phelps argued that the objective of regulatory science is to couple the ethics of societal protection with the science of regulation. He defined public health practice as "the application of the science of preventive medicine, through government, for social ends."

To recap: Why don't we set drinking water standards at levels that would assure total safety? Why not zero mg/L of arsenic and zero coliforms/100mL? Because we simply do not have the resources to eliminate all toxic materials in drinking water. Therefore, the standards are set at the lowest level that is technologically and economically practicable, illustrating the principle of expediency.

In addition to monitoring the safety of our drinking water, the EPA also oversees, and states operate, programs designed to reduce the flow of pollutants into natural watercourses. The limits on all discharges into watercourses are known as *discharge standards*. All industries and communities that discharge wastewater into surface watercourses are required to obtain a NPDES (National Pollution Discharge Elimination System) permit. While some detractors have labeled these "permits to continue pollution," the permitting system has had a major beneficial effect on the quality of surface waters. The intent is to tighten these limits as required in order to enhance water quality.

Tied to the effluent standards are quality-based surface water standards, often called *stream standards*. All surface waters in the USA are now classified according to a system of standards based on their greatest beneficial use. The highest classification is usually reserved for pristine waters, with the best use as a source of drinking water. The next highest include waters that have had wastes discharged into them but that nevertheless exhibit high levels of quality. The categories continue in order of decreasing quality, while the lowest water quality is useful only for irrigation and transport. The categories are often linked to the stream's ability to propagate fish.

The objective is to attempt to establish the highest possible classification for all surface waters and then to use the NPDES permits to turn the screws on polluters and enhance the water quality and increase the classification of the watercourse. Once at a higher classification, no discharge is allowed that would degrade the water to a lower quality level.

The EPA gave each individual state the opportunity to develop its own stream standards, arguing that the engineers and scientists from each state would naturally be most familiar with the details of their surface water and the

TABLE 5.2. Simplified Stream Standards.

Classification	Highest beneficial use	Maximum allowable coliform concentration, number/100 mL
A	Raw water for drinking water treatment	less than 10
B	Swimming and other water contact sports	10 to 1000
C	Non-water contact sports	1000 to 10,000
D	Irrigation of crops	less than 100,000

needs for such standards. Hence the surface water standards are different for each state.

Consider a simplified form of such standards, as listed in Table 5.2, in which only one criterion, coliform organisms (indicators of human pollution) is used to differentiate between the various classifications. In real stream standards, many different measurements of water quality are used. We are making this example (unrealistically) simple to show how the process works.

Consider now a small watershed as shown in Figure 5.21(a). The stream water in these creeks is sampled, and the level of coliform organisms is measured, with the results as shown in the figure. Comparing the measured results with the standards, we see that the highest classifications that can be assigned to these streams is shown in Figure 5.21(b). For example, the uppermost reach has a coliform level of 160 coliforms/100 mL, and since this is between 10 and 1000, the stream is classified as a B stream. All other sections of the streams are classified according to the coliform organism levels measured in the streamwater.

Now the state will use the effluent standard permit system to increase the water quality in the streams. Suppose, for example, a wastewater treatment plant is located as shown in Figure 5.21(c), and the effluent from this plant is responsible for the elevated coliform levels downstream of its discharge, resulting in a stream quality classification of C. The state pollution control people now suggest to the community that if they want to continue to operate this plant, the effluent coliform levels have to be reduced. Suppose the community is responsive to this request (request of course is a euphemism—legal clout is often needed), and the effluent quality is improved. If the effluent coliform level is reduced, resulting in coliform levels as shown in Figure 5.21(c), the state immediately reclassifies this stream to a higher classification [Figure 5.21(d)]. Once at that classification, it can never go back to the lower water quality.

What would happen if an industry, such as a hog farm, wants to locate on the same stream? In order for them to discharge their treated wastewater into the stream they would essentially have to disinfect their discharge (kill off all the coliforms and other microorganisms) before they can obtain a discharge permit, since the now more restrictive level of water quality cannot be compromised. The objective is to eventually be able to reclassify all rivers and streams as Class A or B.

Sometimes the state's efforts are complicated by other societal considerations. Some states have "right-to-farm" legislation that allows farmers to discharge normal farm waste, such as runoff from pastures. Such legislation is then used inappropriately (but legally) by large-scale hog factory farms.

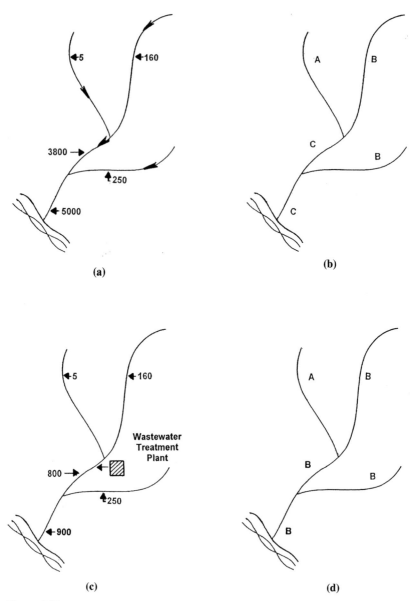

Figure 5.21. Stream standards for a typical watershed. (a) shows the coliform levels at the beginning of the process, (b) are the resulting stream classifications according to the standards in Table 5.2, (c) are the coliform levels after the wastewater treatment plant upgrades its treatment, and (d) are the resulting stream classifications.

The Pittsburgh Case

The powers of dilution were for a long time thought to be the answer to the wastewater disposal problems. The old saw was "The solution to pollution is dilution."

George Fuller, an eminent sanitary engineer observed during the first decade of the 1900s that "the disposal of sewage by dilution is a proper method when by dispersion in water the impurities are consumed by bacteria and larger forms of plant and animal life or otherwise disposed of so that no nuisance results."

But using the waterways as treatment systems was not acceptable to downstream communities, and many of them sued the upstream dischargers on the basis of the riparian water law which states that the water flowing through one's land must not be diminished in quantity or unimpaired in quality. This upset the larger cities, which were the major cause of pollution, and the magazine *City Hall* intoned in an editorial:

> The average city or individual considers a sewage disposal plant a novelty or unnecessary expense and they only build them when a threat of a damage suit arises and then to appease the wrath of the plaintiff, or in the case of some states, to meet with the seemingly useless requirements of the State Board of Health. Then any sort of a plan or device prepared or proposed by an engineer or concern will do, as long as it is not too expensive.

But the state health departments thought otherwise and pushed to have cities construct effective wastewater treatment plants. The conflict between cities that polluted the rivers and the communities that used the rivers for a water supply was never more clearly etched than in the Pittsburgh case.

In 1905 Pennsylvania passed a law forbidding the discharge of untreated sewage into streams or rivers. In 1910, Dr. Samuel G. Dixon, the Pennsylvania Commissioner of Health, required Pittsburgh to submit a comprehensive plan for replacing the city's combined sewerage system with a separate system and a treatment plant. In response, the city hired two well-known sanitary engineers, Allen Hazen and George Whipple, to make recommendations. After a year of investigation, Hazen and Whipple submitted what the journal *Engineering Record* called "the most important sewerage and sewage disposal report made in the United States."

Hazen and Whipple estimated that replacing Pittsburgh's combined sewers with a separate system and building a treatment plant would cost Pittsburgh taxpayers a minimum of $46 million (1912 dollars), but that the 26 towns downstream from Pittsburgh on the Ohio River could provide filtered water for their residents for far less money. It was a no brainer to them that the most efficient option was to have the 26 towns treat the water, and allow the Ohio River to be an open conduit for the removal of human waste. No precedent existed, they argued, "for a city's replacing the combined system by a separate system for the propose of protecting supplies of other cities." Hazen and Whipple concluded that "no radical change in the method of sewerage or of sewage disposal as now practiced by the City of Pittsburgh is now necessary or desirable."

(continued)

(continued)

Engineering opinion overwhelmingly supported the Hazen and Whipple report and viewed the controversy as an issue on "how far engineers are at liberty to exercise their own judgment as to what is best for their clients and how far they must give way to their medical colleagues."

Uncertain of his ability to compel Pittsburgh to build a separate system and treat its own sewage, Dixon retreated, and issued the city a temporary discharge permit. The State Commissioner of Health continued to issue such permits to the city until 1939.

Source: Martin Melrose, *The Sanitary City*, Johns Hopkins Press, Baltimore, 2000, and Tarr, J. A. and F. C. McMichael "Historic turning Points in Municipal Water supply and Wastewater disposal, 1850–1932," *Civil Engineering*, vol. 47, no. 10, October 1977.

The Pigeon River

The Pigeon River flows from the Great Smokey Mountains of North Carolina into Tennessee, passing through Canton, North Carolina, which boasted a large paper mill. Upstream of Canton the Pigeon river was clear and clean, but below the paper mill the river turned a dark brown color and emitted a sulfuric odor. The state of North Carolina for many years tried to get the paper mill to clean up its discharges, but the importance of the mill to the economically depressed area make firm legal action politically impossible. Thus the Pigeon River remained an open industrial sewer emptying into the state of Tennessee.

This did not sit well with the folks in Tennessee, but they could do nothing about it. The water was polluted out of their jurisdiction, and setting high stream standards on the Tennessee portion of the Pigeon River made no sense, because such rules had to be enforced in North Carolina. Finally, the Tennessee water pollution agency pushed the U. S. EPA to take action. The controversy escalated with the discovery of dioxin in the stream water, and the State of Tennessee put up warning signs along the river prohibiting the eating of fish caught from the river.

In 1992, the mill was sold to a consortium of workers, and was renamed Blue Ridge Paper Products. With assistance from numerous sources, the new company was able to fund extensive pollution control measures to the tune of $330 million. With this treatment, the dioxin levels dropped to below the toxic limit, and the water color returned to normal in the Tennessee portion of the Pigeon.

The dilemma illustrates one of the problems in water pollution control in the United States. If each state is allowed to set its own water quality standards, these may be incompatible as a river flows across state lines. The standards set by the downstream state must be enforced by the upstream state, which would not attain any benefit from the money spent for such pollution control.

5.6 SYMBOLS

BOD = biochemical oxygen demand, mg/L

C_o = initial (influent) concentration, mg/L

C = final (effluent) concentration, mg/L

COD = chemical oxygen demand, mg/L

D = dilution of the sample

F = food to a biological treatment system, often expressed as pounds of BOD per day

FDO = final dissolved oxygen of the sample bottle

FDO' = final dissolved oxygen of the seeded dilution water

F/M = food/microorganism ratio

IDO = initial dissolved oxygen in the bottle containing both effluent sample and seeded dilution water

IDO' = initial dissolved oxygen of the seeded dilution water

M = microorganisms concentration in aeration tank, often expressed as concentration of VSS

MLSS = mixed liquor suspended solids

N = nitrogen concentration, often as mg/L

P = production of primary sludge, dry solids, such as kg/day

P = phosphorus concentration, often as mg/L

Q = influent flow rate to the treatment plant

R = percent removal of a pollutant,

SS = concentration of suspended solids, often as mg/L

SVI = sludge volume index

TKN = total Kjeldahl nitrogen, mg/L

VSS = volatile suspended solids

W = BOD loading, such as kg BOD/day

X = mL of seeded dilution water in sample bottle

Y = total mL in the BOD bottle

5.7 PROBLEMS

5.1 Using reasonable values, estimate the sludge production in a 1.0 m³/s wastewater treatment plant. Estimate both sludge volume per time and dry solids per time.

5.2 A 1-L cylinder is used to measure the settleability of 0.5% suspended solids sludge. After 30 min, the settled sludge solids occupy 600 mL. Calculate the SVI.

5.3 What measures of "stability" would you need if a sludge from a wastewater treatment plant were to be

a. placed on the White House lawn.
b. dumped into a trout stream.
c. sprayed on a playground.
d. sprayed on a vegetable garden.

5.4 A wastewater treatment plant is to be designed for a flow of 300 m³/hour. What would the production of raw primary sludge if the recovery of solids by the primary clarifier is expected to be 60% and if the solids concentration of the underflow from the primary clarifier is expected to be 5% solids? Express your answer in both dry kg/day and in m³ wet sludge/day.

5.5 A small treatment plant for a dairy waste has a flow of 10 m³/hour and an influent BOD of 500 mg/L. There are no suspended solids in the influent, so there is no primary clarifier. The wastewater goes directly into an activated sludge system that includes a final clarifier. What is the expected production of waste activated sludge, expressed as kg of dry sludge/day?

5.6 The following data were reported on the operation of a wastewater treatment plant:

Parameter	Influent, mg/L	Effluent, mg/L
BOD	200	20
Suspended solids	220	15
Phosphorus	10	0.5

a. What percent removal was experienced for each of these parameters?
b. What kind of treatment plant might have produced such an effluent? Draw a flow diagram showing what you believe the unit operations might have been.

5.7 Describe the condition of a primary clarifier one day after the raw sludge pumps broke down. Imagine you have to do a report on what you might have experienced. Use all of your senses (well, maybe not all).

5.8 The influent and effluent data for a secondary treatment plant (activated sludge) are shown below. Also shown are the effluent limits (this what the effluent is required to be).

Parameter	Influent, mg/L	Effluent, mg/L	Effluent limits, mg/L
BOD	180	40	30
Suspended solids	200	260	30
Phosphorus	10	8	10

a. What are the percent removal efficiencies?
b. Which of the effluent limits are not met?
c. What do you think is wrong with this plant? What might be happening to cause such an effluent?

5.9 Draw a block diagram of unit operations necessary to treat the following wastes to effluent limits of BOD = 20 mg/L, SS = 20 mg/L, and P = 1 mg/L.

Waste	BOD, mg/L	Suspended solids (SS), mg/L	Phosphorus (P), mg/L
Domestic	200	200	10
Chemical industry	40,000	0	0
Pickling canning	0	300	1
Fertilizer manufacturing	300	300	200

5.10 If the volume of sludge after 30 minutes was 300 mL both before and during a bulking problem, and SVI was 100 and 250 before and during the bulking problem, what was the mixed liquor suspended solids (MLSS) concentration before and during the bulking problem?

5.11 The influent to the wastewater treatment plant has a BOD of 180 mg/L. The primary clarifier removes 40% of this BOD, sending the rest to the activated sludge system. The flow to the plant is 2 million gallons per day, and the aeration basin has dimensions of 20 × 6 × 4 m. The operator runs the aeration basin at a MLSS concentration of 3,000 mg/L. What is the food-to-microorganism ratio for this plant?

5.12 A treatment plant operator is running her MLSS in the aeration basin at a solids concentration of 1000 mg/L. She decides to increase the solids concentration in the aeration basin. How would she go about doing this? What valves would she have to turn and what changes in her operating process would be necessary?

5.13 A transoceanic flight in a Boeing 747 with 430 persons on board takes 7 hours. Assume that it requires 5 liters of water to flush the toilet. Make any other assumptions necessary. What fraction of the total payload (people) would the flush water represent? (Assume an average passenger weighs 70 kg.) How would you reduce the weight of the flush water? (The railroad system of simply discharging the waste into air is for obvious reasons illegal.)

5.14 Download a topographic map of any small town in the United States. (www.usgs.gov) Make a photocopy of this page, and sketch a sewerage system for this community. Sewers have to be available to all buildings; they have to flow downhill; they have to be in straight lines and connected by manholes; and the entire system has to empty into a intercepting sewer. Use pumping stations and force mains if necessary. Clearly indicate with colored pens the direction of flow in the sewers and the force mains.

5.15 A community of 100,000 produces an average dry weather wastewater flow of 500 L per capita per day. What might be the range of expected wet weather flow?

5.16 Using your campus map, design a sewerage system for the campus, placing manholes and pumping stations as needed.

5.17 A treatment plant operator is having a difficult time with his plant. The BOD of the effluent keeps on being too high. He develops an ingenious solution, however. He discovers that if he runs the BOD test in the refrigerator instead of the incubator, he gets much lower effluent BOD values. Why is this funny?

5.18 Using the Internet, investigate the stream standards for any one of the 50 states. Copy them and write a short paragraph stating how the specific state would reasonably be expected to formulate such standards.

5.19 In your own words, define the "principle of expediency," and use one example of how it works to some aspect of civilized life other than the quality of drinking water.

5.20 Suppose Ms. Sanderson (the one of Pennsylvania Coal Company fame) had lived in California. Would she have had greater claim to the quality of the water flowing in the creek, or less claim? Speculate how the trial would have ended if Ms. Sanderson lived in California.

5.21 Using the Internet, investigate how one of the drinking water standards has been set. Read any criteria documents published by the U. S. EPA and research published by others. Do you believe, based on what you have learned about this pollutant, that the present standard is properly set?

5.22 Call your wastewater treatment plant operator and ask what the effluent standards are for this wastewater treatment plant. Why do you believe they are set at these limits? Is there something special about the watercourse into which the plant discharges?

5.23 A lab tech runs BOD tests on a stream water downstream from a wastewater plant discharge. She decides to use non-seeded dilution water, and obtains the results below:

Bottle No.	Sample	Initial DO mg/L	Final (5 day) DO mg/L
1	Dilution water	9.2	9.2
2	Full strength sample	9.1	0.8
3	1:2 dilution	9.0	4.5
4	1:4 dilution	9.0	6.8

a. What BOD_5 should she report?
b. Was her decisions to use a non-seeded dilution water a good one? How do you know?
c. Suppose she had made a mistake and placed the BOD bottles in the refrigerator instead of the incubator. Would the BOD_5 values have been the same, higher, or lower?

CHAPTER 6

Hazardous Waste

HAZARDOUS waste management is a relatively recent field of practice for environmental engineers and scientists. Prior to 1976, numerous hazardous waste sites existed in the U.S.; however, hazardous waste was not officially defined by EPA, nor were the criteria that categorized hazardous waste sites. During the 1970s, various episodes of hazardous waste releases in communities captured the public spotlight, received national attention, and resulted in Congress initiating hazardous waste legislation.

Discussion in this chapter is limited to hazardous waste as defined by federal legislation. Radioactive wastes are regulated under the Atomic Energy Act and are therefore not considered as hazardous wastes. For mixed wastes that contain hazardous and radioactive wastes, EPA, the Nuclear Regulatory Commission (NRC), and the Department of Energy (DOE) have regulatory authority.

One of the most immediate problems today is the clean-up of existing hazardous waste dumps, and Congress has focused attention on the development of legislation to address this problem. As of this writing, over 2200 hazardous waste sites have been deemed eligible for federal assistance for cleanup, and nearly 30,000 sites await evaluation to determine how serious they are.

6.1 HAZARDOUS WASTE LAWS

Congress determined that two lines of action were required to address the hazardous waste problem. Numerous abandoned facilities were known to be releasing toxic pollutants, and currently operating industries were producing (and therefore, disposing of) increasing quantities of hazardous wastes. Arguably, it would seem wise to write laws that would address the abandoned sites that caused so much public concern. However, the first law addressed the

225

Love Canal

It was a grand dream. William T. Love recognized an opportunity for electricity generation from Niagara Falls and the potential for industrial development. To achieve this, Love wanted to build a canal that would also allow ships to pass around the Niagara Falls and travel between the two great lakes, Erie and Ontario. The project started in the 1890s, but soon floundered due to inadequate financing and also due to the development of alternating electrical current, which made it unnecessary for industries to locate near a source of power production. Hooker Chemical Company purchased the land adjacent to the Canal in the early 1990s and constructed a production facility. In 1942, Hooker Chemical began disposal of its industrial waste in the Canal. This was war time in the United States, and there was little concern for possible environmental consequences. Hooker Chemical (now Occidental Chemical Corporation) disposed of over 21,000 tons of chemical wastes including halogenated pesticides, chlorobenzenes, and other hazardous materials into the old Love Canal. The disposal continued until 1952, at which time the company covered the site with soil and deeded it to the city of Niagara Falls, which wanted to use it for a public park. In the transfer of the deed, Hooker specifically stated that the site was used for the burial of hazardous materials, and warned the city that this fact should govern future decisions on the use of the land. Everything Hooker Chemical did during those years was legal and aboveboard.

About this time, the Niagara Falls Board of Education was looking around for a place to construct a new elementary school, and the old Love Canal seemed like a perfect spot. The area was a growing suburb, with densely packed single-family residences on streets paralleling the old canal. A school on the site seemed like a perfect solution, and so it was built.

In the 1960s the first complaints began, and intensified during the early 1970s. The groundwater table rose during those years and brought to the surface some of the buried chemicals. Children in the school playground were seen playing with strange 55-gallon drums that popped out of the ground. The contaminated liquids started to ooze into the basements of the nearby residents, causing odor and health problems. More importantly perhaps, the contaminated liquid was found to have entered the storm sewers and was being discharged upstream of the water intake for the Niagara Falls water treatment plant.

The situation reached a crisis point, and President Jimmy Carter declared an environmental emergency in 1978, resulting in the evacuation of 950 families in an area of 10 square blocks around the canal.

But the solution presented a difficult engineering problem. Excavating the waste would have been dangerous work, and would probably have caused the death of workers. Digging up the waste would also have exposed it to the atmosphere resulting in uncontrolled toxic air emissions. Finally, there was the question as to what would be done with the waste. Since it was all mixed up, no single solution, such as incineration would have been appropriate. The U. S. EPA finally decided that the only thing to do with this dump was to isolate it and continue to monitor and treat the groundwater. The contaminated soil on the school site was excavated, detoxified and stabilized and the building itself was razed. All of the sewers were cleaned, removing 62,000 tons of sediment that had to be treated and removed to a

(continued)

(continued)

remote site. At the present time, the groundwater is still being pumped and treated, thus preventing further contamination.

The cost is staggering, and a final accounting is still not available. Occidental Chemical paid $129 million and continues to pay for oversight and monitoring. The rest of the funds are from the Federal Emergency Management Agency and from the U. S. Army, which was found to have contributed waste to the canal.

The Love Canal story had the effect of galvanizing the American public into understanding the problems of hazardous waste, and was the impetus for the passage of several significant pieces of legislation such as the Resource Conservation and Recovery Act (RCRA), the Comprehensive Environmental Response, Compensation, and Liability Act (CERCLA), and the Toxic Substances Control Act.

Dioxin and Times Beach, Missouri

Times Beach was a popular resort community along the Merrimac River, about 17 miles west of Stylus. With few resources, the roads in the town were not paved and dust on the roads was controlled by spraying oil. For two years, 1972 and 1973, the contract for the road spraying went to a waste oil hauler named Russell Bliss. The roads were paved in 1973, and the spraying ceased.

Bliss obtained his waste oil from the Northeastern Pharmaceutical and Chemical Company in Verona, Missouri, which manufactured hexachlorophene, a bactericidal chemical. In the production of hexachlorophene, considerable quantities of dioxin had to be removed and disposed of. A significant amount of the dioxin was contained in the "still bottoms" of chemical reactors, and the company found that having it burned in a chemical incinerator was expensive. The company was taken over by Syntex Agribusiness in 1972, and the new company decided to contract with Russell Bliss to haul away the still bottom waste without telling Bliss what was in the oily substance. Bliss mixed it with other waste oils, and this is what he used to oil the roads in Times Beach, unaware that the oil contained high concentrations of dioxin (greater than 2000 ppm). Bliss also used the oil to spray the roads in nearby farms, and it was the death of horses at these farms that first alerted the Center for Disease Control, which sampled the soil at the farms. They found the dioxin, but did not make the connection with Bliss. Finally in 1979 the U.S. EPA became aware of the problem when a former employee of the company told them about the sloppy practices in handling the dioxin-laden waste. The EPA converged on Times Beach in "moonsuits," and panic set in among the populace. The situation was not helped by the message from the EPA to the residents of the town: "If you are in town it is advisable for you to leave and if you are out of town do not go back."

The residents of Times Beach were eventually evacuated and their properties were bought by the federal government for $33 million.

After everyone had moved out of Times Beach, the houses were razed and Syntex Corporation was required to build an incinerator for burning the contaminated soil. The Superfund site was eventually decontaminated at a cost of over $200 million and the site now is a beautiful riverside park.

The Kepone Tragedy

The Allied Chemical plant in Hopewell, Virginia, was in operation since 1928 and had produced many different chemicals during its lifetime. In the 1940s the plant started to manufacture organic insecticides, which had recently been invented, DDT being the first and most widely used. In 1949 it started to manufacture Kepone, a particularly potent herbicide that was so highly toxic and carcinogenic that Allied withdrew its application to the Department of Agriculture to sell this chemical to American farmers. It was, however, very effective and cheap to make, and so Allied started to market it overseas.

In the 1970s the national pollutant discharge permit system went into effect and Allied was to list all the chemicals it was discharging into the James River. Recognizing the problem with Kepone, Allied decided not to list it as part of their discharge, and a few years later "tolled" the manufacture of Kepone to a small company called Life Science Products Co., set up by two former Allied employees, William Moore and Virgil Huntofte. The practice of "tolling," longstanding in chemical manufacture, involves giving all the technical information to another company as well as an exclusive right to manufacture a certain chemical—for the payment of certain fees, of course. Life Sciences Products set up a small plant in Hopewell and started to manufacture Kepone, discharging all of its wastes into the sewerage system.

The operator of the Hopewell wastewater treatment plant soon found that he had a dead anaerobic digester. He had no idea what killed his digester, and tried vainly to restart it by giving it antacids. Recall the discussion in Chapter 5 regarding the sensitivity of methane-producing organisms in anaerobic digesters.

In 1975 one of the workers at the Life Sciences Products plant visited his physician, Dr. Chow, complaining of tremors, shakes, and weight loss. Dr. Chow took a sample of blood and sent it to the Center for Disease Control in Atlanta for analysis. What they discovered was that the worker had nearly 8 mg/L of Kepone in his blood! The State of Virginia immediately closed down the plant and took everyone into a health program. Over 75 people were found to have Kepone poisoning. It is unknown how many of these people eventually developed cancer. The Kepone that killed the digester in the wastewater treatment plant flowed into the James River, and over 100 miles of the river was closed to fishing due to Kepone contamination. The sewers through which the waste from Life Science flowed were so contaminated that they were abandoned and new sewers built. These sealed sewers are still under the streets of Hopewell, and serve as a reminder of corporate avarice.

management of currently generated hazardous waste. In retrospect, this had a more profound effect in that it reduced the likelihood that new abandoned facilities would be created. Once a regulatory program for currently generated hazardous waste was in place, Congress then took action on the cleanup of abandoned hazardous waste facilities.

6.1.1 Resource Conservation and Recovery Act

The 1976 *Resource Conservation and Recovery Act* (RCRA) authorized EPA to write regulations that governed businesses that generated hazardous waste. The primary objective of RCRA was to control the generation, transport, storage, treatment, and disposal of hazardous waste. This objective was met by first classifying hazardous wastes.

A waste is classified hazardous if it is either specifically listed in the regulations or if it displays one of the defined characteristics of hazard. Hazardous waste regulations actually include four lists of hazardous wastes: the "F", "K", "P", and "U" lists. The "F" list includes individual chemicals from nonspecific sources, the "K" list is from specific industrial sources, whereas the "P" or "U" lists include commercial products that are either out-of-date or off-specification ("P" denotes acutely hazardous, "U" denotes toxic).

Alternatively, any waste material may be classified as hazardous according to one of the following characteristics:

- ignitable
- corrosive
- reactive
- toxic

An ignitable waste will "flash" or combust below 60°C, a corrosive waste has a pH less than 2.0 or greater than 12.5, a reactive waste reacts violently when exposed to air or water, and toxic wastes contain greater than 100 times the concentration of various chemicals listed by the Safe Drinking Water Act (based on a laboratory test known as the Toxic Characteristic Leaching Procedure).

Undesignated wastes may also be regulated as hazardous if a listed hazardous waste is mixed with a non-hazardous waste. This was meant to discourage generators of hazardous waste from "diluting" hazardous materials with non-hazardous. This classification is known as the *mixture rule* and, if applied, results in a net increase in hazardous waste that must be managed and disposed of. This can be very costly financially and in terms of future liability.

Recognizing that some treatment processes, such as hazardous waste incinerators and landfills produce residues, EPA developed the concept of a *"derived-from"* hazardous waste. Examples of derived from hazardous wastes are incinerator ash and hazardous landfill leachate.

Industries may petition EPA to remove or "delist" a particular waste from the hazardous classification by submitting laboratory analysis that demonstrates either that chemical concentrations in the waste are below regulatory levels or the waste no longer exhibits characteristics of hazardous wastes.

In order to understand the types and magnitude of hazardous wastes that were being produced, EPA instituted a tracking system in which all generators of hazardous wastes were required to obtain an identification number and to complete a document known as a manifest that detailed the type, quantity, transport, storage, treatment and disposal of each waste. All generators of hazardous wastes are required to forward copies of each manifest to EPA, state environmental agencies, and to maintain a file of all manifests. For regulatory purposes, generators were defined as facilities that produced greater than 1000 kg/month of hazardous waste. Generators that produced less than 1000 kg per month were termed *small quantity generators* and were exempted from most regulatory requirements. Also, businesses involved in the treatment, storage, or disposal were required to obtain an identification number and apply for a permit to manage hazardous wastes. First, a RCRA Part A permit would be issued by EPA based on a brief application submitted by a treatment, disposal or storage facility. The Part A permit was temporary. Subsequently, they would have to apply for an RCRA Part B permit. The Part B application was detailed and technical, and often required the services of an environmental engineering consulting firm to complete.

EPA made the generator of hazardous waste ultimately responsible by establishing 'cradle to grave' responsibility. This meant that the generator is responsible for the waste from the moment it is produced, during storage, transport, treatment, and even after disposal. Cradle to grave liability implies that the generator is forever responsible—even if a licensed hazardous waste landfill eventually causes contaminated ground water, all generators that sent waste to the facility can be held liable for the cleanup.

RCRA also imposed specific design and performance requirements on hazardous waste landfills and incinerators. When capacity was reached, the landfill closure procedure included installation of an impermeable cap (to minimize leachate production) and a vegetative cover to limit erosion. Hazardous waste incinerators were required to meet specific efficiency levels known as *destruction removal efficiency*. For most hazardous wastes, incinerators were required to achieve 99.99% destruction, whereas for chemicals on the hazardous waste "P" list, the required destruction was set at 99.9999%.

6.1.2 CERCLA

In 1980, Congress passed CERCLA with the objective of responding to threats to human health or the environment from previously disposed of hazardous wastes. CERCLA is the acronym for the *Comprehensive Environmental Response, Compensation, and Liability Act.* Its name gives an indication of the major issues of this legislation: environmental response pertains to hazardous waste sites that were created from past disposal practices; compensation refers to the financial support necessary to clean up hazardous waste sites, and liability would be assigned to generators of the waste found at sites.

To make sure past disposal of hazardous waste was cleaned up properly, EPA was given authority and a $1.6 billion budget from taxes on major industries, such as chemical manufacturers and petroleum refiners. Because of the contributions from industry, the clean-up fund became known as the *Superfund*, and the sites to be cleaned up became known as *Superfund sites*. Rather than use Superfund money for all sites, EPA's strategy is to identify *potentially responsible parties* (PRPs) and convince them to pay for cleanup of a site. In this manner some of the fund could be used for sites where no PRPs could be identified.

EPA was given leverage to help convince PRPs to clean up sites, by means of the liability assigned by CERCLA. Specifically, generators of hazardous waste at abandoned sites are assigned strict, or joint and several liability. Strict liability means that negligence need not be demonstrated for a party to be liable. As an example, consider a local gasoline station that stores its gasoline in underground storage tanks (this is typical). Suppose a neighbor with a ground water well discovers that the water in the well has a strange odor and tastes bad. The neighbor sues the gasoline station owner for contaminating the well. In his defense, the station owner shows detailed records that show close management of gasoline inventory and daily verification of the amount of gasoline pumped from the underground tanks. The station owner is still found liable based on strict liability. That is, even though the owner was not negligent and had no idea that his tanks were leaking, he is still responsible for the neighbor's contaminated well. By contrast, according to joint and several liability, multiple PRPs at a hazardous waste site can be made to contribute (equally or proportionate to the amount of waste disposed) to the site cleanup. Alternatively, if only one potentially responsible party of a group can be located, the one PRP can be held responsible for the cleanup of the entire site. The leverage of liability plus the ability of EPA to clean up the site of an unwilling PRP followed by a lawsuit for up to three times the actual cost has gone a long way to convince PRPs to clean up contaminated sites.

Another requirement of CERCLA is that EPA is directed to maintain a list of the 400 "worst" sites in the country. To everyone's surprise, the list of 400 "worst" soon became much longer than that and a means for determining the priority for site cleanup became necessary. In order to rank the sites, EPA uses a hazard ranking system known as the Mitre model. The model considers proximity of the site to humans, obvious discharges to the environment, open access to the site, likelihood of contamination of drinking water sources and many other factors. Once acknowledged, each of the factors is assigned a numerical value that is used to calculate the overall ranking of the site. Highly ranked sites are placed on the *National Priorities List*, which is updated annually by EPA.

The list is a part of the *National Contingency Plan*, which also includes the strategy that must be employed to bring a contaminated site to acceptable cleanup levels. The list dictates that first, a *remedial investigation* will be conducted to determine sources of contamination, the pathways that contaminants are taking to leave the site, and the potential adverse effects on human health and the environment. Once the remedial investigation is underway, a feasibility study is initiated to assess alternatives that are applicable in cleaning up the site. The *feasibility study* culminates with the selection of the optimal alternative for site cleanup. This is determined based on engineering criteria, such as applicability, performance, costs, operational requirements, size and availability of equipment. EPA and state regulatory agencies review the plans and issue their record of decision, which details what and where contamination exists and how it will be cleaned up.

6.1.3 Hazardous and Solid Waste Amendments

By 1984, it was recognized that modifications to RCRA were needed. At that point Congress implemented the *Hazardous and Solid Waste Amendments* (HSWA). Based on information from manifests and RCRA Part B applications, EPA learned that the technology of choice for most of the currently generated hazardous waste was landfilling. HSWA sought to discourage landfilling by invoking a land disposal restriction (also known as "the land ban"). EPA was directed to examine within five years the suitability of each listed waste for land disposal. Wastes that were not examined would automatically be banned from land disposal. EPA was also directed to establish treatment standards for specific hazardous chemicals of concern. Some chemicals were assigned concentration limits; whereas others were "technology-specific" (i.e., dioxin contaminated wastes must be incinerated). At that time the Reagan administration campaigned on a program of smaller government and had gutted EPA's budget. These policies made implementing HSWA a daunting task for EPA.

HSWA also established new authority for EPA to force generators of hazardous waste to comply with regulations. Under HSWA, EPA was given authority to enforce corrective action for RCRA facilities in need of cleanup.

In order to bring more generators into the full regulatory requirements, HSWA lowered the small quantity generator exemption to 100 kg/month. That is, those who generated greater than 100 kg/month of hazardous waste were now required to comply with all hazardous waste requirements. Note that the small-quantity generator exemption is unique among countries in the industrialized world. Indeed, some states in the U.S., notably California, provide no exemption to hazardous waste generators, regardless of the amount produced.

In another effort to reduce the possibility of new contaminated sites, HSWA set strict requirements for facilities with underground storage tanks that held products containing hazardous chemicals. This was based on the recognition that there were tens of thousands of *leaking underground storage tanks* causing groundwater contamination. HSWA required that owners of tanks implement a routine leak testing program, repair or replace leaking tanks, and implement new construction requirements for new tanks (cathodic protection, steel in non-corrosive environments), spill and overflow protection, leak detection, and most important, owners must demonstrate financial responsibility for potential tank leakage. This final item essentially forced many private owners of gasoline stations out of business since they could not afford the high insurance premiums required to insure cleanup of soil and groundwater should the tanks leak.

6.1.4 SARA

In 1986, a major piece of legislation was promulgated by Congress to amend CERCLA. The *Superfund Amendments and Reauthorization Act* (SARA) was written to address deficiencies in CERCLA. Most significant was that the initial $1.6 billion allocated for site cleanup was initially considered sufficient to address all contaminated sites within five years! This suggests how little was understood of the complexities of hazardous waste site cleanup, let alone the tremendous legal roadblocks that were established due to liability requirements established by CERCLA.

To facilitate site cleanup, another $8.5 billion was allocated for Superfund sites, and procedures were implemented to emphasize faster cleanups and permanent remedies. Permanent remedies were defined as technologies that would destroy contamination rather then merely transfer it elsewhere (such as landfilling). Furthermore, EPA was directed to favor cleanup alternatives that were known to reduce the mobility, toxicity, and volume of the contamination. EPA was also directed to consider additional environmental legislation when determining cleanup limits, such as the Safe Drinking Water Act, the Clean Air Act, Clean Water Act, and the Toxic Substances Control Act.

Under CERCLA, EPA had no authority to regulate environmental cleanup on federal government sites such as Department of Energy (DOE) and Department of Defense (DOD) locations. Some DOE and DOD sites were known to be among the most highly contaminated sites in the U.S. SARA gave full authority to EPA to require cleanup of these facilities.

Another important revision involved the "innocent landowner" defense. Prior to CERCLA, a recent owner of previously used industrial property could be held liable for contamination. Under SARA, a prudent business person could hire an environmental expert to examine an industrial property of interest prior to sale to determine the likelihood of environmental contamination, and whether a current industrial owner was in compliance with all environmental requirements. If so, a new business person could purchase the property and provide the report of the hired expert in defense, in cases where contamination was found in the future. Such contamination had to be unrelated to current operations.

6.2 HAZARDOUS CHEMICALS

This section is an overview of some of the most commonly-found chemicals at hazardous waste sites and their linkage to particular industrial activities. The majority of hazardous wastes comprise organic chemicals (i.e., chemicals containing carbon, hydrogen, and oxygen other than CO_2, HCO_3^-, or CO_3^{2-}). Examples of inorganic chemical contamination at hazardous waste sites involves various heavy metals. Organic molecules are made up of carbon-hydrogen bonds that are often involved in the sharing of electrons. Chemical bonds with shared electrons are known as covalent bonds. Properties of organic chemicals, such as solubility in water, density, and affinity for other organics are often explained on the basis of chemical structure and bonding.

Organic chemicals that consist of carbon atoms linked in straight or branched chains are termed *aliphatics* whereas chemicals consisting of carbons in ringed structures are known as *aromatics*. Among the aliphatics are alkanes, which are made up of single bonds between carbon atoms, alkenes which have at least one double bond between carbons, and alkynes which have at least one triple bond between carbon atoms. Figure 6.1 illustrates these compounds. Among hazardous waste treatment professionals, organic compounds are often referred to based on traditional names that are used in the chemical process industry. Such names often give little indication of the structure or composition of the compound. To address this, the International Union of Pure and Applied Chemists (IUPAC) developed a systematic approach to the nomenclature of organic compounds. Because of this, some hazardous waste sites may describe chemicals based on traditional names or IUPAC names, or both. For example, perchloroethylene is the traditional name for a chlorinated solvent found at many hazardous waste sites. However, a laboratory report on ground water at such a site may quantify this chemical and refer to it by the IUPAC name, tetrachloroethene. Access to a good text on organic chemistry, or better still, an introductory course in organic chemistry is recommended for

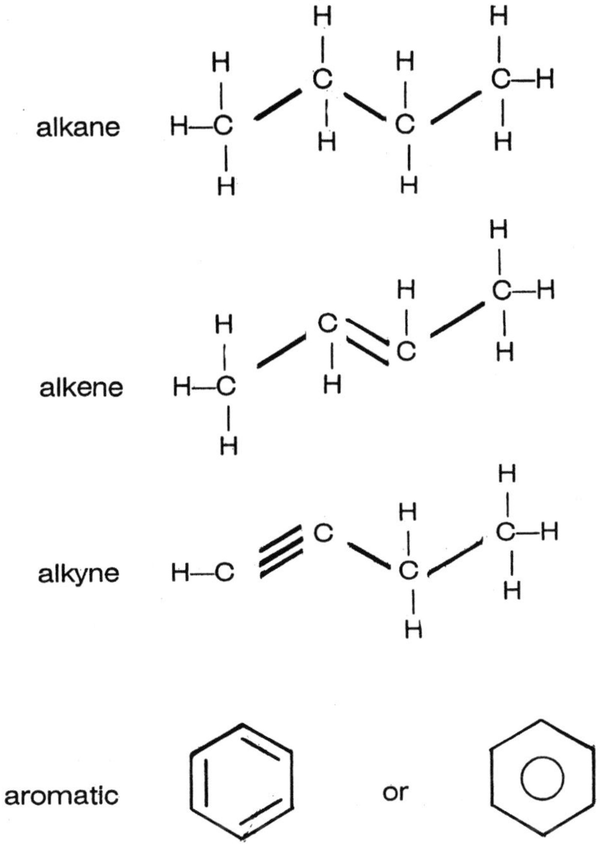

Figure 6.1. Aliphatics and aromatics.

those intending to pursue a career in hazardous waste assessment and remediation.

6.2.1 Links to Industrial Activity

The chemicals benzene, toluene, ethylbenzene and xylene, known collectively as BTEX, are associated with petroleum products such as gasoline, diesel fuels and heating oil. Oil refineries and gasoline stations with leaking underground storage tanks would be expected to exhibit soil and ground water contamination including the BTEX compounds. A diagram of the aromatic BTEX chemicals is shown in Figure 6.2, where it is understood that a carbon atom exists at each node on the diagrams. Note from Figure 6.2 that xylene describes three chemicals with identical numbers of carbon and hydrogen atoms, but with slightly different structures. Chemicals with the same chemical formula but different structures are termed isomers. Chemical isomers exhibit somewhat different properties due to differences in structure.

Methyl (tertiary) butyl ether, better known as MTBE, was added to gasoline to promote more complete combustion in automobile engines. Since it was added to much of the gasoline sold nationally and since there are numerous leaking underground storage tanks, MTBE is frequently found as a groundwater contaminant. It has proven to be a technical challenge to those involved in site cleanup, since it is highly soluble in water and is resistant to biodegradation.

Numerous chemical solvents have been employed by many industries for various applications such as degreasing of engine and fabricated metal parts, dry cleaning of clothing, printed circuit boards, and even removing caffeine from coffee! Many of these chemicals are either known or suspected human carcinogens and pose environmental problems; hence, their use has diminished widely and less aggressive alternatives are in use or under consideration. Chlorinated solvents often found at hazardous waste sites include derivatives of methane (chemicals with one carbon atom) such as chloroform, carbon tetrachloride, and methylene chloride. Others are derivatives of ethane (chemicals with two carbon atoms) such as trichloroethene (TCE), perchloroethene (PCE), and dichloroethene (DCE). Chlorinated solvent contamination is sometimes difficult to associate with a particular industrial activity due to the widespread application of these chemicals. However, military bases frequently have these chemicals on site due to their usefulness in

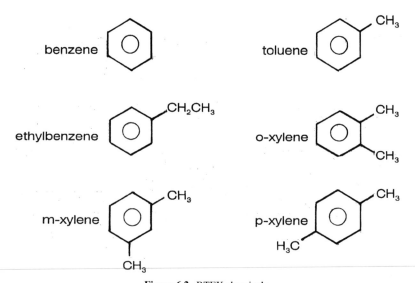

Figure 6.2. BTEX chemicals.

From Air Pollution to Groundwater Pollution

MTBE has been used in U.S. gasoline at low levels since 1979 to replace lead as an "anti-knock" additive. Engine "knocking" occurs when a car quickly accelerates and is sometimes described by the sound of marbles dropping into a coffee can. The reason for engine knock is often low octane gasoline and this can be remedied by an octane enhancer such as MTBE. The 1990 Clean Air Act Amendments required cleaner auto emissions, and MTBE was labeled as beneficial to achieve this purpose. Specifically, MTBE was used as a fuel additive in motor gasoline to raise the oxygen content of gasoline. Oxygen helps gasoline burn more completely, and thus reduces tailpipe emissions. In this application, MTBE serves as an "oxygenate." To fulfill this need, MTBE was produced in large quantities (over 200,000 barrels per day in the U.S. in 1999). Because MTBE is highly soluble in water it is highly mobile once released into the environment. MTBE has also proven resistant to biodegradation, and its health effects are currently under investigation. As a result, the State of California banned its use as a gasoline additive and EPA is expected to rule similarly in the near future. This is an example of how solutions to some pollution problems may cause others and illustrates the need for a holistic viewpoint on environmental protection.

---

degreasing mechanical components during maintenance. Figure 6.3 includes diagrams of several chlorinated solvents.

Polyaromatic hydrocarbons, known as PAHs, are large molecules constituted by rings of carbon atoms. PAHs were widely used in the preservation of wood products such as railroad ties and telephone poles. Creosote is a chemical formulation (of up to 16 different PAHs), which may be familiar to some who have observed a dark syrup-like substance oozing from telephone poles. To protect exposed wood from insect attack and degradation, facilities would inject chemicals, such as PAHs, under pressure to insure the penetration deep into the wood fibers. So-called "pole treating plants" were small businesses and not very well managed from an environmental standpoint. As such, these facilities number among the top industrial activities responsible for hazardous waste sites in the U.S.

Another chemical that serves as a wood preservative is pentachlorophenol (PCP). PCP is a single benzene ring surrounded by five chlorine atoms and a hydroxyl group. Refer to Figure 6.4 for a diagram of some PAHs and PCP. As an aside, the wood products industry has recently provided residential and commercial carpenters with pressure treated wood containing chromated copper arsenic (CCA). Whereas CCA-treated wood is easier to handle than that treated with PAHs or PCP, recent concerns over arsenic toxicity for young children who play on popular wood playground structures have resulted in considering other ways to make wood resistant to attack (and to find other materials for playgrounds).

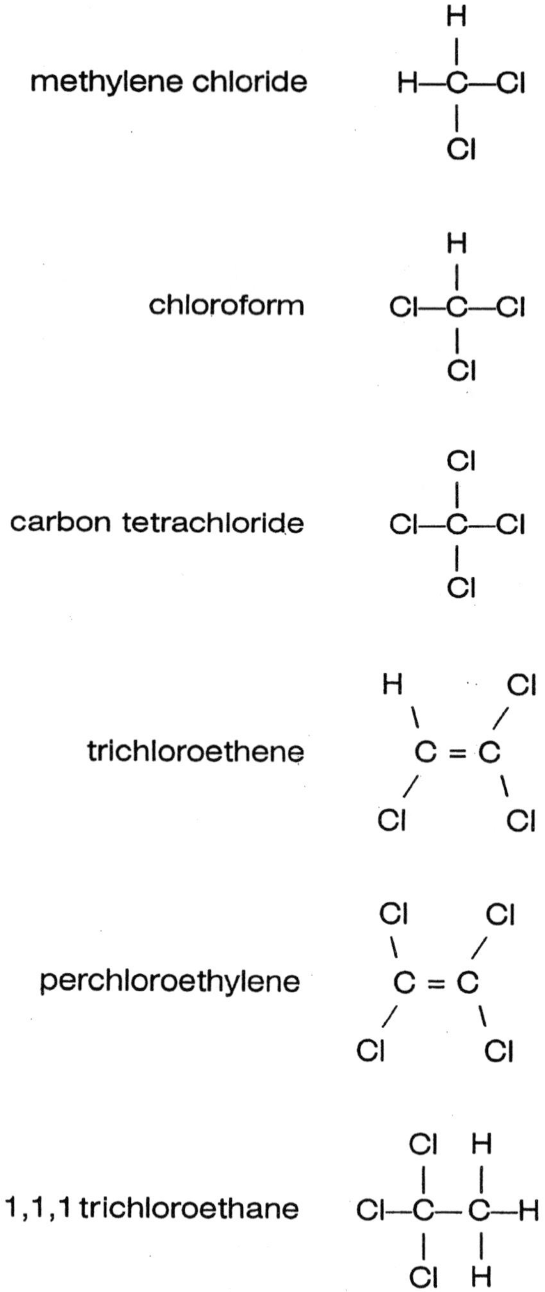

methylene chloride

chloroform

carbon tetrachloride

trichloroethene

perchloroethylene

1,1,1 trichloroethane

Figure 6.3. Chlorinated solvents.

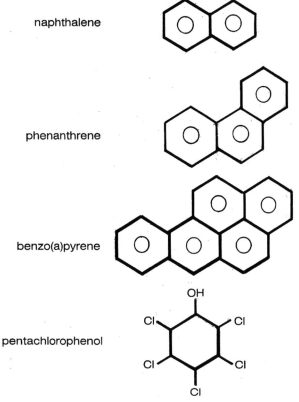

naphthalene

phenanthrene

benzo(a)pyrene

pentachlorophenol

Figure 6.4. Some PAHs and PCP.

Middletown Airfield Site

During the Korean War, Olmsted Air Force Base in south central Pennsylvania served as a facility for the maintaining, overhauling, and testing of aircraft. The solvent TCE was regularly used to remove grease and grime from mechanical parts. "Dirty" or spent TCE was accumulated in 55-gallon drums. Filled drums were disposed of on site in an excavated pit. During the mid-1960s, the site was purchased by private entities and the state, and encompassed Harrisburg International Airport, a Penn State University campus, dormitory quarters, and several industrial properties. In 1983, contamination in water supply wells resulted in the Middletown Airfield Site being placed on the Superfund List with the Pennsylvania Department of Transportation and the U.S. Air Force named as responsible parties. A water treatment system was installed to treat extracted water and return it to potable quality. The water supply on the site now provides water to about 3,500 full-time users, as well as to airline travelers and industrial users. Approximately 19,500 people obtain drinking water from wells within 3 miles of the site. EPA now considers the site safe for its intended use, and has deleted the site from the National Priorities List.

Another group of chlorinated aromatics is the polychlorinated biphenyls, more often referred to as PCBs. PCBs were widely used in industrial transformers and capacitors due to their ability to serve as a cooling medium while offering great resistance to flammability. Unfortunately, during maintenance or replacement of transformers, leaks of oil (that contained the PCBs) occurred, and soil and ground water contamination resulted. Because PCBs are known to be toxic to human health and the environment, the use of PCBs has ceased and limits were set during the 1980s to designate transformers that must be replaced and those that would be routinely monitored for leaks.

Viewers of cartoons will recall the escapades of Wiley E. Coyote and his frequent (mis)use of TNT in trying to nab the elusive Roadrunner. TNT refers to trinitrotoluene, an explosive that is found near chemical production facilities that manufacture or used TNT in practice ranges for the military. Imagine being called in to assist in the remediation of a site containing waste piles of TNT and picric acid (another explosive) and you understand a totally new meaning associated with hazardous waste cleanup!

Various chlorinated pesticides have been developed by the chemical industry to improve food crop yield and provide such benefits as making our lawns weed free (one measure of a wealthy society). The use of dichlorodiphenyltrichloroethane, better known as DDT, is banned in the U.S. but used widely elsewhere. DDT's image was initially tarnished by Rachel Carson in her book *Silent Spring*. More recently, debates over DDT's use in African countries have considered the benefits of reduced mosquito populations (and therefore, lowered incidence of malaria) versus the human and environmental toxicity of the chemical. Other chlorinated pesticides of concern that may be found in some household products are methoxychlor, lindane, aldrin, dieldrin, and endosulfin.

Dioxin is an aromatic contaminant that is generated as a byproduct of synthesis of chlorinated organics such as trichlorophenol. Dioxin is also generated during incineration of chlorinated wastes. Initial evaluations suggested that dioxin is extremely toxic to human health. More recently, dioxin is a suspected endocrine disruptor (a substance that can interfere with reproduction and thyroid hormones) and is therefore under further evaluation.

During the 1990s, perchlorate emerged as a contaminant of concern due to its persistence in the environment and solubility in water. Perchlorate is an inorganic anion with a chemical formula ClO_4^- and was employed as propellant of missiles and rockets. Perchlorate has a limited shelf life and must be replaced periodically. Since its initial use in the 1940s, the inventory of missiles and rockets in the U.S. required rinsing and replacement of perchlorate; improper management of the rinse water has resulted in ground water contamination in many states. Health risks are still under evaluation; however there are concerns over the role of perchlorate in inhibiting mental development, causing cancer, and acting as an endocrine disruptor.

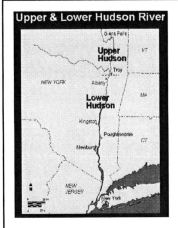

Upper & Lower Hudson River

PCB Contamination of the Hudson River

PCBs, or polychlorinated biphenyls, when they were invented by Monsanto in 1929, were the perfect chemicals. They had all the desirable properties to make them useful as heat transfer fluids in electrical transformers. They were apparently non-toxic, did not biodegrade, were chemically stable, were difficult to burn, had great heat transfer properties, and most of all, were inexpensive to manufacture. Most electrical transformers for almost 50 years contained PCBs.

The first concern with PCBs occurred in 1968, when a villager in Yusho, Japan, became ill after eating PCB-contaminated fish and the toxic nature of PCBs was revealed. Finally in 1977, Monsanto ceased the manufacture of PCBs, and by 1979, the U. S. EPA had determined that PCBs were human carcinogens and banned their use.

The General Electric Corporation facilities along the Hudson River north of New York City used PCBs in the manufacture of electrical equipment and from 1947 to about 1976 dumped more than one million pounds of PCBs into the river. Being non-biodegradable, the PCBs are now deposited along a 31-mile stretch of the Hudson River from Fort Edward southward to Troy, New York. The concentration of PCBs in the sediments range as high as 3,000 ppm, whereas a level of 50 ppm is considered toxic by the EPA and must be treated as a hazardous waste. Because PCBs do not readily dissolve in water, the PCB levels in the river are low; however, because the amount was immense, the estimated PCBs transferred from the sediments to the lower Hudson was about 8,000 pounds per year in the 1970s and about 1,000 pounds annually today.

The PCB levels in the water resulted in contaminated fish, and this became a public health problem. Finally the EPA banned all fishing in the upper Hudson, and this devastated a booming industry. Since the 1970s, the levels of PCBs in fish have been dropping, and the EPA has allowed limited fishing in the river. Still, epidemiological studies among the human population have shown that people who eat a modest diet of Hudson River fish have a 1,000 times greater cancer risk than the general population, and thus eating Hudson River fish is not a very good idea.

Other top-of-the-food chain species such as owls and eagles have been found to have high PCB concentrations. Even low PCB concentrations in such birds as the bald eagle and land species such as minks can cause reproductive failure.

The State of New York environmental agencies finally determined that General Electric was responsible for the PCB contamination and got the company to agree to conduct tests and to continue monitoring the situation. Most importantly, the deal absolved GE of further liability for cleanup. The argument was that any cleanup, no matter how conducted and with whatever good intentions, would result in much greater potential harm to humans and the environment than just allowing

(continued)

(continued)

the PCBs to remain in the sediment and slowly dissipate. By doing nothing, the PCBs would be expected to reach acceptable levels by the year 2040.

But the U. S. EPA would not let the matter drop. It designated the upper Hudson River a Superfund site, and determined that dredging the sediments and burning them to remove the PCBs was the only option. This project is expected to take about 11 years to complete. People living along the upper Hudson are opposed to the plan, fearing disruption and potential environmental damage, while people in the lower valley see this as the fastest route to full restoration of the river. As of this writing the EPA is planning to start the dredging operations.

6.2.2 Chemical Properties

The behavior of hazardous chemicals in water and in association with aquifer material is important in predicting a chemical's movement once it is released to the environment. The magnitude of solubility of a chemical in water gives an indication of its propensity to dissolve in and travel with water in an aquifer or stream. The maximum concentration of a chemical that will dissolve in water is dependent on its structure and water temperature.

Some general rules of thumb illustrate the relative solubility of chemicals:

- Larger molecules are less soluble than smaller molecules. Compare the structure of one of the PAH compounds shown in Figure 6.4 with chloroform (Figure 6.3). For comparison, the solubility (at 20°C) of pentachlorophenol is 14 mg/L, whereas it is 8,000 mg/L for chloroform. [1]
- Highly polar chemicals are highly soluble. Water is a polar molecule, which means it exhibits a positive charge at one end (around hydrogen) and a negative charge at the other (around oxygen). Therefore, chemicals that are polar associate with water to a greater extent and this results in increased solubility. The phrase "like dissolves like" is sometimes used to remember that polar chemicals will dissolve in a polar solvent such as water. Some examples of polar chemicals are methanol, ethanol, acetone, and ethylene glycol. Likewise, non-polar chemicals, such as hexane, and highly chlorinated species (discussed previously), are sparingly soluble.
- Halogens (i.e., chlorine, bromine and fluorine) decrease solubility. For example, TCE has three chlorines, and its solubility is approximately 1000 mg/L at 20°C. By contrast, substituting a fourth chlorine for hydrogen on TCE results in PCE and a decrease in solubility to approximately 200 mg/L. Consider this as a guideline to be used when comparing the solubility of a list of chemicals. That is, the magnitude of

solubility can be projected based on the relative number of halogen atoms on the molecule.

Chemicals are referred to as "sinkers" or "floaters" based on their density in water. The density of a chemical is determined by its mass (in grams or kilograms) per volume (in liters). Specific gravity is a comparison of the density of a chemical with that of water. As a point of reference, water (at 20°C) has a density of 1 kg/L. Chemicals with a specific gravity greater than 1 are said to be denser than water and will therefore sink in water. "Sinkers" are also referred to as dense non-aqueous phase liquids, or DNAPLs. Likewise, chemicals with a specific gravity less than one will float on water and are also referred to as light non-aqueous phase liquids (LNAPLs). Chemicals associated with petroleum products (BTEX) are examples of LNAPLs whereas chlorinated solvents (PCE, TCE, and CF) are DNAPLs.

Partition Coefficients

The propensity of chemicals to move through the aquifer under a hazardous waste site is calculated using *partition coefficients*. The *octanol-water* (K_{ow}) partition coefficient estimates how much a chemical in question wants to stay with the organic fraction of the soil and how much it wants to be with water. Octanol, an organic solvent, is used to estimate how much the chemical wants to adhere to organics.

If octanol is placed in a container with water, two distinct phases will be observed. If a chemical in question is placed in the container and the container is mixed and then allowed to come to equilibrium, the concentration of the chemical in octanol can be compared to the concentration of the chemical in the water. The calculation is then

$$K_{ow} = C_o/C_w \tag{6.1}$$

where

K_{ow} = octanol-water partition coefficient
C_o = concentration of the chemical in the octanol
C_w = concentration of the chemical in water

A high K_{ow} suggests that the chemical has a greater propensity for organic solvents, and hence the organic faction of soils, than it has for water, and therefore this chemical would stick to the soil matrix and not move very rapidly through saturated soil.

The *soil distribution coefficient, K_d,* similarly is a measure of how readily the chemical adheres to soil surfaces as opposed to moving with the groundwater, and the *organic carbon (K_{oc})* partition coefficient relates to the propensity of the chemical to stick to the organic fraction of soil. These three partition coefficients typically agree in that a specific chemical moving through the aquifer would have low values for all three and hence would move rapidly through the aquifer (it likes to be with the water), or the chemical might have high partition coefficient values and it would therefore not move very rapidly with the water, desiring to stay attached to the organics associate with the soil.

Partition coefficients can be used in a relative sense to estimate the potential behavior of multiple chemicals at a hazardous waste site. If several chemicals have been seeping into the groundwater, some will move faster than others with the water in the aquifer, and the relative movement of the chemicals can be estimated from the partition coefficients.

Octanol-water coefficients range from 10^{-3} to 10^5 for most chemicals. Because of this wide span of values, textbooks often list the logarithm (base 10) of K_{ow}, or $log\ K_{ow}$. High values of $log\ K_{ow}$ for a chemical means that the chemical is highly hydrophobic (water fearing) and would prefer to associate with organics in the soil as opposed to flow with the water. Conversely, chemicals with low $log\ K_{ow}$ would like to associate with water, would move more readily with groundwater and would be less likely to be retained by the soil.

Example 6.1

Three chemicals, perchloroethylene (PCE), trichloroethylene (TCE), dichloroethylene (DCE), and vinyl chloride (VC) are moving from a hazardous waste site into the aquifer. These three have $log\ K_{ow}$ values as shown below:

Perchloroethylene (PCE)	2.60
Trichloroethylene (TCE)	2.38
Dichloroethylene (DCE)	1.90
Vinyl chloride (VC)	1.38

What would a plume look like in the aquifer? Which chemicals would be expected to move the fastest with the groundwater through the aquifer?

The $log\ K_{ow}$ value for perchloroethylene is the highest, so it is expected to adhere most tightly to the soil matrix. Vinyl chloride on the other hand has the lowest $log\ K_{ow}$ and it would be expected to move the fastest through the soil. The plume from such a hazardous waste site would be similar to that shown in Figure 6.5.

Keep in mind that $log\ K_{ow}$ values are shown, so a difference between 2 and 1 is highly significant. Note from the figure that PCE, the chemical with the greatest K_{ow} migrates with the water to a lesser extent than VC, the chemical with the least K_{ow}. Furthermore, the relative movement of each chemical is predicted by the magnitude of K_{ow}. Chemicals with relatively high K_{ow} are less hydrophilic (water-loving), will associate more strongly with aquifer material and will not move as readily with groundwater.

The soil distribution coefficient is a measure of the degree to which a

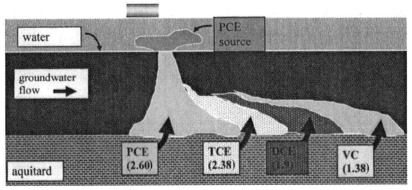

Figure 6.5. Spatial distribution of chemicals based on log K_{ow}.

contaminant is distributed between the sorbed (soil) phase and aqueous (water phase). Because soil organic matter is the most common sorbent in an aquifer, K_d may be based on organic carbon content of the soil. This leads to the organic carbon K_{oc} partition coefficient which is expressed based on the mass sorbed to soil organic carbon per mass dissolved in water. Furthermore, K_d may be estimated from K_{oc}, if the fractional content of organic carbon in a soil sample is known. The soil distribution coefficient may be calculated from K_{oc} from Equation 6.2:

$$K_d = K_{oc} f_{oc} \tag{6.2}$$

where

f_{oc} = the fraction of organic carbon in soil (dimensionless)

Empirical relationships have been determined between the partition coefficients. For example for PAHs and chlorinated hydrocarbons, the following relationship has been observed: [2]

$$K_{oc} = 0.63 \, K_{ow} \tag{6.3}$$

The partition coefficients enable an environmental engineer or scientist to estimate relative chemical behavior in the subsurface in a quantitative manner. It is possible to quantify the movement of a chemical relative to water in an aquifer by calculating the *retardation factor*. The retardation factor (R) is the relative contaminant velocity and is expressed as shown in Equation 6.4:

$$R = (\text{ground water velocity})/(\text{contaminant velocity}) \tag{6.4}$$

The retardation factor may be estimated based on Equation 6.5.

$$R = 1 + (\rho_B / n) K_d \tag{6.5}$$

where

ρ_B = soil bulk density (g/cm^3)
n = effective porosity (dimensionless)
K_d = soil distribution coefficient

Example 6.2

Determine the retardation factor for PCE, TCE, DCE, and VC and quantify the relative rate of movement of each chemical in groundwater. Assume a soil bulk density of 1.9 g/cm^3, an effective porosity of 0.25, a fractional organic content (foc) of 0.015 and that equation 6.2 is valid for these chemicals.

Begin with the log K_{ow} values for each chemical. Use Equation 6.2 to calculate K_{oc} from K_{ow}, then use equation 6.1 to calculate K_d. Next, calculate R from Equation 6.5 and list all results in a table as follows.

	log K_{ow}	K_{ow}	K_{oc}	K_d	R
PCE	2.60	398	251	3.77	29.7
TCE	2.38	240	152	2.28	18.3
DCE	1.90	79	50	0.75	6.7
VC	1.38	24	15	0.23	2.7

The interpretation, based on comparison of retardation factors, is that PCE would move through this aquifer more than 10 times slower than VC, and nearly 30 times more slowly than water.

Note that, from a qualitative standpoint, the movement of a chemical in an aquifer, relative to other chemicals, can be predicted by comparing their partition coefficients, K_{ow}, K_d, and K_{oc}. The chemical with the highest partition constant such as K_{ow} moves slowest, and the chemical with the lowest K_{ow} moves fastest. The same is true for K_{oc} and K_d.

Volatilization

Volatilization is the transfer of a chemical from the solid or liquid phase to the gaseous phase. Hazardous chemicals that are highly volatile are relatively easy to remove from water but, in doing so, may become hazardous air pollutants. A chemical's *vapor pressure* or *Henry's Law constant* indicates volatility. Vapor pressure is the pressure exerted at the interface of a pure chemical and the atmosphere. High vapor pressure corresponds to high volatility of a pure chemical (or mixture of chemicals). An example of a mixture of chemicals with high vapor pressure would be if you spill some gasoline on the driveway during a hot, sunny day. You would smell the gasoline quite rapidly, and this would be an indication that the chemicals in gasoline have volatilized from the liquid phase (gasoline) to the gaseous phase (air).

Henry's Law Constant describes the equilibrium of a dilute contaminant (one that is dissolved in water). In a closed system, the concentration of chemical in solution is in equilibrium with the concentration in the overlying gas phase (also known as "head space"). A high Henry's Law constant denotes that a chemical exhibits high volatility when dissolved in water. To illustrate this, envision a bottle half-filled with water. Add a few drops of gasoline to the water and close the container. Shake the container to achieve equilibrium (i.e., the gasoline will partition between the water and head space according to its Henry's Law constant). Then remove the bottle top and you would expect to smell gasoline, which would indicate that the chemicals have volatilized from the water to the air. As with the partition coefficients discussed previously, one may compare the Henry's Law constants from a list of chemicals to determine which are more likely to volatilize from water to air. Henry's Law constant is an important design parameter that engineers consider when designing an air stripper, a technology used to remove volatile chemicals from water.

6.3 SITE CHARACTERIZATION

To appropriately characterize a hazardous waste site, engineers and scientists follow a methodical procedure. Characterization begins with a site inspection by a team of professionals, which may be comprised of engineers, hydrogeologists, and toxicologists. The team would look for evidence of "releases" of contaminants to the environment in the form of discharges to waterways, seepage from soil, and strange odors emanating to the atmosphere. If discharges were observed, the team would work to identify the source of such releases and determine what immediate action is needed to stop the contamination from leaving the site. In anticipation of a subsequent risk assessment, site access would be evaluated to determine if nearby residents could inadvertently or deliberately enter the site and become exposed to hazardous waste.

To facilitate on-site inspection, the team would review historical records to determine site use, chemicals and wastes produced, waste management processes on site, and the status of environmental permits with federal and state regulatory agencies. If construction plans are available, they would be studied carefully to determine the location of waste discharge and treatment facilities. At some facilities, former waste disposal sites, such as lagoons and excavated pits, were filled and covered and may not be visible to the inspection team. Based on previous site use, knowledge of chemicals produced and waste disposed would be documented.

Next, a sampling program would be designed that specifies expected locations of contamination, chemicals and media. Sampling locations would be specified based on statistical analysis; however, the number of samples to be collected is determined based on a compromise between wanting the most information possible and the cost of mobilizing sampling equipment, placing a team on-site, and the price of laboratory analysis.

Before a team enters a site for sampling, a health and safety plan must be developed to protect the team from exposure during sample collection, excavation, and installation of sampling wells. The health and safety plan details procedures for sampling, safety equipment that must be worn and used by all staff, training for workers, monitoring procedures, and decontamination procedures for workers and equipment leaving the site.

To determine site conditions, various geophysical methods may be employed to identify location of ground water, existence of buried drums, and waste pits. A drill rig is often employed to install wells into the subsurface to allow collection of subsurface soil and groundwater samples. As the drill bit is advanced, samples of soil can be collected, and this allows a hydrogeologist to determine the composition of the aquifer, and the existence of clay and silt layers and bedrock formations. With wells appropriately located, samples can be employed to characterize ground water with respect to chemicals and concentrations, and a continuous cross-sectional map can be developed to estimate the location of aquifers and the groundwater gradient.

6.4 TOXICOLOGY

Toxicology considers the toxic effects to humans and other living organisms caused by exposure to hazardous chemicals. As stated in Chapter 2, we limit our discussion of toxicology to the toxic effect of chemicals on human health. *Toxicology* is a relatively new (ca. 30 years) interdisciplinary science that is included here to facilitate an appreciation of the basis for many critical decisions in hazardous waste management. Some issues to realize before we embark on details are that for most hazardous chemicals, our knowledge of toxicology is based largely on inferences from laboratory data. Specifically, laboratory rodents are exposed to chemicals in carefully controlled experiments, and the effects of these chemicals are quantified and documented. Also, in order to insure that an adverse effect will be observed, scientists will often administer a relatively high dose of the chemical to a rodent. By comparison, human exposure at a hazardous waste site often involves relatively low doses of chemicals. Despite such inferences and the limitations that arise, toxicology is still considered to be an essential tool on which to base the assessment of risk of a hazardous waste site.

To promote an ill effect in the human body, a chemical must reach a critical location known as a target organ. For many chemicals, the dose must be sufficiently high and contact time must be of sufficient duration. The means by which a chemical gains access to the body is known as *exposure route*. There are three possible exposure routes: *inhalation, ingestion,* and *dermal contact.* For many chemicals, inhalation is considered the most serious exposure route since chemicals gain quick access to the circulatory system via the lungs. Ingestion is considered a somewhat more protective exposure route; whereas a chemical can gain access to the circulatory system, there is the possibility of transformation of a chemical in the digestive tract. The skin is considered to be an effective barrier to many chemicals and therefore dermal contact is thought to be the most protective of the exposure routes. Keep in mind that even if a chemical gains access to the body, natural protection mechanisms exist. For

ALICE HAMILTON

Alice Hamilton (1869–1970) graduated from both the Fort Wayne College of Medicine in Indiana and University of Michigan School of Medicine. Following her medical degree in 1893 she did internships in Munich and Leipzig (being allowed to sit in the class of all men as long as she did not make herself conspicuous).

Some of her early experiences were working in a settlement house in Chicago, where she started educational programs and clinics for the destitute. In 1902 she recognized the connection between waste disposal and typhoid fever, and was able to initiate change in the city health department. In her work with the poor, she noted the connection between unsafe conditions at work and the health of the workers, and in 1910, became the director of the new Occupational Disease Commission.

From there she moved in 1919 to the faculty of the Harvard Medical School where she founded the program in occupational medicine (and was the only female member of the faculty, and was appointed only on the condition that she not join the faculty club!). She was a leading participant in two occupational controversies, the leaded gasoline debate and the health of the radium dial painters (known as the "radium girls"). In the leaded gasoline debate, she showed how lead can accumulate in the bones and fought against the industry claims that there is a natural threshold of lead in the human body. She fought unsuccessfully against the introduction of lead into gasoline in the 1920s, and her work was not vindicated until the 1970s, when lead in gasoline was finally banned (ironically for air pollution control reasons, not human health). In the radium dial painters controversy, Hamilton's epidemiological studies showed how radiation exposure to women painting glow-in-the-dark watch dials was causing a high incidence of cancer.

Hamilton is acknowledged to be the founder of occupational medicine, and during her long lifetime received many honors and awards. In 1944 she was listed in *Men of Science*, which must have caused her to chuckle.

Pesticides in Central America

For many years both Shell Oil and Dow Chemical supplied a pesticide containing dibromochloropropane (DBCP) to Standard Fruit Company for use on its banana plantations, even though Shell Oil knew since the 1950s that DBCP causes sterility in laboratory animals. Even after it was shown that DBCP also causes sterility in humans and it was banned in the United States, Shell continued to market the pesticide in Central America.

In 1984 banana plantation workers from several Central American countries filed a class action suit against Shell, claiming that they became sterile and faced a high risk of cancer. In response, Shell claimed that it was inconvenient to continue the case because the workers were in Costa Rica, a claim that was quickly thrown out of court. Shell finally settled out of court with the Costa Rican workers and paid $20 million in damages to the 16,000 claimants. A scientist from Shell is quoted as saying: "Anyway, from what I hear they could use a little birth control down there."

Quote from: David Weir and Constance Matthiessen "Will the Circle be Unbroken?" *Mother Jones*, June 1989.

---

example, certain chemicals may be exhaled, or may be eliminated through the sweat glands, urine, or feces. Alternatively, some chemicals may be biotransformed to a non-toxic form, whereas other chemicals may be stored in the fat cells. An example of a beneficial biotransformation is conversion of benzene, a confirmed human carcinogen, to phenol, a polar chemical that associates with water and may therefore be more readily eliminated.

To illustrate the concept of the fate of chemicals in the body, consider that toxicologists refer to three doses of chemical exposure: the *administered dose* is what has been applied to the body (in air, liquid, or on the skin); the *intake dose* is the amount that crosses a body barrier (lung cells, stomach lining, or skin), and the *target dose* is the amount that eventually reaches a target organ. Due to natural protection mechanisms, the target dose would be less than the intake dose, which is less than the administered dose. Factors that affect chemical absorption by the body include the medium that contains the chemical (i.e., soil, solvent or air), properties of the chemical and its affinity for water versus target organ tissue, the exposure route, and the susceptibility of individuals.

Data from laboratory evaluations are represented in the form of *dose-response* graphs. Such graphical relationships correlate the frequency of specific toxic effect as a function of dose. An example dose-response curve is shown in Figure 6.6. From the figure, as the dose increases, the severity of response increases. Because such evaluations are time consuming and expensive, and it is desired to observe an effect, often a relatively high chemical dose, known as the *maximum tolerable dose*, is applied, a dose high enough to elicit an ill effect but not cause death.

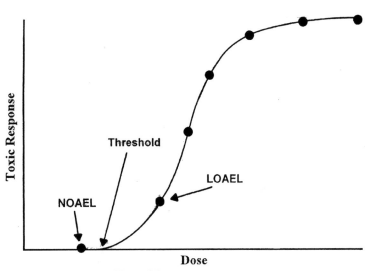

Figure 6.6. Dose-response curve.

Responses observed from toxicology studies result in chemicals being classified as *carcinogens* (those that promote the onset of cancer) or *non-carcinogens* (those that promote all other ill effects, such as organ damage or developmental disorders). For non-carcinogens, dose-response curves may reveal a *threshold*, which is defined as a maximum chemical dose that results in no adverse effect. Most often, the threshold is not an applied dose but is determined graphically and exists between the maximum applied dose which caused no ill effect and the minimum applied dose that did cause an adverse effect. Figure 6.6 illustrates this concept and identifies the *No Observed Adverse Effect Level* (NOAEL) and the *Lowest Observed Adverse Effect Level* (LOAEL).

The *Acceptable Daily Intake* (ADI) value for humans is determined by reducing the NOAEL dose by a factor of 100. The ADI is thus more conservative than the NOAEL value. The rationale for reducing the NOAEL value is as follows: the NOAEL is divided by 10 to account for extrapolation of results from laboratory animals to humans; the NOAEL is divided again by 10 to take into account an expected variation of responses among a population of humans exposed to a chemical. However, the ADI value may not always be considered the presumed "safe" dose of a chemical at a hazardous waste site. EPA employs a factor known as the *reference dose* (RfD), as the value considered to be safe for human exposure. The reference dose may be equivalent to the ADI or may be more conservative depending on the conditions of the laboratory data. If the ADI for a specific chemical is based on a shorter-term laboratory study (sub-chronic) as opposed to long-term (chronic), the ADI will be reduced by a factor of 10. Furthermore, to determine the reference dose, the ADI may be reduced by another factor of 10 depending on the quality of laboratory data that was employed (this is a professional judgment issue). Finally, if the NOAEL was not identified and the LOAEL was employed, the ADI would be reduced by another factor of 10. In summary, for a long-term, superior quality laboratory test, ADI and reference dose would be identical. In other cases, the reference dose, which is the value used by EPA in determining the safe dose, may be less than ADI by up to three orders of magnitude. For non-carcinogens, the bottom line is, if the exposure dose is less than the reference dose, EPA considers this a safe condition.

Regarding carcinogens, current procedures stipulate that no safe level of exposure exists. Support for such a policy is based on the theory that one molecule of a carcinogenic chemical is sufficient to initiate the development of a tumor. Therefore, a threshold value is not determined and "safe" concentrations are not estimated from NOAEL or LOAEL values. Instead, the slope of the dose-response curve is used to determine a *slope factor*. The slope factor is employed to estimate the *incremental risk* of developing cancer as a direct result of chemical exposure. Incremental risk is due solely to the chemical exposure, whereas total or overall risk is attributed to all exposures and activities that could lead to an ill effect. An example of background risk would be the incidence of cancer by people who have not been exposed to a particular hazardous chemical, whereas incremental risk is the incidence rate of cancer due only to the exposure to a specific chemical. The sum of incremental and background risk is overall risk. As pointed out in Chapter 2, the background risk often exceeds the incremental risk from hazardous chemicals by a wide margin. Why then are most people so concerned with the health effects of hazardous wastes? The answer has to do with free will. If people willingly choose to engage in an activity (smoking cigarettes or bungee jumping) they are also willing to assume the risks inherent in such activity. By contrast, exposure to or proximity to hazardous waste sites is considered to be an unwanted activity. The concern for risk is thus magnified accordingly. Somehow, the risk associated with hazardous wastes must be compared to a standard that is considered acceptable.

6.5 SYMBOLS

BTEX = benzene, toluene, ethylbenzene, xylene
CERCLA = Comprehensive Environmental Response, Compensation, and
Liability Act
C_o = concentration of a chemical in octanol (mg/L)
C_w = concentration of a chemical in water (mg/L)
DCE = dichloroethylene

DDT = dichlorodiphenyltrichloroethane
DOD = Department of Defense
DOE = Department of Energy
f_{oc} = fraction of organic carbon in soil
HSWA = Hazardous Solid Waste Amendments
K_d = soil distribution partition coefficient
K_{oc} = organic carbon partition coefficient
K_{ow} = octanol-water partition coefficient
MTBE = methyl tertiary butyl ether
n = effective porosity
PAH = polyaromatic hydrocarbon
PCE = perchloroethylene
PCP = pentachlorophenol
PRP = potentially responsible party
R = retardation factor
RCRA = Resource Conservation and Recovery Act
RfD = reference dose
SARA = Superfund Amendment and Reauthorization Act
TCE = trichloroethylene
VOC = volatile organic compound
ρ_B = soil bulk density

6.6 REFERENCES

1. Schnoor, J.L. *Environmental Modeling*. John Wiley & Sons, New York. 1996.
2. Karickhoff, S.W., Brown, D.S., and Scott, T.A. *Sorption of hydrophobic pollutants on natural sediments*. Wat. Res. 13. 1979.

6.7 PROBLEMS

6.1 Discuss the decision of Congress to first address hazardous waste production rather than abandoned hazardous waste sites. Prepare a table that states the advantages and disadvantages of this approach.

6.2 Describe by examples that have nothing to do with hazardous waste the legal concepts of (a) strict liability and (b) joint-and-several liability.

6.3 Define the four characteristics of hazardous waste. Use these to evaluate three common household chemicals. Which ones do you think might be classified as hazardous and which toxic characteristic(s) would they exhibit?

6.4 Describe by example (a) the mixture rule, (b) derived-from hazardous waste, and (c) delisted hazardous waste.

6.5 What are the advantages of a potentially responsible party in agreeing to clean up a hazardous waste site as opposed to refusing to cooperate and relying on EPA to complete the cleanup? If you are a CEO of a small company that has discovered a hazardous waste dump on your land (a dump that you did not contribute to) what action would you take?

6.6 Argue the advantages and disadvantages of the small-quantity generator exemption from the perspective of a business owner and from the standpoint of an environmentalist.

Search EPA's website to obtain supportive statistics for your argument.

6.7 Discuss the benefits of the "innocent landowner" defense from a land use perspective. Consider the impact on previously occupied industrial land and rural land if the innocent landowner defense had not been developed.

6.8 Describe the industrial activity associated with the following chemicals: TCE, PAHs, PCBs, PCP, dioxin, DDT, methylene chloride, trinitrotoluene. Do you believe any of these are produced in your home town? Why or why not?

6.9 Draw a sketch that describes the behavior of a DNAPL or LNAPL when leaked from an underground storage tank and identify some chemicals that are DNAPLs and others that are LNAPLs.

6.10 Describe the inferences that are often made with toxicological data. Do these inferences lead people to have less or more confidence in the accuracy of the data?

6.11 Define exposure route, list the exposure routes, and rank the exposure in order of (typical) severity for three chemicals. You can choose any chemicals, even if they are not hazardous.

6.12 Describe the rationale EPA uses in funding the clean-up of Superfund sites.

6.13 Why does the EPA define the removal of hazardous waste by the destruction removal efficiency instead of just insisting that zero waste be discharged? Use the principle of expediency in your answer.

6.14 A contaminated aquifer contains methylene chloride, carbon tetrachloride, and chloroform in approximately equal amounts. Based on relative mobility of these chemicals in ground water, which compound would be (a) at the leading edge of the contaminant plume (i.e., moving fastest) (b) which compound in the middle (c) which one at the trailing edge (i.e., moving slowest)? The log K_{ow} for methylene chloride, carbon tetrachloride, and chloroform is 1.30, 2.64, and 1.97, respectively. (d) Determine the relative velocity of these chemicals in an aquifer by calculating the retardation factors as described in Example 6.1.

CHAPTER 7

Management of Hazardous Waste

THE previous chapter is all about chemicals that can be considered hazardous, and about the laws that have been passed to help solve the problem of these wastes. CERCLA is intended to clean up old hazardous waste sites, and RCRA is responsible for management of currently generated hazardous materials. In this chapter we discuss the justification and the technology of reducing the risk of harm from hazardous waste substances.

7.1 RISK ASSESSMENT OF HAZARDOUS WASTE

Risk assessment is used in hazardous waste management for decision-making and to understand health and environmental hazards associated with contaminated sites. Quantitative risk assessment results in numerical values that may be used to determine if a site requires remediation, and if so, to what level must contaminants be reduced to consider the cleanup effort acceptable. The application of quantitative risk assessment to determine acceptable contaminant concentrations may be a surprising realization. The point is that sites are not cleaned up to a pristine level.

The first step in conducting a risk assessment occurs during the site visit described in section 6.3. Media (soil, ground water, surface water, air) that are contaminated are identified, potential release points for chemicals are itemized, and potential populations that could be exposed are identified. Once laboratory analysis identifies and quantifies the contaminants at a site, the risk assessment proceeds to determine all potential *exposure pathways* for chemicals. An exposure pathway describes, in sequential fashion, chemical migration from the point of release to the point of reaching a receptor (for human health risk assessment, a receptor is a human being). Figure 7.1 illustrates hypothetical exposure pathways for chemicals.

Figure 7.1 suggests that some chemicals on site must be volatile and may partition to the gaseous phase (the atmosphere) and be transported downwind

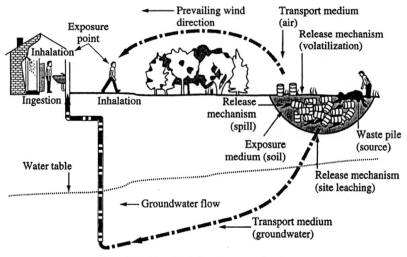

Figure 7.1. Hypothetical environmental pathways.

to an unsuspecting pedestrian. Once the chemical reaches the pedestrian, the pathway has been determined. Entry of the chemical to the human body is not a part of exposure pathway. Recall that chemical entry occurs by one of three exposure routes.

Once exposure pathways are documented, potential exposure concentrations of chemicals must be calculated based on knowledge of chemical reactions, dilution in air or water, sorption, and volatilization. Appropriate exposure routes are determined. Then the average and maximum daily intake rates are calculated and normalized according to body weight as discussed in Chapter 2.

Calculated average daily intake values are next compared to EPA's reference dose (*RfD*), for non-carcinogens, or slope factor (*SF*) for carcinogens (see p. 251) Equation 7.1 illustrates calculation of the average daily intake (*I*).

$$I = \frac{M_{abs} D}{WT} \qquad (7.1)$$

where

I = average daily intake (mg/kg-d)
M_{abs} = mass of chemical absorbed per day (mg/d)
D = total days exposure occurred (d)
W = body weight (kg)
T = averaging time (d) = 70 years for carcinogens or total exposure
  duration for non-carcinogens

Once the average daily intake (*I*) is calculated, it may be compared to the *RfD* to assess safety as shown by Equation 7.2.

$$HI = \frac{I}{RfD} \qquad (7.2)$$

where

HI = hazard index (dimensionless).

If the average daily intake is less than the RfD, HI will be less than one and this is considered to be safe. Alternatively, I may be compared to the slope factor (SF) for carcinogens as shown by Equation 7.3.

$$\text{Risk} = I \times SF \qquad (7.3)$$

If the risk is less than 1×10^{-6} (one in a million), EPA considers the risk acceptable. It is important to use the appropriate exposure route (inhalation, ingestion, or dermal contact) when choosing the reference dose or slope factor. Updated reference for RfD and SF values may be obtained from EPA at http://www.epa.gov/iris/. General guidance is that, for chemicals with no dermal contact data, ingestion is used.

Examples 7.1 and 7.2 illustrate risk assessment calculations for non-carcinogens and carcinogens, respectively.

Example 7.1

A homeowner has a private well that contains 120 µg/L m-dinitrobenzene. Consider that the homeowner consumes 2 liters per day of water for 10 years, weighs 70 kg, and EPA's reference dose for m-dinitrobenzene is 1×10^{-4} mg/kg-d. Determine if the homeowner's well is safe to drink. If not, determine what concentration of m-dinitrobenzene would be considered safe.

Calculate the average daily intake dose by the homeowner:

$$I = \text{average daily intake} = \frac{(0.12 \text{ mg/L})(2 \text{ L/d})(365 \text{ d/y})(10 \text{ y})}{70 \text{ kg}(10 \text{ y})(365 \text{ d/y})}$$

$$I = 3.4 \times 10^{-3} \text{ mg/kg} - d$$

$$\text{Hazard Index} = \frac{3.4 \times 10^{-3} \text{ mg/kg} - d}{1 \times 10^{-4} \text{ mg/kg} - d} = 34 > 1 \quad \text{not safe!}$$

$$\text{Safe Dose} = RfD = \frac{(C_{safe})(2 \text{ L/d})}{70 \text{ kg}} = 1 \times 10^{-4} \text{ mg/kg} - d$$

$$C_{safe} = 3.5 \text{ µg/L}$$

Based on this evaluation, EPA would consider the homeowner's well to be safe if it contained 3.5 µg/L or less of m-dinitrobenzene (bear in mind that this concentration is near the detection limit of current instrumentation).

Jersey City Chromium

Jersey City, in Hudson County, New Jersey, was once the chromium processing capital of America and over the years, twenty million tons of chromate ore processing residue was sold or given away as fill. There are at least 120 contaminated sites which include ball fields and basements underlying homes and businesses. It is not uncommon for brightly colored chromium compounds to crystallize on damp basement walls, and to "bloom" on soil surfaces where soil moisture evaporates creating something like an orange hoar frost of $Cr(VI)$. A broken water main in the wintertime resulted in the formation of bright green ice due to the presence of $Cr(III)$.

The companies that created the chromium waste problem no longer exist, but liability was inherited by three conglomerates through a series of takeovers. In 1991, Florence Trum, a local resident, successfully sued Maxus Energy, a subsidiary of one of the conglomerates, for the death of her husband, who loaded trucks in a warehouse built directly over a chromium waste disposal site. He developed a hole in the roof of his mouth and cancer of the thorax and it was determined by autopsy that his death was caused by chromium poisoning. While the subsidiary company did not produce the chromium contamination, the judge ruled that they knew about the hazards of chromium.

The State of New Jersey initially spent $30 million to locate, excavate, and remove some of the contaminated soil. But the extent of the problem was overwhelming and they stopped these efforts. The director of toxic waste clean-up for New Jersey admitted that even if the risks of living or working near chromium were known, the state did not have the money to remove it. Initial estimates for site remediation are well over a billion dollars. Citizens of Hudson County are angry and afraid. Those sick with cancer wonder if it could have been prevented. Mrs. Trum perceived the perpetrators as well-dressed business people who were willing to take chances with other peoples' lives. "Big business can do this to the little man. . . ." she said.

The contamination in Jersey City is from industries that used chromium in their processes, including metal plating, leather tanning, and textile manufacturing. The deposition of this chrome in dumps has resulted in chromium contaminated water, soils, and sludge. Chromium is particularly difficult to regulate because of the complexity of its chemical behavior and toxicity, which translates into scientific uncertainty. Uncertainty exacerbates the tendency of regulatory agencies to make conservative and protective assumptions, the tendency of the regulated to question the scientific basis for regulations, and the tendency of potentially exposed citizens to fear potential risk.

Chromium exists in nature primarily in one of two oxidation states—$Cr(III)$ and $Cr(VI)$. In the reduced form of chromium, $Cr(III)$, there is a tendency to form hydroxides, which are relatively insoluble in water at neutral pH values. $Cr(III)$ does not appear to be carcinogenic in animal and bioassays. Organically complexed $Cr(III)$ has recently become one of the more popular dietary supplements in the United States and can be purchased commercially as chromium picolinate or with trade names like Chromalene to help with proper glucose metabolism, weight loss and muscle tone.

(continued)

258

(continued)

When oxidized as Cr(VI), however, chromium is highly toxic. It is implicated in the development of lung cancer and skin lesions in industrial workers. In contrast to Cr(III), nearly all Cr(VI) compounds have been shown to be potent mutagens. The U.S. EPA has classified chromium as a human carcinogen by inhalation, based on evidence that Cr(VI) causes lung cancer. However, by ingestion, chromium has not been shown to be carcinogenic.

What confounds the understanding of chromium chemistry is that under certain environmental conditions, Cr(III) and Cr(VI) can interconvert. In soils containing manganese, Cr(III) can be oxidized to Cr(VI). Given the heterogeneous nature of soils, these redox reactions can occur simultaneously. While organic matter may serve to reduce Cr(VI), it may also complex Cr(III) and make it more solubleùfacilitating its transport in ground water and increasing the likelihood of encountering oxidized manganese present in the soil.

Clean-up limits for chromium are still undecided, but through the controversy, there have evolved some useful technologies to aid in resolution of the disputes. For example, analytical tests to measure and distinguish between Cr(III) and Cr(VI) in soils have been developed. Earlier in the history of New Jersey's chromium problem, these assays were not reliable and would have necessitated remediating to soil concentrations based on total chromium. Other technical/scientific advances include remediation strategies designed to chemically reduce Cr(VI) to Cr(III) in order to reduce risk without excavation and removal of soil designated as hazardous waste. The establishment of clean-up standards is anticipated, but the proposed endpoint based on contact dermatitis is controversial. While some perceive contact dermatitis as a legitimate claim to harm, others have jokingly suggested regulatory limits for poison ivy, which also causes contact dermatitis. The methodology by which dermatitis based soil limits were determined has come under attack by those who question the validity of skin patch tests and the inferences by which patch test results translate into soil Cr(VI) levels.

The frustration with slow cleanup and what the citizens perceive as doubletalk by scientists finally culminated in the unusual step of amending the state constitution to provide funds for hazardous waste cleanups. State environmentalists depicted the constitutional amendment as a referendum on then Gov. Christine Todd Whitman's (R) environmental record, which relaxed enforcement and reduced cleanups. (Whitman became President George W. Bush's new administrator of the U. S. Environmental Protection Agency and subsequently resigned before completion of Bush's first term in office).

Example 7.2

Is it safe to pump gas?

A motorist fills the tank of his car once per week over the course of ten years. It takes 5 minutes to fill the tank and, during this time, he is exposed to 1.5 μg/L benzene in the air. The man breathes 14 L/min of air and 50% of the benzene breathed is expelled during exhalation. Is this safe? The slope factor for inhalation of benzene is 0.029 $(mg/kg-d)^{-1}$.

From EPA's IRIS database, benzene is a known human carcinogen. Therefore, no safe level can be determined. We can only determine if the risk is acceptable.

$$I = \frac{(1.5 \times 10^{-3} \text{ mg/L})(14 \text{ L/min})(5 \text{ min/event})(52 \text{ event/y})(0.5)(10 \text{ y})}{70 \text{ kg } (70 \text{ y})(365 \text{ d/y})}$$

$$= 1.53 \times 10^{-5} \text{ mg/kg} - \text{d}$$

Risk $= (0.029 \text{ kg} - \text{d/mg})(1.53 \times 10^{-5}) = 4.43 \times 10^{-7} < 10^{-6}$ acceptable risk

Note that the averaging time is 70 years in the denominator of the equation. It is standard procedure to use a 70-year averaging time for carcinogens to reflect the long-term nature of studies used to determine slope factors.

7.2 PUMP-AND-TREAT SYSTEMS

Initial efforts to remediate ground water contaminated with hazardous chemicals employed technologies like those described in Chapters 4 and 5 for water and wastewater treatment. To use these technologies, contaminated groundwater must be pumped to the surface, where various treatment processes are placed. This is the general description of what is currently termed a *pump-and-treat system*; groundwater-pumping wells are installed strategically to collect a plume of contaminated ground water and pump it to an above-ground treatment system. Pump-and-treat systems may employ typical water and wastewater treatment processes such as:

• physical separation processes to separate dense non-aqueous phase liquids (DNAPLs) and light non-aqueous phase liquids (LNAPLs) from water (clarifiers for gravity settling or dissolved air flotation (DAF) systems to float oils and LNAPLs, sand filters for removal of fine particulates and biological suspensions),
• chemical processes to adjust the pH of acidic or basic waters or to promote desirable reactions such as precipitation of calcium, ion exchange for reduction in hardness,
• and biological processes for removal of biodegradable organics.

Other technologies, such as air stripping and activated carbon are often used to clean up contaminated ground water. Soil vapor extraction is a process that is employed to remove volatile organics from the subsurface in the *unsaturated zone* (i.e., above the water table).

Advantages of pump-and-treat systems are that they are proven, and many professionals have experience with them, and they are relatively easy to construct and operate. In addition, properly installed pumping wells can serve as hydraulic barrier and thus prevent migration of contaminants to an adjacent property owner. Containment of a plume of contamination is a considerable advantage in that it minimizes the need to utilize a neighbor's land for installation of pumping wells or treatment systems. Figure 7.2 illustrates the benefit of a pumping system in achieving a hydraulic barrier.

Disadvantages of pump-and-treat systems are dependent on the contaminants and composition of the subsurface at a particular site. For example, pump-and-treat systems installed at sites containing chemicals with high K_{ow} values (relatively non-polar) would have to operate for a considerable time period (decades or possibly, centuries) to achieve desired cleanup alternatives. Remember that non-polar chemicals "prefer" to associate with organic material in an aquifer and would therefore be more strongly retained. Such chemicals would very slowly desorb from the aquifer into water and take considerably long to achieve cleanup. Realize that with subsurface contamination, we have very little control over how quickly a chemical will desorb into ground water. Other than adding chemicals that may promote dissolution of contaminants into the water, the only way to stimulate chemical removal is to pump significant amounts of water to the surface. Contaminated water extracted from an aquifer will be replaced with clean water, which will then serve as the medium to encourage continued release of sorbed chemicals. And so it continues; the more strongly sorbed chemicals (high K_{ow}) require

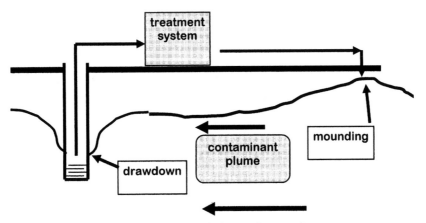

Figure 7.2. Hydraulic barrier by pumping ground water.

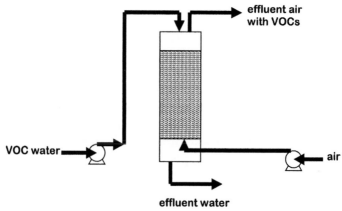

Figure 7.3. Air stripper.

greater volumes of groundwater to be pumped to achieve adequate cleanup and this translates into extended time periods to achieve cleanup goals. Long pumping-times increase operational costs. In addition, the degree of treatment (i.e., final concentrations in an aquifer) may not meet regulatory goals for certain contaminants. Nevertheless, pump-and-treat systems are still widely chosen by engineers to prevent groundwater migration and to achieve cleanup.

Some of the more popular processes employed in pump-and-treat systems are: air stripping, granular activated carbon, and biological treatment processes.

7.2.1 Air Stripping

A diagram of a counter-current *air stripper* is shown in Figure 7.3. Water with volatile organic compounds (VOC) is discharged at the top of a vertical column while air is pumped from bottom to top of the column. The column is filled with porous packing that promotes downward water movement in a random pattern and as a thin film. Thin-film water flow exposes a high surface area to the air moving in an upward direction and this promotes the transfer of chemicals from the water to the air. Naturally, an air stripper is not applicable for all chemicals. Based on our previous discussion on chemical properties, it should be understood that such a process is only viable for chemicals that are volatile in water. Therefore, chemicals with relatively high Henry's Law constants are candidates for air stripping. Once chemicals are transferred from the water to the air, effluent liquid may be directed to subsequent treatment, or discharged to a receiving stream or aquifer, whereas effluent air may or may not be treated prior to discharge.

The fact that an air stripper achieves cleaner water by merely transferring contaminants from water to air can be disturbing. From a holistic viewpoint, how can such a process be considered beneficial when cleaner water is achieved at the expense of air quality? Regulatory agencies reconcile this based on assessment of risk. At some sites, elevated concentrations of contaminants may be determined to pose unacceptable risk to those who may use the water. For non-carcinogenic chemicals, this would mean that chemical concentrations are greater than EPA's reference dose. For carcinogenic chemicals, the incremental risk would be greater than one in a million. By comparison, volatile chemicals that are transferred to the air would be significantly diluted to the point that a risk assessment on breathing air in the region of the air stripper would be considered completely acceptable. However, in other instances, treatment of effluent air from an air stripper would be required. Processes that are often employed to treat gaseous effluent from a stripper are activated carbon (described in the next section) or thermal combustion that destroys the chemicals.

Air strippers are advantageous due to relatively low installation and operational costs, and ease of operation. Air strippers are also a proven and readily available technology. Another factor that is particularly advantageous in remote locations is that an air stripper requires only intermittent checking by a technician every few days. This reduces the operational budget and reduces the burden on those responsible for maintaining the process. Potential disadvantages are that this technology is limited in application to volatile chemicals and that air stripping is a *mass transfer technology*. Mass transfer technologies merely transfer contamination from one environmental medium to another.

Depending on the water's condition, pre-treatment or post-treatment of contaminated water may be required. If water with volatile organic compounds (VOCs) also contains significant iron or hardness concentrations, these constituents must be removed prior to air stripping. If you've ever had water from a well, perhaps you remember a taste of iron even though the water looked clear. Obviously, the iron, is present in the water, but it must be dissolved. Groundwater is often low in dissolved oxygen, and iron will exist in a reduced form, which is soluble. In an air stripper, water is oxygenated as it moves counter-current to air. This causes the dissolved iron to oxidize and precipitate on the packing in the air stripper. If left to continue, rust will form on the packing and restrict the movement of water through the column. Cleaning or replacement of packing represents a very inconvenient maintenance expense and should be avoided. Therefore, the iron needs to be oxidized and allowed to precipitate before the air stripper. Likewise, water with high hardness should be treated to remove calcium and magnesium from the water prior to air stripping to avoid precipitation and clogging of packing. Figure 7.4 illustrates the use of oxidation to remove iron and ion exchange to remove calcium and magnesium before removal of VOCs in an air stripper.

To determine the viability of air stripping for a particular site, engineers

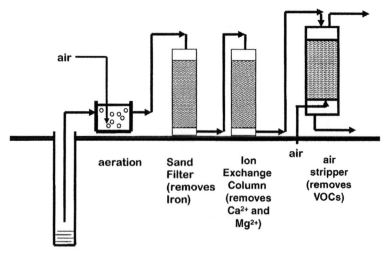

Figure 7.4. Air stripper with iron removal.

consider the Henry's Law constant of chemicals in the ground water. Chemicals with Henry's Law constants ranging from 10^{-5} to 10^{-3} atm-m^3/mol are good candidates for air stripping. Chemicals with Henry's Law constants less than 10^{-5} atm-m^3/mol volatilize slowly from water, and their removal in an air stripper is relatively slow and incomplete. Table 7.1 lists Henry's Law constants for some of the frequently detected hazardous chemicals.

From the table, one can infer that benzene, toluene and chlorinated solvents would be candidates for removal with an air stripper, whereas groundwater with the benzo(a) pyrene (a PAH) would require use of an alternate process to achieve removal (phenanthrene would be marginal). Henry's Law constants are provided in engineering and science handbooks in various dimensions. If dimensionless Henry's Law constants are used, chemicals with values greater than or equal to 0.01 are considered to be good candidates for air stripping.

Besides chemical volatility, a key factor that determines the extent of air stripping is the air-to-water ratio, which is calculated by dividing the air flow by

TABLE 7.1. Henry's Law Constants for some chemicals (20°C) [1].

	H (atm-m^3/mol)
PCE	8.3×10^{-3}
TCE	1×10^{-2}
Benzene	5.5×10^{-3}
Toluene	6.6×10^{-3}
Phenanthrene	3.5×10^{-5}
Benzo(a) pyrene	5.75×10^{-6}

the expected water flow through the stripping column. Air-to-water ratios ranging from 5 to 200 are feasible. It should make sense that, for a constant water flow, chemical removal will increase as the air flow is increased. However, there is a price to pay for operating an air stripper at a very high air-to-water ratio; operational costs will increase due to running the air pump at a high rate. Furthermore, the upper limit of air flow will eventually be realized in that too much air will actually prevent the water from moving downward through the column. This condition is known as "flooding" and must be avoided. Typical full-scale air strippers exist in diameters ranging from 0.5 m to 3 m and heights of 2 m to 20 m.

7.2.2 Granular Activated Carbon

It is becoming somewhat typical that households in America use some form of activated carbon to treat drinking water in the home. If you have an under-the-sink arrangement of pipes, tubing, and canisters, or you pour tap water through a pitcher that contains a sealed canister, chances are good you are relying on *granular activated carbon* (GAC) to purify your water. In the home, GAC is used to remove residual chlorine and low concentrations of organics. At hazardous waste sites, GAC is often used solely or as the final process in a series to achieve low effluent concentrations prior to discharge to receiving streams or groundwater.

GAC can remove organic chemicals from water because it offers numerous "sites" of *adsorption* that resemble a sponge if observed through a scanning electron microscope. Adsorption is described as an electrochemical attraction that carbon has for various chemicals. A carbon source (such as coconut husks, coal, peanut shells) is "activated" by exposing the material to high temperature and pressure; this activation process opens microscopic pores in the carbon and readies it for adsorption on chemicals.

Adsorption should not be confused with absorption, which could be envisioned if soil is used to "dry up" a puddle of waste oil in a factory. The absorbed oil will be released if adequate pressure is applied to soil (such as in a landfill), whereas chemicals that are adsorbed to GAC can only be released from the carbon by "regenerating" the GAC. Regeneration involves processing used GAC in a high-temperature process that effectively oxidizes the chemicals on the carbon. Regenerated GAC may be returned to a site for use; however, there is some loss of carbon in the regeneration process, so supplementation with virgin carbon may be necessary.

Adsorption occurs because a chemical has a specific affinity for activated carbon, and often a relative lack of affinity for water. To determine how well carbon will work at particular site, engineers run laboratory procedures known as *isotherm* tests. Such tests are completed by preparing about 10 bottles, each with increasing amounts of GAC (known as the *adsorbent*) and adding a fixed volume of contaminated water with known concentrations of contaminants (known as the *adsorbate*). The bottles are stoppered and mixed for a few days to allow equilibrium to be reached (i.e., when no further chemical removal occurs). Then, bottles are opened and samples are collected and analyzed to determine the final concentration of chemicals. Isotherm data are useful in determining the *extent* of adsorption at equilibrium. Rough estimates are made of carbon usage and sizes of carbon columns that may be required. However, a more thorough analysis and design would include operation of columns of GAC connected in series. Such a *column study* yields a better indication of how quickly carbon will be used and how many full-scale GAC columns will be required at a particular site. Figure 7.5 is a schematic of GAC columns connected in series.

Activated carbon is particularly advantageous in producing a high-quality effluent. This means that GAC is often employed to meet very stringent discharge limits; as such, GAC is the final process in a treatment system. Also, a GAC system offers the advantages of being easy to operate, needs infrequent checking, and is applicable for many organics in water. Activated carbon is a proven technology and is readily available. Disadvantages are that activated carbon is a mass transfer process that does not achieve destruction of contaminants; they are merely transferred from the aqueous phase to the solid phase (GAC). Finally, the cost of activated carbon can be considerable, so its use is often confined to low-concentration waste streams. The rate of carbon use is proportional to chemical concentration, so low concentrations permit longer life of GAC. This is why other processes are employed at the "front end" of a treatment system where concentrations are higher.

One final problem that must be considered is the need for pretreatment of groundwater prior to activated carbon. Due to the cost of carbon, its use for ground water cleanup should be limited to removal of chemicals that are

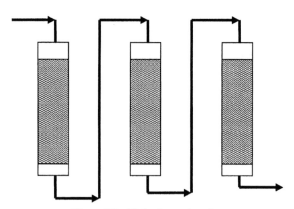

Figure 7.5. GAC columns in series.

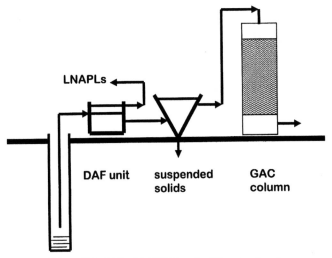

Figure 7.6. GAC with LNAPL and solids pretreatment.

dissolved. At some sites, water may contain oils (from LNAPLs) or suspended solids (from upstream processes), and both of these constituents must be removed before the water reaches the GAC columns. Settling tanks are often employed to remove suspended solids, and DNAPLs, dissolved air flotation (DAF) units will float LNAPLs and separate them from water, and sand filters are sometimes used to further remove suspended solids upstream of GAC systems. Figure 7.6 is a schematic of a GAC system with LNAPL and suspended solids pretreatment.

Applications of activated carbon in hazardous waste treatment are not limited to groundwater. Vapor phase activated carbon (VPC) is sometimes placed on the gaseous effluent from an air stripper to remove contaminants that were transferred from groundwater to the air in a stripper column. VPC is manufactured specifically for pollutants in the gaseous phase and the rate of chemical removal is more rapid that that of GAC.

7.2.3 Bioremediation

Bioremediation is the application of microorganisms to transform hazardous chemicals into non-hazardous forms. Microorganisms may biotransform a chemical to another form and, in doing so, render the chemical non-hazardous. Some chemicals are oxidized to CO_2, and this represents a specific type of biotransformation known as *mineralization*. Some chemicals are readily metabolized by microbes, whereas others are *recalcitrant*, meaning that they are resistant to biodegradation. Recalcitrant chemicals persist in the environment due to specific chemical properties. Such chemicals were created for industrial purposes, they do not exist naturally and are termed *xenobiotic*. Chlorinated chemicals are not found in a pristine environment and, as such, microbes have had to "learn" or adapt to be able to biodegrade these compounds.

In contrast to air stripping and activated carbon, bioremediation offers the advantage of complete destruction of a hazardous chemical. Consider the potential benefits of this if you are managing (or owner of) a business that generates hazardous waste. Specifically, if bioremediation of your hazardous waste was employed, your business could save considerably in environmental management costs, and long-term liability would be reduced. Another advantage of bioremediation is cost. Cost is a site-specific issue that is dependent on the contaminants to be removed and volume to be treated, but often, biological treatment of hazardous waste is cheaper than air stripping or granular activated carbon. However, since biological processes involve living microorganisms, such systems may take more time and effort to start up at a hazardous waste site. Furthermore, operation of a biological process requires more sophisticated knowledge; therefore more intense operator training and daily monitoring are required compared to air stripping and activated carbon.

To maintain a successful bioremediation process, engineers and operators of a treatment facility must make sure that microbial requirements are met. Just as we have specific nutritional requirements, bacteria (which are responsible for most bioremediation) have requirements in order to survive and biodegrade material that we consider to be hazardous. Before we consider these requirements, realize that bacteria exist naturally in the environment. They are often found in abundance at hazardous waste sites in aquifers and soil, and in many instances, they have adapted to hazardous chemicals that have been discharged and have developed the biochemical "machinery" to use xenobiotic chemicals as a food source. This is a very fortunate occurrence for the human race. Furthermore, bacteria in nature exist in an environment of starvation. They are therefore very opportunistic such that, when a chemical is "recognized" as a beneficial food source, biodegradation commences.

All living beings, bacteria included, require an energy source. In many cases, the xenobiotic chemical may serve as the energy source, but this is not always the case (examples follow). In addition, a carbon source must be available to provide for growth of more bacteria. As with the energy source, the carbon source may be satisfied by the xenobiotic chemical (if not, a different chemical must fulfill this requirement). To enable use of an energy source, bacteria must be supplied with a chemical to "breathe," as humans use oxygen. Aerobic bacteria use oxygen, whereas anoxic bacteria use nitrate (NO_3^-), and anaerobic bacteria rely on either carbon dioxide (CO_2) or sulfate (SO_4^{2-}). Most bacteria are able to use only one of these; however a few can exist on either oxygen or nitrate. The diverse capability of various bacteria lends itself to the application of bioremediation for a variety of chemicals in many different microbial environments. The choice of which type of bacteria to employ depends primarily on the chemicals present at a site. For example, aromatic compounds, such as the benzene, toluene ethylbenzene, and xylene (BTEX) chemicals, are more readily biodegraded by aerobic microbes. For such cases, either air or pure oxygen is supplied to an aquifer. By contrast, the chlorinated solvents such as perchloroethene (PCE) and trichloroethene (TCE), are known to biodegrade more readily under anaerobic conditions.

Additional bacterial requirements are nitrogen and phosphorous and micronutrients such as sulfur, potassium, magnesium, calcium, and sodium. For groundwater remediation, it is typical that nitrogen and phosphorous must be supplied, whereas the micronutrients exist naturally in sufficient quantities.

Various environmental factors must be considered to insure successful biodegradation of chemicals at hazardous waste sites. Most microorganisms used in bioremediation prefer a pH near 7.0. Therefore, engineers and scientists must have knowledge of the chemical products of a bioremediation process and must understand what occurs if the pH changes during biodegradation. To preclude a change in pH, additives may be supplied (such as bicarbonate) to resist pH changes. Microorganisms are sometimes classified by the temperature range in which they live. Most bioremediation projects use mesophilic organisms, which thrive in the 20–45°C range; however, there have been application of psychrophilic microbes (1–20°C) in composting of contaminated soils in cold regions. Use of thermophilic microbes (>55°C) in pollution abatement is normally confined to applications involving industrial wastewater treatment and municipal solid waste treatment. Moisture content is obviously not a concern for treatment of contaminated groundwater. However, for bioremediation of contaminated soil, a minimum moisture content of approximately 40 percent should be maintained.

Finally, environmental engineers and scientists must know the inhibitory concentrations of xenobiotic compounds. Many xenobiotics are biodegradable; however, bioremediation can be inhibited if the concentration of the xenobiotic chemical is too high. Actual concentrations are specific to the chemical being degraded.

Microbes have adapted to many chemicals that have been discharged to the environment and have developed the ability to use these chemicals and biotransform them into another form. Factors that influence if and how quickly biodegradation can occur often depend on molecular structure and chemical solubility. Also, chemicals that have other constituents substituted for hydrogen are more difficult to degrade. For example benzene, an aromatic chemical (C_6H_6) is more readily degraded than chlorobenzene (C_6H_5Cl), which has chlorine substituted for hydrogen. In addition, chemical solubility affects biodegradability since all substrates for microbes must exist in the aqueous phase to be absorbed and metabolized.

Most chemicals are biodegraded by bacteria because they benefit the organism. Often they serve as an energy or carbon source. Others may provide a source of oxygen. The point is that the organism benefits from the biodegradation—the organism gains energy and fulfills the purpose of bacteria—to grow more bacteria. Surprisingly, it has been determined that some xenobiotic compounds can be degraded and not confer any apparent benefit to the organism. Biodegradation of a compound with no apparent benefit is known as *cometabolism*. As the name implies, more than one compound must be degraded simultaneously; while a primary substrate that serves as an energy source is being degraded, a secondary substrate (or cometabolite) may be degraded concurrently. Methanotrophic bacteria in a pristine environment biotransform methane to carbon dioxide and water under aerobic conditions. Therefore, they use methane as a source of energy and carbon, and because they are aerobic organisms, they use oxygen. If methanotrophic bacteria are provided with the primary substrate (methane) and oxygen, they can simultaneously biodegrade TCE, which is termed a secondary substrate. In this instance, TCE biodegradation confers no apparent benefit on the methanotrophic bacteria. They have the ability to degrade TCE at low concentrations typically found at hazardous waste sites (less than 0.1 mg/L). However, TCE concentrations that approach 1 to 10 mg/L are known to inhibit biodegradation.

Technologies that use biodegradation of contaminants in groundwater are often identical to those used in the wastewater treatment industry. Smaller-scale activated sludge or trickling filter units are fabricated at an industrial facility and then delivered by truck to a hazardous waste site. At other locations, anaerobic reactors similar to those used in industry have been used. Pre- or post-treatment processes are installed and connected to the biological treatment system. Pretreatment processes may include sedimentation or flotation tanks to remove DNAPLs or LNAPLs, air stripping to remove recalcitrant volatile organic compounds (VOCs). Post-treatment processes include sand filters to remove suspended solids and granular activated carbon (GAC) to achieve low effluent concentrations.

7.2.4 Treatment Systems and Life Cycle Design

Typically, hazardous waste sites are contaminated with multiple chemicals, so pump-and-treat systems often include a number of processes connected in series. It is the responsibility of the engineers designing the treatment system to determine the most appropriate processes and to arrange them in proper order.

Example 7.3

Contaminated groundwater has been sampled and the following constituents were detected: biodegradable and recalcitrant VOCs and LNAPLs. Draw a process-flow diagram that shows the necessary processes to achieve low effluent limits.

To complete this assignment, first prepare a table that lists each type of contaminant and the technologies that could be used. Then, determine the appropriate order of each process and identify any pre- or post-treatment processes that may be necessary.

Constituent	Potential Process	Order of Process	Recommended Process	Pre- or Post-treatment processes
Biodegradable VOCs	Biological— Activated sludge (maybe) Biological— anaerobic process (maybe)	2 (maybe)	Depends on contaminant. Assume (for this example) that they are aerobically degradable, therefore choose activated sludge	
Recalcitrant VOCs	Air Stripping GAC	3	Air stripping (will likely be cheaper to operate than GAC).	4. GAC to achieve low concentrations
LNAPLs	Dissolved air flotation	1	Dissolved air flotation	

Figure 7.7 is a schematic (also known as a flow diagram) of the recommended process. Dissolved air flotation precedes the air stripper to remove LNAPLs that would otherwise "coat" the packing of the air stripper. If the recalcitrant VOCs are not inhibitory, a biological process may be installed before air stripping. Alternatively, the air stripper may be designed to remove all VOCs, and biological treatment would not be used. GAC is the final process to produce effluent with very low concentrations of contaminants.

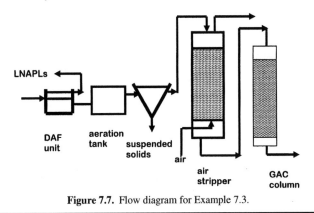

Figure 7.7. Flow diagram for Example 7.3.

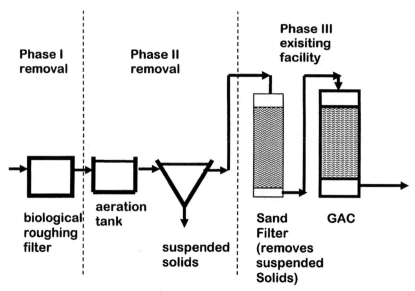

Figure 7.8. Life-cycle design of pump-and-treat system.

Lifecycle design is the recognition that contaminant concentrations from an extraction well system will decrease in with time. This implies that, as a pump-and-treat project progresses, some processes may become unnecessary due to lower concentration influent. The effect of decreasing concentrations is actually beneficial for an air stripper or GAC system. However, a biological process may eventually exhibit problems in maintaining enough biomass to allow good settling properties in a secondary clarifier. With this in mind, a treatment system that initially has a biological "roughing" filter (to reduce biodegradable organics), followed by traditional activated sludge, sand filtration, and finally activated sludge, would be represented by Figure 7.8. This block-flow diagram depicts a pump-and-treat system that is similar to one used at a pole treating facility near Seattle, Washington.

Once the concentration of biodegradable organics becomes lower, the roughing filter can be removed. Eventually, if concentrations decrease further, the activated sludge system could be eliminated, and only filtration and activated carbon would be needed for the remainder of the project life.

7.3 *IN SITU* TREATMENT

In situ is a Latin phrase meaning "in place," and it is used in the context of site cleanup to refer to processes that effect subsurface soil remediation or ground water remediation with no extraction of water or excavation of soil. In effect, *in situ* remediation processes use the aquifer or subsurface as the "reactor" to reduce concentrations of contaminants.

7.3.1 *In situ* Bioremediation

Figure 7.9 is a schematic of an *in situ* bioremediation process implemented by the U.S. Department of Energy at its Savannah River Site in Aiken, SC. This DOE system is an example of cometabolism of trichloroethene (TCE) by methanotrophic bacteria that are indigenous to the contaminated aquifer. Because methanotrophs are aerobic organisms, oxygen is required, and the compressed air fulfills this requirement. Also, methane supplementation satisfies the requirement for a carbon source and energy source. Once the energy source, carbon source, and oxygen are provided (plus, possibly macronutrients), the methanotrophs will cometabolize TCE.

Initially, most applications of *in situ* bioremediation involved aerobic and anoxic systems contaminated with BTEX or petroleum hydrocarbons. However, as laboratory findings are transferred to field-scale applications, *in situ* remediation of chlorinated aliphatics such as PCE and TCE are becoming more common. Whereas TCE may biodegrade under aerobic or anaerobic conditions, PCE biodegradation has been demonstrated only under anaerobic conditions. This can be rationalized, in that PCE is a highly oxidized molecule and bacteria that mediate the biodegradation of PCE have been described as "breathing on PCE" [2]. Figure 7.10 depicts the anaerobic biodegradation of PCE, where it is shown to be sequentially dechlorinated, in that each step achieves the removal of chlorine and its substitution with hydrogen. To

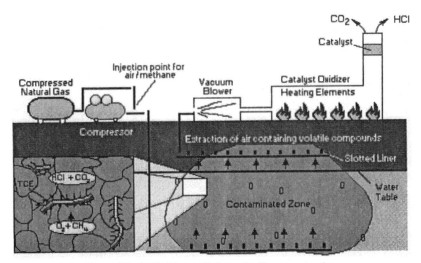

Figure 7.9. *In situ* bioremediation and soil vapor extraction.

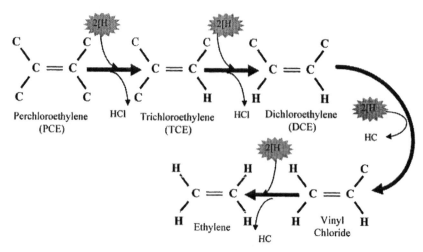

Figure 7.10. PCE sequential dechlorination to ethane.

facilitate this process, an energy source must be supplied. From laboratory research and field-scale tests, many organic chemicals have been shown to serve as energy sources, such as methanol, acetate, and glucose, as well as commonly used substances such as sugar, non-fat dried milk and even various vegetable oils! The challenge in achieving success with *in situ* bioremediation is achieving uniform delivery of additives, such as energy sources, oxygen sources, carbon sources, and macronutrients. In an aquifer, we do not have the luxury of being able to mix the contents of a reactor to make sure all additives are distributed uniformly throughout the process. Work is therefore progressing to improve methods of delivery of additives to the subsurface.

Monitored Natural Attenuation (MNA) is a form of in situ bioremediation that involves no supplementation. During the 1990s, increasing evidence suggested that, at some sites, chemicals were being transformed naturally and with no intervention. Many owners of contaminated sites became interested in determining if MNA is occurring on their property. Whereas MNA has sometimes been labeled as a "Do Nothing" treatment alternative, EPA has placed the burden of proof on PRPs that wish to use MNA to clean up their site. Based on knowledge of how specific chemicals may biodegrade, engineers and scientists have learned what indications to look for to determine if MNA is occurring. A positive indication that MNA is occurring would be detection of a chemical that has not been discharged at a site that is a known biological product of a chemical that was disposed of.

Compared to a pump-and-treat system, *in situ* bioremediation offers the potential for faster and cheaper clean-up. Recall that a pump-and-treat system relies on the extraction of significant quantities of groundwater to encourage desorption of chemicals from an aquifer. This could result in a long-term pumping project lasting for decades, and possibly, centuries [3]. By contrast, indigenous microbes have direct access to contaminants such that an active *in situ* process that supplied necessary additives would require less equipment on site and quicker cleanup times. Better still would be if MNA was occurring to a significant degree and at a vigorous rate to preclude migration of contamination to adjacent properties. In such instances, no supplementation would be required. However, long-term monitoring would be necessary to insure that expected results were occurring. Another precaution to be aware of with MNA is that knowledge of the products of xenobiotic biodegradation must be completely understood. Some sites with PCE contamination have displayed partial MNA by dechlorinating PCE to vinyl chloride. From a toxicology standpoint, vinyl chloride is more potent than PCE, so it is important to analyze samples for products of biodegradation in addition to monitoring disappearance of the initial contaminant. In addition to detection of products of a specific biodegradation reaction, other indicators of MNA at a PCE site would be elevated levels of chloride ions (refer to Figure 7.10), and spatial distribution of less chlorinated product, due to different retardation factors, as shown previously in Figure 6.5.

7.3.2 Soil Vapor Extraction

At many sites, contamination exists above the water table. A simple example of this would be the numerous underground storage tanks at gasoline stations that have developed leaks over decades. In such cases, soil above the water

---

Natural Attenuation in Toronto

During the late 1980s at a railroad switching station in Toronto, Ontario, a nearby aquifer was found to be contaminated with PCE. Historical investigation of site use revealed that rail tank cars full of PCE or methanol were brought into the switching station and then "off-loaded" to tank trucks for local delivery. Commercial hose with industrial fittings was used to transfer the chemicals from the rail cars to the tank trucks. When the transfer was completed, the hose was disconnected, and its contents were left to drain onto the ground and seep into the soil.

Groundwater monitoring wells were drilled, and water was sampled to determine the extent of contamination. Laboratory analysis confirmed the presence of PCE, but also indicated that the groundwater contained TCE, cis-DCE, vinyl chloride, and ethene. It was known that these less chlorinated chemicals were not discharged at this site, so results were puzzling as to how the chemicals found their way into the aquifer. The subsequent discovery that anaerobic bacteria, once supplied with an energy source (methanol), could dechlorinate PCE to ethene solved the puzzle and suggested that natural attenuation was occurring at this site.

table becomes saturated with chemicals from gasoline, such as BTEX and MTBE. If the leak continues for a long time these chemicals eventually reach the water table and contaminate groundwater. If the leak is recognized soon enough, only the soil above the water table must be remediated. Based on economics, the two most likely actions are either to excavate the soil and transport it to a sanitary landfill or to employ *soil vapor extraction* (SVE).

Figure 7.9, shown previously as an example of *in situ* bioremediation at the Savannah River site, also includes an SVE system to remove TCE in soil above the water table. The soil vapor extraction system comprises a series of small-diameter wells that are drilled into and around the periphery of the contaminated area. Above-ground pipes connect the wells in a manifold, which is routed to a vacuum pump. The pump causes negative pressure in the wells and this results in movement of soil gas through the contaminated zone and into the wells. Such a system is analogous to a water pump-and-treat system except, in this case, soil gas is being pumped instead of water. Movement of soil gas promotes the volatilization of chemicals to the soil gas, where they are extracted by the vacuum pump and discharged to a treatment system. VOC-laden gas is typically treated using vapor-phase carbon or destroyed in a thermal combustion unit.

Obviously, soil vapor extraction is applicable only at sites with VOC contamination that exists above the water table. Non-volatile chemicals would not be affected by movement of soil gas. Care must be taken in operating soil vapor extraction wells close to the water table; vigorous pumping rates can result in water being brought into the extraction wells, in a condition known as "upwelling." Finally, chemical volatility is evaluated for soil vapor extraction based on vapor pressure rather than Henry's Law constant. Consider the descriptions of each chemical property given previously to rationalize the use of vapor pressure instead of Henry's Law constant for soil vapor extraction.

7.3.3 Barrier Technology

To prevent migration of subsurface contamination, a *barrier* that is resistant to water and contaminant movement is sometimes installed. Such a barrier is referred to as a *slurry wall* and comprises a mixture of soil and bentonite. Bentonite is a clay that expands when wet and imparts very low water permeability to the slurry wall. Figure 7.11 shows a schematic of a slurry wall system installed at the Love Canal site to prevent contaminant migration and allow pumping of contaminants. The slurry wall has been installed around an area of contamination and to a depth that reaches an *aquitard*, which is a geological formation that does not permit appreciable water movement. In this case, clay is the aquitard; however, bedrock could fulfill this purpose at another site. To prevent groundwater movement under the slurry wall, it is "keyed" into the aquitard.

Slurry walls on their own are considered a containment system that resists movement of contamination. However, no containment system has the ability to prevent the movement of water and contaminants permanently. This realization may have led to the development of slurry walls that direct contamination through a zone of active *in situ* treatment, known as a "funnel-and-gate" system. Alternatively, some slurry walls have been impregnated with reactive materials such as zero valent iron, which has been shown to promote chemical transformation of chlorinated organics such as PCE and TCE.

7.4 INCINERATION

Incineration is a process that can achieve destruction of hazardous organic chemicals by combustion. In an incinerator, organic chemicals serve as fuel and sufficient oxygen must be provided to support complete oxidation to carbon dioxide and water. For some chemicals (such as chlorinated solvents) with relatively lower heating values, auxiliary fuel must be provided to support combustion. To calculate the amount of oxygen necessary to oxidize organics in hazardous waste, a chemical reaction that represents all constituents can be developed. However, since the complete composition of all waste materials is difficult to know, oxygen must be provided in excess to preclude significant incomplete combustion, which results in carbon monoxide production and various organic air pollutants.

End products from the combustion process include water, CO_2, and residual solids known as "incinerator ash." Depending on the components in the waste, metals will either accumulate in the ash or be oxidized and become part of the gaseous discharge from the incinerator. Air pollution equipment must therefore be provided to remove metals, particulates, and acid gases prior to discharge. Recall that incinerator ash from hazardous waste combustion would be classified as a "derived from" hazardous waste, and the owner of the process would need to submit a delisting petition to request that the ash be managed as a non-hazardous material.

Some types of incinerators include liquid injection types (for liquids and sludges), and multiple hearth and rotary kiln for solid wastes. Regardless of the incinerator type, all incinerators are designed based on the "three Ts" of incineration: temperature, time, and turbulence. Each of these factors must be provided in sufficient quantity to facilitate combustion.

As of 1994, EPA reported that 164 hazardous waste incinerators were either permitted or under interim status in the U.S. Of the total, 21 were permitted commercial facilities [4]. The remainder of incinerators were either under permit appeal or non-commercial processes.

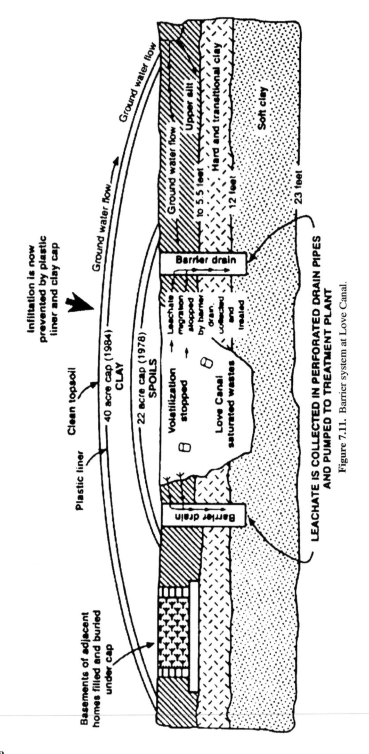

Figure 7.11. Barrier system at Love Canal.

The Drake Chemical Company Superfund Site

The Drake Chemical Company of Lock Haven, PA was a major producer of chemicals during the Second World War and continued to provide employment opportunities to the economically depressed town after the war. One of its waste chemicals which the company disposed of in an open pit was beta-napathylamine, a compound used as dye. Unfortunately, beta-napathylamine is also a potent carcinogen, having been found to cause bladder cancer. In 1962 the State of Pennsylvania banned the production of this chemical, but the damage to the groundwater had already been done with the disposal of beta-napathylamine into the uncontrolled pit. The order from the state caused Drake to stop manufacturing beta-napathylamine, but it continued to produce other chemicals, without much concern for the environment or the health of the people in Lock Haven. Finally in 1981 the U. S. EPA closed the facility and took control of the property. They discovered several unlined lagoons and hundreds of often unmarked drums of chemicals stored in makeshift buildings. After removing the drums and draining the lagoons, they found that the beta-napathylamine had seeped into nearby property and into creeks, creating a serious health hazard. The EPA's attempt to clean the soil and the water was, however, met with public opposition. Much of the public blamed the EPA for forcing Drake Chemical, a major local employer, to close the plant. In addition, the best way to treat the contaminated soil was to burn it in an incinerator, and the EPA made plans to bring in a portable unit. Now the public, not at all happy with EPA being there in the first place, became concerned with the emissions from the incinerator. After many studies and the involvement of the U. S. Army Corps of Engineers, the incinerator was finally allowed to burn the soil, which was then spread out and covered with 3.5 feet of topsoil. The groundwater was pumped and treated, and this continued until the levels of beta-napathylamine levels reached background concentrations. The project was not completed until 1999 with the EPA paying for the legal fees of the lawyers who argued against the cleanup.

7.5 LANDFILLS

Landfills represent the final disposal alternative for hazardous wastes. Because we are not able to completely eliminate or destroy hazardous wastes, and because some treatment processes (incineration, GAC, biological) produce residues that may be hazardous, landfills remain a necessary technology.

RCRA enabled EPA to set new standards for hazardous and municipal solid waste landfill design and operation. It is generally accepted that, with time, landfills will leak. Therefore, RCRA focused on minimizing the potential of landfills to create future Superfund sites.

As waste is placed in a landfill, precipitation and moisture from the waste begins to accumulate and move through the buried waste. This liquid becomes contaminated as it makes its way from top to bottom through the landfill. If not contained, this liquid (known as *leachate*) will migrate to groundwater and cause contamination.

To prevent the discharge of leachate and groundwater contamination, landfills are typically located away from sources of drinking water. Furthermore, landfills are often sited in areas with geological formations that restrict the movement of leachate. Clay and glacial till are examples of geological formations that restrict leachate movement. Alternatively, a clay barrier is installed and compacted. To further contain leachate, landfills are constructed with two liners of high-density polyethylene under the waste. Figure 7.12 shows that the liner system begins with the placement of a liner on the compacted clay. Next a drainage layer consisting of fine gravel or synthetic material is placed on top of the liner. Perforated pipes are placed at low spots in the drainage layer to enable collection and subsequent treatment of leachate. Another liner is placed on top of the lower drainage layer, and another (top) drainage layer is installed [4]. A filtration layer is then added, and then waste may be deposited.

As leachate is produced in the landfill, it migrates downward and is eventually collected in the top drainage layer and piping system, known as the primary leachate collection system. The topmost liner restricts the downward movement of the leachate. However, due to flaws that occur during the installation process, or localized liner failure, some leachate does find its way to the bottom drainage layer and piping system. This is known as the secondary leachate collection system. Landfills are required to monitor the daily production of primary and secondary leachate, with the requirement that, if secondary leachate production reaches a specific percentage of primary leachate, action must be taken to locate and repair the leak in the primary (topmost) liner.

Groundwater monitor wells are installed upgradient and downgradient of the landfill. These wells are sampled and analyzed routinely to determine if leachate has been released from the secondary liner system.

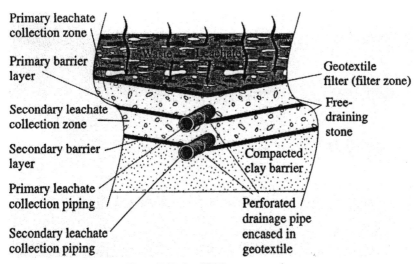

Primary leachate collection zone

Primary barrier layer

Secondary leachate collection zone

Secondary barrier layer

Primary leachate collection piping

Secondary leachate collection piping

Geotextile filter (filter zone)

Free-draining stone

Compacted clay barrier

Perforated drainage pipe encased in geotextile

Figure 7.12. Landfill liner cross section.

To minimize the entrance of excess water to the landfill, a "cover" is installed over the waste, once a section of landfill reaches its permitted height. The cover consists of another layer of impermeable high-density polyethylene followed by a sloped drainage layer to prevent pooling of precipitation. The drainage layer is covered with top soil, and grass is planted to reduce erosion.

Landfills are permitted to accept either hazardous waste or municipal solid waste. Municipal solid waste is household refuse, yard waste, garbage, commercial waste from office buildings and schools, and construction and demolition materials. These solid wastes are not hazardous and may be commingled as they are disposed of, buried, and compacted at a landfill. By contrast, solid waste landfills that are permitted to accept hazardous waste are required to maintain an inventory of all waste that is accepted, and to organize the disposal of compatible wastes into isolated cells of the landfill. Isolation of compatible waste insures that undesirable or violent chemical reactions do not occur once the waste is disposed of.

7.6 SYMBOLS

BTEX = benzene, toluene, ethylbenzene, xylene
D = total days exposure (d)
DCE = dichloroethylene
DNAPL = dense non-aqueous phase liquid
GAC = granular activated carbon\
H = Henry's Law constant
HI = hazard index (dimensionless)

I = average daily intake (mg/kg-d)
LNAPL = light non-aqueous phase liquid
K_d = soil distribution partition coefficient
K_{oc} = organic carbon partition coefficient
K_{ow} = octanol-water partition coefficient
M_{abs} = mass of chemical absorbed per day (mg/d)
MNA = monitored natural attenuation
MTBE = methyl tertiary butyl ether
PCE = perchloroethylene
RCRA = Resource Conservation and Recovery Act
RfD = reference dose
SF = slope factor
SWE = soil water extraction
T = average time (d) = 70 years for carcinogens or total exposure duration for non-carcinogens
TCE = trichloroethylene
VOC = volatile organic compound
VPC = vapor phase activated carbon
W = body weight (kg)

7.7 REFERENCES

1. Schnoor, J. L. *Environmental Modeling*. John Wiley & Sons, New York.1996.
2. McCarty, P. L. *Breathing with chlorinated solvents*. Science. 276. June1997.
3. Macdonald, J. A. and Kavanaugh, M. C. *Restoring contaminated groundwater: an achievable goal?* Environ. Sci. Technol. 28.1994.
4. http://www.epa.gov/epaoswer/hazwaste/combust/general/universe.txt.

7.8 PROBLEMS

7.1 Describe qualitatively how the risk assessment procedure is used to determine if contamination poses an unacceptable hazard due to non-carcinogens or carcinogens.

7.2 Define RfD and SF and describe how they are used in risk assessment.

7.3 Calculate the hazard index and the risk to a 9-year old child who ingests chloroform (CF) in water at summer camp. Assume the following: body weight = 29 kg, water ingestion = 2 L/d, resides at camp = 90 day/yr for 3 years, RfD for CF = 0.01 kg/mg-d. SF for CF = 0.0061 (kg/mg-d)$^{-1}$.

7.4 Describe verbally and graphically how a packed-tower air stripper works. Include in your description the relevant fluid flows and contaminant transfer.

7.5 Describe under what conditions air stripping is appropriate, what contaminants air stripping is most likely to remove, and what pre- and post-treatment technologies may be necessary.

7.6 Describe the purpose of granular activated carbon and for what conditions and contaminants granular activated carbon is appropriate, and (verbally and graphically) what pre- and post-treatment technologies may be necessary.

7.7 Describe verbally, and via schematic, how air stripping, granular activated carbon, and vapor-phase carbon may be employed collectively in a groundwater treatment system and what types of contaminants each technology removes.

7.8 Define recalcitrant and xenobiotic.

7.9 Discuss the advantages and disadvantages of biological treatment of hazardous waste

7.10 Give an example of cometabolism and specifically , name the xenobiotic compound being degraded and energy source, oxygen source, and carbon source.

7.11 Draw a flow diagram of a complete pump-and-treat system with appropriate technologies to remove VOCs, biodegradable organics with low Henry's Law constants, DNAPLs and LNAPLs, and dissolved iron. The system must meet very stringent effluent limits.

7.12 Describe the importance of understanding the intermediate compounds that may be produced during bioremediation. Give an example of how a problem can occur by only measuring the disappearance of the initial xenobiotic compound.

7.13 Define monitored natural attenuation (MNA),

7.14 Discuss the advantages of a successful bioremediation system versus air stripping or GAC from the standpoint of EPA and from the standpoint of a PRP (potentially responsible party).

7.15 Sketch a cross-section of a landfill, including the buried waste, primary and secondary liner and leachate collection systems, and upgradient and downgradient monitor wells. Include the water table and show the direction of groundwater. Use hypothetical laboratory analysis of groundwater to illustrate a landfill that is causing contamination. Also, consider a case where groundwater contamination exists but due to a source other than the landfill. How would laboratory analysis from the monitor wells support this scenario?

CHAPTER 8

Non-Hazardous Solid Waste

CITIES in ancient times were horrible places to live. Human waste of all kinds was simply thrown into the most convenient corners, and the people lived among this waste and squalor. Only when the social discards became dangerous for defense was action taken. In Athens, in 500 B.C., a law was passed to require all waste material to be deposited more than a mile out of town because the piles of rubbish next to the city walls provided an opportunity for invaders to scale up and over the walls. Rome had similar problems, and eventually developed a waste collection program in 14 A.D.

The cities in the Middle Ages in Europe were characterized by unimaginable filth. Pigs and other animals roamed the streets, and wastewater was dumped out of windows on unsuspecting passersby. In 1300 the Black Death, which was to a great degree accelerated by the filth, reduced the populations in cities and alleviated the waste problems until the industrial revolution in the mid-1800s brought people back to the cities.

The living conditions of the working poor in 19th century European cities have been graphically chronicled by Charles Dickens and other writers of that period. Industrial production and the massing of wealth governed society, and human conditions were of secondary importance. Water supply and wastewater disposal were by current standards totally inadequate. For example, Manchester, England, had on average one toilet per 200 people. About one sixth of the people lived in cellars, often with walls oozing human waste from adjacent cess pools. People often lived around small courtyards where human waste was piled, and which also served as the children's playground.

In the United States, conditions in many of the cities were similarly appalling. Waste was disposed of by the judicious method of throwing it into streets where rag pickers would try to salvage what had secondary value. Animals would devour foodstuffs. In 1834, Charleston WV enacted a law protecting garbage-eating vultures from hunters. One of the greatest problems in cities during the turn of the last century was the waste created by horses.

Early Solid Waste Management in Washington DC

Although Georgetown (now a part of Washington DC) was the first city in the new United States to pass a solid waste ordinance (do not dump in the streets), there were no municipal removal operations, and each person had to hire a "carter" to take the refuse. John Adams, the first president to live in the White House, was also the first to have a personal carter. In most cases, the carters brought the refuse to a small incinerator within the city limits and burned it, taking the residuals to a dump. The smell from these incinerators apparently was overwhelming.

The story is that Thomas Jefferson, while riding through the streets of Washington, was sickened by the smell of burning refuse and returning to the White House, instructed his secretary to discuss with the presidential carter a plan for collecting the refuse from all government buildings and removing it to a dump outside the city. Thus President Jefferson initiated the first contract for the collection of wastes from federal buildings.

Modern refuse management in American cities had a rocky start, mostly because of corruption in local governments. One of the first truly effective programs was started in New York City by George E. Waring, but the real impetus for many municipal solid waste and street cleaning programs came from women's organizations.

Today, solid waste professionals recognized that issues related to managing solid waste must be addressed with a holistic approach. For example, if more waste is recycled, this can have a negative financial impact on the landfill

The Horse as a Source of Solid Waste

At the turn of the 19th century, the prime mover in cities was the horse. The jobs they did ranged from pulling streetcars to collecting refuse, and their numbers were incredible—Chicago alone in 1880 had over 82,000 horses and there were over 3 million horses in all cities in the United States. A horse produces between 15 and 30 pounds of manure daily, so the problem of storing and removing this waste was staggering. The working conditions for these poor creatures was terrible, and their life span was only about two years, resulting in a huge number of carcasses to be removed. This removal was done by scavengers who used the carcasses for various purposes. The automobile replaced an immense pollution problem in the cities.

GEORGE E. WARING, JR.

This is the same George E. Waring who revolutionized the drainage of cities by constructing small-diameter sewers in Memphis. Although these innovations were important in the development of urban areas, Waring's greatest contribution was in solid waste management.

The appointment of Colonel George E. Waring as street cleaning commissioner of New York City in 1895 in many ways signals the beginning of modern refuse management the United States. Col. Waring tackled the problem in New York with his characteristic energy and chutzpah. First he reorganized the street cleaning department, getting rid of political deadwood, and hired men with education and military training to take over. His greatest accomplishment was the organization of the separation at the household level and the use of separate containers for these materials which were then collected and disposed of separately. In addition, he designed and operated a plant for separating mixed refuse. These two operations we know today as recycling and recovery.

He totally revolutionized street cleaning, including issuing white coats to the street sweepers, suggesting to all the importance of this civil activity in terms of maintaining public health. He also organized a 500-strong cadre of youngsters who participated in street cleaning, his thinking being that if he could get the young people to take seriously the cleanliness of streets, they would then have an effect on their parents, who might have different views.

There is no doubt that Waring's programs were spectacularly effective, but continuing them after he left office proved difficult. As the administration in City Hall changed, new priorities developed, and New York street cleanliness suffered. Waring's tenure on the job was, however, the crowning achievement of his career, and illustrated what a combination of political will, engineering knowhow, and unbounded enthusiasm can accomplish.

because less refuse is landfilled. Since many landfill costs are fixed (there is a cost regardless whether any refuse is landfilled), a drop in the incoming refuse can have severe economic ramifications. The various methods of solid waste management are therefore interlocking and interdependent.

Recognizing this fact, the U. S. EPA has developed a national strategy for the management of solid waste, called the *Integrated Solid Waste Management* (ISWM). The intent of this plan is to assist local communities in their decision-making by encouraging strategies that are the most environmentally

Municipal Housekeeping

During the Progressive Era (the early part of the 20th century), the impetus to clean up the cities often originated with the banding together of women who were dissatisfied with the condition of the streets and parks. One of the earliest champions of this cause was Mildred Chadsey, commissioner of sanitation in Cleveland, who called the cleaning up of cities "municipal housekeeping," and went on to argue that the art of housekeeping in the home is not materially different from keeping a city clean. Women's clubs in many cities became active in the fight for maintaining a cleaner urban environment, often pressing the elected officials to action. In Reading, Pennsylvania, for example, women's clubs banded together, picked up brooms, and went on a street-cleaning mission. The aldermen were sorely embarrassed, and political action followed the well-publicized event.

acceptable. The EPA ISWM overall strategy suggests that the list of the most to least desirable solid waste management strategies should be:

- Reducing the quantity of waste generated
- Reusing the materials
- Recycling and recovery of materials
- Combusting for energy recovery
- Landfilling

That is, when an integrated solid waste management plan is implemented for a community, the first means of attacking the problem should be reducing the waste at the source. This action minimizes the impact of natural resource and energy reserves.

Reuse is the next most desirable activity, and this also reduces the impact on natural resources and energy. Recycling is the third option, and should be undertaken when most of the waste reduction and reuse options have been implemented. Unfortunately, the EPA confuses recycling with recovery, and groups them together as meaning any technique that results in the diversion of waste from landfills. As previously defined, *recycling* is the collection and processing of the separated waste, ending up as new consumer product. *Recovery* is the separation of mixed waste, also with the end result of producing new raw materials for industry.

The decade of the 1990s witnessed dramatic growth in recycling/recovery in the United States. For example, the State of North Carolina was recycling about

---

Garbage Archeology

A long-running project at Arizona State University, headed up by William Rathje, has been studying the lifestyle of human civilizations based on what they choose to throw away. Their most telling work is done by digging into our own landfills and correlating the objects they find with the social values.
Some findings are:

- The amount of food waste in refuse depends on the availability of the food, but surprisingly this is an inverse relationship. For example, in 1975 there was a sugar shortage in the United States, and the results of the 1976 landfill sampling showed high amounts of sugar as people finally threw away the sugar they had been hording during the shortage.
- Higher-income families buy larger packages and have less packaging waste.
- Much of the waste from lower-income households is related to the automobile, while middle income households produce larger quantities of lawn waste.
- The volume occupied by plastic has not changed much over the years in spite of the fact that plastic use has increased. The plastic is thinner and is more efficiently used.

And finally, a most interesting conclusion:

- The degradation in landfills is so slow that newspapers from over 50 years ago are still readable.

10% of residential and commercial solid waste in 1988, and ten years later was running at about 26% recycling rate. In 1988 there were only three curbside collection programs in North Carolina, and ten years later there were 260 such programs serving 3.4 million people, or 45% of the population of the state. In addition, 120 yard trimmings composting facilities process over 700,000 tons of material a year.

The fourth level of the ISWM plan is solid waste combustion, which really should include all methods of treatment. The idea is to transform solid waste into a non-polluting product. This conversion may be by combustion, but other thermal and chemical treatment methods may eventually prove just as effective.

Finally, if all of the above techniques have been implemented and/or considered, and there is still waste left over (which there will be), the final solution is landfilling. At this time there really is no alternative to landfilling (except disposal in deep water—which is now illegal) and therefore every community must develop a landfilling alternative.

While this ISWM strategy is useful, it can lead to problems if taken literally. Communities must balance the above strategies to fit their local needs. This is where judgment comes into play, and where the solid waste manager really earns his/her salary. All the options have to be juggled and any unique conditions integrated into the decision. The economics, history, politics, and aspirations of the community are important in developing the recommendations.

8.1 REDUCTION OF SOLID WASTE

Waste reduction can be achieved in three basic ways: (1) reduction in the amount of material used per product without sacrificing the utility of that product, (2) increasing the lifetime of a product, and/or (3) eliminating the need for the product.

Waste reduction in industry is called *pollution prevention*—an attractive concept to industry, because in many cases the cost of treating waste is greater than the cost of changing the process so that the waste is not produced in the first place. Any manufacturing operation produces waste. As long as this waste can be readily disposed of there is little incentive to change the operation. If, however, the cost of waste disposal is great, the company has an incentive to seek improved manufacturing techniques that reduce the amount of waste. Pollution prevention as a corporate concept was pioneered by such companies as 3M and DuPont, and has as its driving force the objective of reducing cost (and hence increasing the competitive advantage of the manufactured goods in the market place). For example, automobile manufacturers for years painted new cars using spray enamel paint. The cars were then dried in special ovens that gave them a glossy finish. Unfortunately, such operations produced large amounts of volatile organic compounds (VOCs) that had to be controlled, and control measures were increasingly expensive. The manufacturers then developed a new method of painting, using dry powers applied under great pressure. Not only did this result in better finishes, but it virtually eliminated the problem with the VOCs. Pollution prevention is the process of changing an operation in such a manner that pollutants are not even emitted.

Reduction of waste on the household level is called *waste reduction*, sometimes referred to as *source reduction* by the U. S. EPA. Typical alternative actions that result in a reduction of the amount of municipal solid waste being produced include refusing to accept bags at stores, using laundry detergent refills instead of purchasing new containers, bringing own bags to grocery stores, stopping junk mail deliveries, and using cloth diapers. Unfortunately, the level of participation in source reduction is low compared to recycling activities. Even though source reduction is the first solid waste alternative for EPA (see above) and eight states have source reduction goals which range from no net increase of waste per capita to 10% reductions, few people participate in

---

The Coffee Cup Debate

What is better for the environment—using paper coffee cups or foam plastic coffee cups? Many people with environmental awareness insist the paper cup is superior and will often even disdain the use of foam plastic cups. Are they right? A study to evaluate the environmental impact of each container can be useful. In one study, the air and water emissions were estimated, with the following results (all numbers are pounds per 10,000 cups):

Type of container	Atmospheric emissions	Waterborne discharges	Industrial waste	Postconsumer solid waste
Foam polystyrene	11.8	2.1	18.6	120
Plastic-coated paper	18.1	2.9	54.3	218
Wax-coated paper	21.8	4.5	71.0	266

Based on this comparison, the plastic cup is looking good. But the picture is not that simple. The polystyrene cup is made from oil, a non-replenishable resource, and when it goes into a landfill, it never decays, taking up a large volume of landfill space. The paper cup is made from trees, a replenishable resources, and it (or at least its paper part) is biodegradable. On the other hand, when the paper cup decays in the landfill, it creates methane, a global warming gas, and unless this methane is captured, it can have a much greater global effect than the plastic cup.

So what is the answer? Don't use either disposable cup, but use a ceramic mug instead!

---

such programs. There is some evidence, however, that where communities initiate disposal fees based on volume or weight of refuse generated, the amount of refuse is reduced by anywhere between 10% and 30%. Public information programs can significantly help in reducing the amount of waste generated. A study of 250 homes in Greensboro, North Carolina, found that a 10% waste reduction can be achieved following a public information program.

8.2 REUSE OF SOLID WASTE

Reuse is an integral part of society, from church rummage sales to passing down children's clothing between siblings. Many of our products are reused without much thought given to ethical considerations. These products simply have utility and value for more than one purpose. For example paper bags obtained in the supermarket are often used to pack refuse for transport from the house to the trash can or to haul recycleables to the curb for pickup. Newspapers are rolled up to make fireplace logs, and coffee cans are used to hold nails. All of these are examples of reuse.

8.3 RECYCLING OF SOLID WASTE

The process of recycling requires that the owner of the waste material first segregate the useful fraction, so that it can be collected apart from the rest of the solid waste. Many of the components of municipal solid waste can be recycled for remanufacturing and subsequent use, the most important being paper, steel, aluminum, plastic, glass, and yard waste.

Theoretically, vast amounts of materials can be recycled from refuse, but this is not an easy task, regardless of how it is approached. In recycling, a person about to discard an item must first identify it by some characteristics and then manually separate it into a storage bin. The separation relies on some readily identifiable characteristic or property of the specific material that distinguishes it from all others. This characteristic is known as a *code* and this code is used to separate the material from the rest of the mixed refuse using a *switch*.

In recycling, the code is simple and visual. Anyone can distinguish newspapers from aluminum cans. But sometimes confusion can occur, such as not differentiating aluminum cans from steel cans, for example, or newsprint from glossy magazines, especially if the glossy magazines are intermingled with the Sunday paper. The most difficult operation in recycling is the identification and separation of plastics. Because mixed plastic has few uses, plastic recycling is more economical if the different types of plastic are separated from each other. Most people, however, cannot distinguish one type of plastic from another. The plastics industry has responded by marking most consumer products with a code that identifies the type of plastic, as shown in Figure 8.1. Plastics that can be recycled are all common products used in everyday life, some of which are listed in Table 8.1.

Theoretically, all a person about to discard an unwanted plastic item has to do is to look at the code and separate the various types of plastic accordingly. In fact, there is almost no chance that a domestic household will have seven different waste receptacles for plastics, nor are there enough of each of these plastics to be economically collected and used. Typically only the most common types of plastic are recycled, including PETE (polyethylene terephthalate), the material out of which the 2-liter soft drink bottles are made, and HDPE (high-density polyethylene), the white translucent plastic used for milk bottles.

Taking into account transportation and processing charges, it still appears that the economics for curbside recycling and materials recovery facilities in

PETE   HDPE   PVC   LDPE   PP   PS   OTHER
Figure 8.1. Symbols used on plastic to aid in recycling.

TABLE 8.1. Common Types of Plastics that May be Recycled.

Code Number	Chemical Name	Abbreviation	Typical uses
1	Polyethylene terephthalate	PETE	Soft drink bottles
2	High-density polyethylene	HDPE	Milk cartons
3	Polyvinyl chloride	PVC	Food packaging, wire insulation and pipe
4	Low-density polyethylene	LDPE	Plastic film used for food wrapping, trash bags, grocery bags and baby diapers
5	Polypropylene	PP	Automobile battery casings and bottle caps
6	Polystyrene	PS	Food packaging, foam cups and plates, and eating utensils
7	Mixed plastic		Fence posts, benches, and pallets

metropolitan areas (close proximity to refuse and markets) are quite favorable. One proof of this is the impressive number of materials recovery facilities in operation or under construction. The success of recycling programs has occurred despite the severe obstacles our present economic system places on the use of secondary materials. Some of these obstacles are:

Location of wastes. The transportation costs of the waste may prohibit the implementation of recycling and recovery. Secondary materials have to be shipped to market and if the source is too far away the cost of the transport can be prohibitive.

Low value of material. The reason an item is considered waste is that the material (even when pure) has little value. For example the price paid for secondary polystyrene has fluctuated from a positive payment of $100 per ton to a negative payment of $300 per ton.

Uncertainty of supply. The production of solid waste depends on the willingness of collectors to transport it, the cooperation of consumers to throw things away according to a predictable pattern, and the economics of marketing and product substitution, which may significantly influence the availability of a material. Conversion from aluminum to plastic beverage containers, whether by legislation, marketing options, or consumer preferences, will significantly change the available aluminum in solid waste. The substitution of a high-value material, aluminum, for a low-value material, plastic, will adversely affect recycling. Potential solid waste processors thus have little control over their raw materials.

Administrative and institutional constraints. Some communities are unwilling to pay the additional cost to implement curbside recycling programs.

Other cities may have labor or contractual restrictions preventing the implementation of resource recovery projects. For example, many yard waste composting facilities are prohibited by land use ordinances from accepting sludge or food waste, both of which would increase the value of the compost.

Legal restrictions. Some cities, such as San Diego, are prohibited from charging their residents for solid waste service, thus making it difficult to implement curbside recycling.

Uncertain markets. Recovery facilities must depend on the willingness of customers to purchase the end products—materials or energy. Often such markets are fickle, being either small, fragile operations or large, vertically integrated corporations that purchase the products on margin so as to satisfy unusually heavy short-duration demand.

8.4 RECOVERY FROM SOLID WASTE

Recovery is defined as the process in which the refuse is collected without prior separation and the desired materials are separated at a central facility. A typical materials recovery facility (MRF) is shown in Figure 8.2.

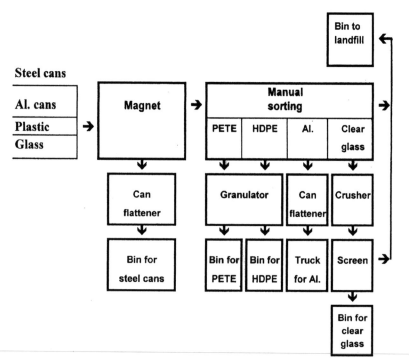

Figure 8.2. A typical materials recovery facility.

The various recovery operations in a MRF have a chance of succeeding if the material presented for separation is clearly identified by a code and if the switch is then sensitive to that code. Currently, no such technology exists. It is impossible, for example, to mechanically identify and separate all of the PETE soft drink bottles from refuse. In fact, most recovery operations employ *pickers*, human beings who identify the most readily separable materials such as corrugated cardboard and HDPE milk bottles before the refuse is mechanically processed.

Most items in refuse are not made of a single material, and to be able to use mechanical separation, these items must be separated into discrete pieces consisting of a single material. A common "tin can", for example, contains steel in its body, zinc on the seam, a paper wrapper on the outside, and perhaps an aluminum top. Other common items in refuse provide equally challenging problems in separation.

One means of producing single-material pieces and thereby assisting in the separation process is to decrease the particle size of refuse by grinding up the larger pieces. Grinding will increase the number of particles and achieve many clean (single-material) particles. The size reduction step, although not strictly materials separation, is employed in some materials recovery facilities, especially if refuse-derived fuel is produced. Size reduction is followed by various other processes like air classification (which separates the light paper and plastics) and magnetic separation (for iron and steel).

The recovery of materials, although it sounds terribly attractive, is still a marginal option. The most difficult problem faced by engineers designing such facilities is the availability of firm markets for the recovered product. Occasionally, the markets are quite volatile, and secondary material prices can fluctuate wildly. One example is the secondary paper market.

Paper industry companies are *vertically integrated*, meaning that the company owns and operates all of the steps in the papermaking process. They own the lands on which the forests are grown, they do their own logging and take the logs to their own paper mill. Finally, the company markets the finished paper to the public. This is schematically shown in Figure 8.3.

Suppose a paper company finds that it has an annual base demand of 100 million tons of paper. It then adjusts the logging and pulp and paper operations to meet this demand. Now suppose there is a short-term fluctuation of 5 million tons that has to be met. There is no way the paper company can plant the trees necessary to meet this immediate demand, nor are they able to increase the capacity of the pulp and paper mills on such short notice. What they do then is to go to the secondary paper market and purchase secondary fiber to meet the incremental demand. If several large paper companies find that they have an increased demand they will all try to purchase secondary paper and suddenly the demand will shoot the price of waste paper up. When either the demand decreases or the paper company has been able to expand its capacity it no

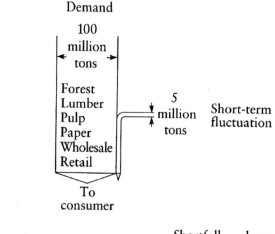

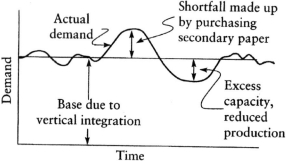

Figure 8.3. The paper industry treats secondary fiber (recycled paper) on the margin.

longer needs the secondary paper and the price of waste paper plummets. Because paper companies purchase waste paper *on the margin*, secondary paper dealers are always in either a boom or bust situation, and the price is highly variable. When paper companies that use only secondary paper to produce consumer products increase their production (due to the demand for recycled or recovered paper), these extreme fluctuations are dampened out.

One of the objectives of many well-meaning organizations is the complete recycling of all manufactured materials so that we have no net extraction of natural resources from the earth. Unfortunately, this is not an easy task. We use numerous materials that cannot, either technically or economically, be recycled. For example, packaging materials, solvents, refrigerants, many plastics, paints, pesticides, explosives, and many, many more materials cannot be recycled.

Even though the U. S. EPA is pushing states to increase their recycling rate, and has set a goal of 50% recycling, this goal has been demonstrated to be unreasonable by a study done by another EPA office. Their final report showed that as the recycled fraction of a material (e.g., aluminum, steel, glass) increases, the cost in energy and adverse environmental impact (measures as the emission of global warming gasses) also increases. This effect is not linear, and greatly increases as the rate of recycling approaches 25%. This argument is quite reasonable. Consider the recycling of glass, for example. Some fraction of glass is easy and inexpensive to recycle, such as the voluntary curb-side programs in many communities. But such programs recover only a small fraction of the total glass available for recycling. If the next step is to recover glass bottles out of mixed refuse in a materials recovery facility (MRF), the cost per ton is far higher than glass from curbside collection. Even this will not result in all glass being recovered, and any other program, such as collection of glass from street and highway cleanup, will cost even more. And finally, beverage bottles are not the only form of glass in refuse. It should be clear that as the fraction captured increases the unit cost has to increase. What the U. S. EPA did not count on, however, was that the cost of recycling common materials jumps significantly at around 25%. Unless we are willing to accept the dramatically increased costs and adverse environmental impacts, it seems unreasonable to expect recycling rates to go much beyond 25%.

But there are presently municipalities, particularly in California, that are claiming success in reaching 50% recycling. How are they doing this?

Whenever a recycling program is initiated in a community the leaders want to know how successful it has been. Success can be measured in many ways, including:

- Fraction of refuse components diverted from the landfill, tons/week.
- Fraction of households participating in the program. Participation can be defined in many ways, such as having set out recyclables at least once a month.
- Fraction of households participating on any given week.
- Profit from the sale of recyclables, $/week.

Any of these methods are useful, but they have to be well defined and used consistently. The danger is that the program can use misleading indicators. Even if the first method of evaluating program success is used, there are a number of ways of doing the calculation. In all cases, the equation used is the same:

$$R = \frac{D}{G} \times 100 \qquad (8.1)$$

where

R = recycling rate, %
D = refuse diverted to materials recovery, tons/week
G = refuse generated, tons/week

Example 8.1

A community has average generation rates of refuse and other wastes as shown in the table below. All of this material goes to the community sanitary landfill.

Material	Generation rate (tons/week)
Paper	300
Glass bottles	20
Aluminum	15
Other refuse	400
Yard waste	100
Construction Rubble	250

The community now initiates a curbside collection program. It also creates a special landfill for construction rubble, and decides that it wants to start composting yard waste. The curbside recycling results are as follows:

Material	Collected and sent to secondary materials industry (tons/week)
Paper	150
Glass	10
Aluminum	5

What is the recycling rate for this community?

1. If we used only the three items; paper, glass and aluminum, the total generated is 335 tons/week, and the community was able to divert 165 tons per week to materials recovery, the recycling rate is

$$R = \frac{D}{G} \times 100 = \frac{165}{335} \times 100 = 49\%$$

2. If we use all of the refuse, the recycling rate is

$$R = \frac{165}{735} \times 100 = 22\%$$

3. If we use all of the solid waste generated by the community in the equation, including the yard waste and the construction rubble that used to go to the landfill, the recycling rate is

$$R = \frac{165}{1085} \times 100 = 15\%$$

4. If we count the yard waste composting operation as recycling, and if we also count the diversion of the construction rubble from the landfill, the recycling rate is

$$R = \frac{165 + 100 + 250}{1085} \times 100 = 48\%$$

So the way the recycling rate is calculated can have a great effect on the outcome. The objective in this example is to show that when recycling rates are advertised, the calculation has to be specified if the result is to be meaningful.

8.5 DISPOSAL OF SOLID WASTE

The disposal of solid wastes is a misnomer. Our present practices amount to nothing more than hiding waste well enough so it cannot be readily found. The only two realistic options for storing waste on a long-term basis are in the oceans (or other large bodies of water) and on land. The former is forbidden by federal law and is becoming similarly illegal in most other developed nations. Little else needs to be said of ocean disposal, except perhaps that its use was a less than glorious chapter in the annals of public health and environmental engineering.

The placement of solid waste on land is called a *dump* in the USA and a *tip* in Great Britain (as in *tipping*). The dump is by far the least expensive means of solid waste disposal, and thus was the original method of choice for almost all inland communities. The operation of a dump is simple and involves nothing more than making sure that the trucks empty at the proper spot. Volume is often reduced by setting the dumps on fire, thus prolonging dump life.

Rodents, odor, air pollution, and insects at the dump, however, can result in serious public health and aesthetic problems, and alternate methods for disposal were necessary. Larger communities can afford to use a combustor for volume reduction, but smaller towns cannot afford such capital investment. This has led to the development of the sanitary landfill. The *sanitary landfill* differs markedly from open dumps in that the latter were simply places to dump wastes, while sanitary landfills are engineered operations, designed and operated according to acceptable standards. The basic principle of a landfill operation is to prepare a site with liners to deter pollution of groundwater, deposit the refuse in the pit, compact it with specially built heavy machinery with huge steel wheels, and cover the material with earth at the conclusion of each day's operation (Figure 8.4). Siting and developing a proper landfill requires planning and engineering design skills.

Even though the tipping fees paid for the use of landfills is charged on the basis of weight of refuse accepted, landfill capacity is measured in terms of volume, not weight. Engineers designing the landfills first estimate the total volume available to them and then estimate the density of the refuse as it is deposited and compacted in the landfill. The density of refuse increases markedly from the time it is first generated in the kitchen until it is finally placed into the landfill. The *as generated* bulk density of MSW is perhaps 60 to 180 kg/m^3 while the compacted waste in a landfill exceeds 700 kg/m^3. Landfills are required to have daily covers. The more soil placed on the refuse reduces the volume available for the refuse itself. Commonly, engineers estimate that the volume occupied by the cover dirt is 1/4th of the total landfill volume.

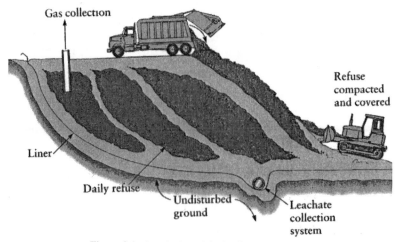

Figure 8.4. A typical municipal solid waste landfill.

Example 8.2

Imagine a town where 10,000 households each fill up one 0.4 m³ container of refuse per week. The density of the refuse as collected is 120 kg/m³ and the density of the refuse when compacted in a landfill is 720 kg/m³. Assume that 10% of the volume is occupied by the cover dirt. What landfill volume is necessary in order to dispose of this refuse?

This problem is solved using a mass balance. Imagine the landfill operation as a black box, and the refuse goes from the households to the landfill.

$$Mass\ out = Mass\ in$$

$$V_L D_L = V_P D_P$$

where V and D are the volume and bulk density of the refuse, and subscripts L and P denote loose and packed refuse. Bulk density is simply the mass (kg) of refuse in a given volume (m³). If you stuff refuse into a 1 m³ box and weigh the refuse you put in, the bulk density is the mass divided by the volume, or kg/m³. Thus

$$(10,000\ households)(0.4\ m^3/household)(120\ kg/m^3) = (720\ kg/m^3)V_P$$

$$V_P = 667\ m^3/week$$

If 10% of the total volume occupied is taken up by the cover dirt, then the total landfill volume necessary to dispose of this waste is

$$T = 667 + 0.10(T)$$

where T is the total volume. Thus $T = 740$ m³/week

Sanitary landfills are not inert. The buried organic material decomposes anaerobically, producing various gases such as methane and carbon dioxide, and liquids that have extremely high pollutional capacity when they enter the groundwater. Liners made of either impervious clay or synthetic materials such as plastic are used to try to prevent the movement of this liquid, called *leachate*, into the groundwater. Figure 8.5 shows how a synthetic landfill liner is installed in a prepared pit. The seams have to be carefully sealed, and a layer of soil placed on the liner to prevent landfill vehicles from puncturing it.

Synthetic landfill liners are useful in capturing most of the leachate, but they are never perfect. No landfill is sufficiently tight that groundwater contamination by leachate is totally avoided. Wells have to be drilled around the landfill to check for groundwater contamination from leaking liners, and if such contamination is found, remedial action is necessary. And of course the landfill never disappears—it will be there for many years to come, limiting the use of the land for other purposes.

Modern landfills also require the gases generated by the decomposition of the organic materials to be collected and either burned or vented to the atmosphere. The gases are about 50% carbon dioxide and 50% methane, both of which are greenhouse gases. In the past, when gas control in landfills was not practiced, the gases caused problems with odor, soil productivity, and even explosions. Larger landfills use the gases for running turbines for the production of electricity for sale to the power company, while smaller landfills burn the gas and vent to the atmosphere. The fact that landfills produce methane

Figure 8.5. Installation of a liner in a landfill.

The Wilmington, North Carolina, Landfill

In the 1970s sanitary landfills were just being introduced to most of the United States, and the level of knowledge about how landfills operated was primitive. Sanitary landfills in those days were all unlined, and consisted of simple pits or trenches where the refuse was buried. The largest and most famous landfills were in Southern California, and much of the information on landfill design and operation came from these facilities.

In the early 1970s the City of Wilmington, North Carolina, a coastal port at the mouth of the Cape Fear River, was searching for a solution to its solid waste disposal problem. The city engineer came to the State of North Carolina asking for permission to construct a landfill just outside of the city along the banks of the Cape Fear River. The young engineer in charge of such applications recognized that this was a terrible site on which to build a landfill. First of all, the soil was all sand, and directly next to the landfill site were a number of private wells. The City of Wilmington depended on that same aquifer for its drinking water. The site was also in the flood plain of the river and thus susceptible to flooding. Given all these problems, the young engineer rejected the application to approve the landfill.

The engineer's boss, an older engineer with questionable credentials, argued that landfills do not produce leachate and that this would be an ideal location for the landfill. His information on landfills not producing leachate came from the Southern California landfills, which were deep canyon landfills at locations that received almost no rain. He simply extrapolated this knowledge to the wet sandy environment in Wilmington without thinking it through, and ordered the young engineer to approve the application, which he did.

A year after the landfill went into operation, the water from the private wells began to taste bad, and tests showed the water to be contaminated. Instead of recognizing the disaster about to happen, the city kept it quiet and offered to pipe city water to the affected people. But as the number of contaminated wells grew, the magnitude of the problem became too large to ignore. The state sent in a team of hydrologic experts and soon discovered a huge plume of leachate headed directly to the city and the city water supply wells. Emergency interception wells were drilled into the sand and the water was piped out at a rate sufficient to intercept the leachate plume. By this time the amount of solid waste in the landfill was such that it would have been much too expensive to dig it out, so the only solution was to keep pumping and treating the water from the intercepting wells. As of this writing, the wells are still producing polluted water.

The most interesting part of this story, however, is the young engineer—the one who signed the agreement to allow Wilmington to construct the landfill. He has not been heard from again. He just disappeared. Did he get out of engineering? Or out of the state? It would be interesting to ask him what he now thinks about this situation, and if he had it to do over again, would he stand up to his boss who was asking him to approve something that he knew was wrong and would cause great environmental harm? Most of all, how did this incident change his professional life?

The Manila Dump Collapse

The victims were all poor people who lived in an area called Payatas, the location of the solid waste dump for the city of Manila in the Philippines. This was not a landfill, but a dump. Garbage was brought to the site and simply piled up. In some locations, the pile of refuse was seven stories high. The squatters who built their shanties in the shadow of the dump would scavenge through the dump, looking for things to sell. The pile was stable during dry weather, but a typhoon in July of 2000 brought heavy rains, soaked the compacted garbage, and destabilized it to the point that a huge wall of it came crashing down on top of the shantytown, and then burst into flame, igniting from the stoves in the shanties. Rescuers worked for days digging into the trash, hampered by the stench and instability of the refuse, and pulled more than 60 people out alive. The death toll is still not known accurately, but latest estimates are that over 100 people perished when the refuse collapsed.

gas has been known for a long time, and many accidental explosions have occurred when the gas has seeped into basements and other enclosed areas where it could form explosive mixtures with oxygen. Modern landfills are required to collect the gases produced in a landfill and either flare them or collect them for subsequent beneficial use.

The first system for collecting and using landfill gas was placed in operation

The Biodegradable Plastic Bag Caper

In 1988 the competition in the sale of plastic trash bags was fierce, and Mobil Oil Company, the maker of the Hefty brand of trash bags, decided to boost sales by advertising its bags as "biodegradable," a claim they were never able to prove scientifically. In fact, a company spokesperson in 1989 admitted that:

> "[Biodegradable bags] are not an answer to landfill crowding or littering. . . . [it] is just a marketing tool. . . . We're talking out of both sides of our mouths because we want to sell bags. I don't think the average consumer even knows what degradability means. Customers don't care if it solves the solid waste problem. It makes them feel good."

Several states initiated lawsuits against Mobil and the word "degradable" was eventually removed from the labels. The scam did not hurt the sale of Hefty bags, however, which had record sales in 1989.

In retrospect, it is not telling a lie (saying the bags are degradable when they aren't) but rather the cavalier attitude about telling the lie that is most troubling about the way some corporations operate. They assume that the public is accepting of corporate lies and does not care. Truth is no longer important.

in California after a number of gas-extraction wells were driven around a deep landfill to prevent lateral migration of gas. These wells burned off about 1000 ft³/min (44 m³ /min) of gas without the need for auxiliary fuel. Based on this experience, to capture and use this gas instead of just wastefully burning it off seemed a reasonable alternative. Huge landfills (such as Palos Verdes, California, 15 million tons in place) could produce gas with a total heating value of 10 trillion Btu.

8.6 ENERGY CONVERSION

An alternative to allowing refuse to biodegrade and form a useful fuel is the combustion of refuse and energy recovered as heat. The potential for energy recovery from solid waste is significant. Of the 196 million tonnes of waste generated annually in the United States, over 80% is combustible, yielding a heat value equivalent to about 1 million barrels of oil per day. This figure is equivalent to about 4.6% of all the fuel consumed by all utilities, 10% of all the coal consumed by all utilities, and about 20% of the electrical energy demand of the private sector of a municipality. Although these figures must obviously be adjusted to reflect the losses incurred in producing electricity, the use of refuse as a source of energy clearly has tremendous potential.

Refuse can be burned as is (in so called *mass burn combustors*), or processed to produce a *refuse-derived fuel*. As shown in a cutaway drawing of a waste-to-energy facility in Figure 8.6, the refuse is dumped from the collection trucks into a pit that serves to mix and equalize the flow over the 24 hour period since such facilities must operate around the clock. A crane lifts the refuse from the pit and places it in a chute that feeds the furnace. The grate mechanism moves the refuse, tumbling it and forcing in air from the bottom as well as the top as the combustion takes place. The hot gases produced from the burning refuse is cooled by a bank of tubes filled with water. The hot water goes to a boiler where steam is produced. This steam can be used for heating and cooling, or for producing electricity in a turbine. The cooled gases are then cleaned by dry scrubbers, baghouses, electrostatic precipitators and/or other control devices and discharged through a stack.

The more the solid waste is processed prior to its combustion the better the heat value and usefulness as a substitute for a fossil fuel. Such processing removes much of the non-combustible materials such as glass and metals, and reduces the size of the paper and plastic particles so they burn more evenly. Refuse that has been so processed is called a *refuse-derived fuel* (RDF). The simplest form of RDF is shredding the solid waste to produce a more homogeneous fuel, without removing any of the metals or other non-combustibles. If the metals and glass are removed, the fuel is improved in its heat value and handling. This organic fraction can be further processed into pellets that are an excellent addition to coal in coal-fired furnaces.

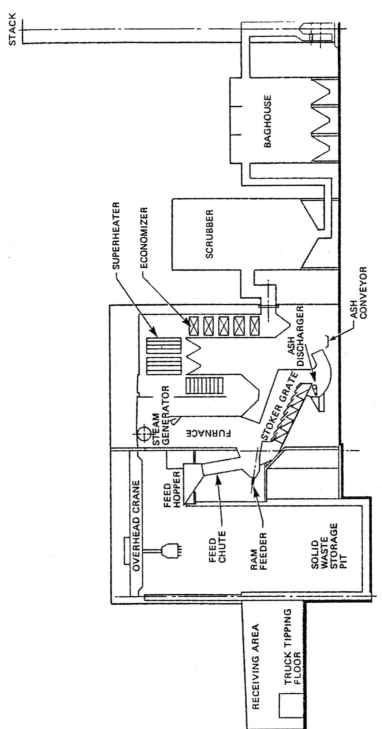

Figure 8.6. A cross section of a typical mass-burn facility.

305

One reason waste-to-energy facilities have not found greater favor is the concern with their emissions, but this seems to be misplaced. Studies have shown that the risk of municipal solid waste combustion facilities on human health are minimal, and with modern air pollution control equipment there should be no measurable effect of either the gaseous or particulate emissions.

The refuse combustion option is, however, complicated by the problem of ash disposal. Although the volume of the refuse is reduced by over 90% in waste-to-energy facilities, the remaining 10% still has to be disposed of. Many special wastes such as old refrigerators cannot be incinerated and must either be processed to recover the metals or landfilled. A landfill is therefore necessary even if the refuse is combusted, and a waste-to-energy plant is not an ultimate

---

The Gargage Barge

Awareness of municipal solid waste problems was greatly heightened by the saga of the "garbage barge." The year was 1987, and the barge named *Mobro* had been loaded in New York with municipal solid waste and found itself with nowhere to discharge the load. With disposal into the ocean being illegal, the barge was towed from port to port, with six states and three countries rejecting the captain's pleas to offload its unwanted cargo.

The media picked up on this unfortunate incident and trumpeted the "garbage crisis" to anyone who would listen. Reporters honed their finest hyperbole, claiming that the barge could not unload because all our landfills were full and the United States would soon be covered by solid waste from coast to coast. Unless we did something soon, they claimed, we could all be strangled in garbage.

The story of the hapless *Mobro* is actually a story of an entrepreneurial enterprise gone sour. An Alabama businessman, Lowell Harrelson, wanted to construct a facility for converting municipal refuse to methane gas and recognized that baled refuse would be the best form of refuse for that purpose. He purchased the bales of municipal solid waste from New York City and was going to find a landfill somewhere on the East coast or in the Caribbean where he could deposit the bales and start making methane. Unfortunately, he did not get the proper permits for bringing refuse into the municipalities, and the barge was refused permission to offload its cargo. As the journey continued, the press coverage grew, and no local politicians would agree to allow the garbage to enter their ports. Poor Harrelson finally had to burn his investment in a Brooklyn incinerator.

The "garbage crisis" never developed, of course. Large waste disposal corporations constructed huge landfills in remote areas and now most of the municipal solid waste in the Eastern seaboard is being trucked or rail-hauled to these landfills, which are now competing for business.

disposal facility. Refuse combustion ash concentrates the inorganic hazardous materials in refuse and the combustion process may create other chemicals that may be toxic. Most solid waste ash is placed in municipal landfills, but other facilities have constructed their own landfills exclusively for ash disposal (so-called *monofills*). Ash has been increasingly used as a raw material for the production of useful aggregate for producing building materials, as an aggregate for road construction, or other applications where a porous, predictable (and inexpensive) material is useful.

8.7 LITTER

Litter is a special type of solid waste. It is distinct from other types of solid waste in that it is not deposited in proper receptacles. We usually think of litter as existing in public places, but litter could be on private premises as well. Although litter is usually considered to be a visual affront only, it may also be a health hazard. Broken glass and food for rats are but two examples. It is also a drain on economic resources, because the public must pay to have it collected and removed when it is on public property.

The collection of litter is of secondary importance to a community because it does not represent a critical public service, as does police and fire protection, water treatment, or collection of refuse from residences and commercial establishments. And litter removal is expensive, costing U.S. municipalities millions annually.

The composition of roadside litter can vary considerably from place to place, as can the method of data collection. One major problem with any litter data analysis is that the reports fail to specify the guidelines used in the collection and identification of litter, and seldom specify the way in which the percentages of the various components were calculated. For example, a broken bottle can be counted as either one item or many items, depending on the guidelines. Similarly, the results can be calculated as a percent of the total of each of the following items:

- Items by actual count.
- Total weight of litter.
- Volume of litter.
- Visible items by actual count.

Because of the problem of not having a standard counting technique, the following guidelines for conducting litter studies are suggested:

1. Count as one item all pieces larger than 2.5 cm (1 in). This count includes removable tabs from beverage cans.

2. Do not count rocks, dirt, or animal droppings.
3. Count as one item all pieces of any item clearly belonging together, such as a broken bottle. Otherwise count each piece of glass, newspaper, and so on, singly.
4. Do not count small, readily decomposable material, such as apple cores.
5. For roadside litter surveys, measure all items within the officially designated right-of-way.
6. Empty liquids out of all bottles and cans before collecting them.

The litter survey, if conducted along a road, should be started by driving along the road at a slow speed and having the passenger record the visible items into a tape recorder for future transcription. Next, the litter is identified, recorded, and manually collected. The items should be separated during collection into as many components as feasible. The collected items are then weighed and the volume measured.

For community litter surveys, the photometric technique developed for Keep America Beautiful, Inc., has found wide acceptance. The block faces of a community are first numbered, and a preliminary sample size established. About 5% of the blockfaces are usually adequate. Using the random number technique, the block faces and the locations on those block faces to be measured are selected. A marker is located in the front center of the survey area, and a chalkboard is used to identify the location and date. To facilitate the counting of litter from the developed photographs, a picture is taken of a clean pavement laid out with white marking tape in a 1-ft grid, 6 ft wide, 16 ft long, parallel to a street curb. A transparency of the grid is prepared and the resulting 96-square grid is placed on top of each litter photograph. The litter is then counted and classified using a magnifying glass. The first photographs are used for establishing the baseline litter conditions. The litter rating (L) is calculated for each picture (location) as the squares containing some litter compared to the total number of squares (%).

Litter can theoretically be controlled by cognitive, social, and technological means: a cognitive solution would be convincing people not to litter; a social solution would be depriving the public of items that might become litter, or fining them heavily if they are caught; and a technical solution would be cleaning up after littering has occurred.

Other litter psychology studies have been directed at finding out what motivates people not to litter. One reason people would not litter is self interest. The Oregon Bottle Law, now adopted in many states, places an artificial tax on the beverage container, and it is advantageous to take the bottle in for a cash refund instead of throwing it in the street.

The second method of litter control is to prevent items that might become litter from ever reaching the consumer. For example, in the example of the hot dog wrapper, it would seem reasonable to suggest that 100% litter-free results could be obtained by not giving customers a paper wrapper around their hot dog. The banning of tear away metal tabs on beer and soft drink cans is a practical means of controlling this type of litter.

The third method of litter control is to clean up the mess once it has occurred. This system is commonly used in sports stadiums and other public areas where no effort is made to ask people to properly dispose of their waste. For roadside litter, it seems that the most economical litter control alternative is actually frequent cleanup, and attempts have been made at designing mechanical litter collection machines. One towed device has proven both inexpensive and effective. It works by having a series of rotating plastic teeth which fling the litter into a collection basket, much like a leaf collector connected to a lawn mower. A more sophisticated and ambitious unit, developed by a major manufacturer of beverage containers, uses a vacuum arm on a truck which sucks up the roadside litter while cruising at highway speed.

Who Litters?

Who are the people who litter? Is it possible to draw some kind of profile of these people so that litter control measured can be most effectively directed at them?

Psychologists have been trying to get an handle on that. In one study, the actions of 272 persons were observed when they bought a hot dog wrapped in paper. Of interest was the final deposition of the wrapper. Ninety-one people chose to dispose of the wrapper improperly (they littered). The probability of any one person littering, based on this sample, could be calculated as

$$E = 0.019 + 0.414(A) + 0.165(B) + 0.153(C)$$

where E is the probability that a person would litter; and A, B, and C = 0, except that A = 1 if the person is 18 years old or younger, B = 1 if there are no trash cans conveniently located, and C = 1 if the area is already dirty with litter. From the study, it is clear that age is quite important, with younger people being much more likely to litter than are older persons. There was no statistical difference between 19- to 26-year-olds and persons older than 26 years. Economic status and race were found not to be independently significant. Similarly, gender was found to be statistically insignificant. Because the study was conducted in 1973, its validity to today's urban populations may be questionable. Intuitively, however, the role of younger persons as the major contributors to urban litter remains valid.

Source: Finnie, W. C., "Field experiments in litter control," *Environment and Behavior*, v. 5, n. 2, 1973.

Litter in the Oceans

While we usually think of litter where it is visible to us, the worst effect of litter is on non-human animals far out of sight and often far from our consciousness. One such problem is the presence of plastics in the oceans.

Only rough estimates are available on the amount of plastics entering the oceans, but a credible number is about 1 million tons per year, or about 100 tons per hour. One of the major problems is the presence of lost or discarded fishing nets that are made of plastics and will not biodegrade. These nets may drift in surface waters and continue to entangle fish and turtles until they become too heavy with the decaying carcasses that they sink to the bottom. Most of the plastic debris has come from ships, but human carelessness on the beaches is a major source as well.

Marine organisms consume plastics by mistake, believing it to be food. While it is impossible to know how many organisms have died as a result of such consumption, we do have some data on the effect of plastic bags on sea turtles. The turtles mistake the bags for jellyfish, which is one of their main food sources, and the bags block up their gastrointestinal tract. On Long Island, eleven dead leatherback turtles washed ashore and autopsies showed that all of them had ingested plastic bags.

The sea turtle hospital on Topsail Island, North Carolina has treated nearly 100 sea turtles over the past six years and kept track of their injuries, many of which were due to the ingestion of plastics from the oceans. One case is a loggerhead turtle named Abbot. This turtle struggled for days entangled in a plastic net, and the more she struggled the tighter the net became. Circulation was cut off from on one of the flippers so that when the turtle was found on the shore, rotten flesh was falling off the bones. The sea hospital staff amputated the flipper and had enough healthy skin to cover the stump. The turtle was named Abbott, in honor of a major league baseball player with one arm who pitched a no-hitter, Jim Abbott. He never gave up, and neither did the turtle. Several weeks after coming to the hospital she passed a huge piece of net, testament to her desperate struggle to free herself. A year later she was released back to the ocean after a truly amazing recovery. More information on the work at the Sea Turtle Hospital can be found at www.seaturtlehospital.org.

Source: Jean Beasley, Director, Topsail Turtle Project, Karen Beasley Sea Turtle Rescue & Rehabilitation Center, Topsail Island, NC

8.8 SOLID WASTE LAW

Prior to the 1970s the only federal legislation that addressed solid waste was the 1899 *Rivers and Harbors Act*, which prohibited the dumping of large objects into navigable waterways. The federal government was not involved in solid waste matters, except as a major producer of solid waste, much of which was managed by the Department of Interior. Municipal solid waste was commonly thrown into unlined open dumps which were intentionally set on fire to reduce volume. In larger communities, solid waste was sent to incinerators, which had minimal air emission controls and did a poor job of reducing the volume of waste. In Durham North Carolina, for example, the city incinerator in the 1950s was called the "Durham Toaster" in the newspapers, because apparently the organic matter that emerged from the incinerator was barely singed.

The first federal legislation intended to assist in the management of solid waste was the 1965 *Solid Waste Disposal Act*, which provided technical assistance to the states through the U. S. Public Health Service. The emphasis in this legislation was the development of more efficient methods of disposal, and not to protect human health.

In 1976, the Congress of the United States passed the *Resource Conservation and Recovery Act* (RCRA). With its 1984 amendments, RCRA is a strong piece of legislation that mainly addressed the problem with hazardous waste, but also specifies guidelines for non-hazardous solid waste disposal. Subtitle D in this act is the municipal solid waste section, and landfills that fall under these requirements are commonly called *Subtitle D landfills*. In 1991 under Subpart D, the U. S. EPA adopted regulations to establish minimum national landfill criteria for all solid waste landfills. A key component of the standard was to require landfills to install composite liner systems consisting of a plastic liner on top of compacted clay. Specification for the composite liner were also included in the regulation. Existing landfills had two years to comply with these standards.

The combustion of solid waste is controlled by the 1970 *Clean Air Act* (with subsequent amendments). With this legislation began the process of closing burning dumps and uncontrolled incinerators. In every case, the federal agencies involved (mostly the U. S. EPA) required the individual states to set up local guidelines that adhere to the federal standards, which are then approved by the U. S. EPA. The actual enforcement of those requirements is left to the states.

The siting of solid waste facilities is further complicated by local opposition, which often drags out the process. In some states it now takes over ten years to site a new landfill, even in the absence of local opposition. Municipalities that traditionally managed their solid waste locally are now shipping waste across state borders to remote regional landfills just to avoid having to go through the process of developing a local solution to the problem. Communities are increasingly unwilling to go through the process of siting a new landfill, transfer station, or combustor. In the end, the engineering challenges are minor compared to the regulatory and political challenges of siting new solid waste facilities.

Most states have passed strong legislation encouraging and promoting recycling. In the 1990s over 40 states established recycling goals. For example, California mandated that 25 percent of the waste be diverted from landfills by 1995 and that 50 percent be diverted by 2000. In Pennsylvania, every community over 5000 population is mandated to set up a recycling program. Often, local problems in siting new landfills drive the recycling effort. If a substantial amount of solid waste can be diverted from the landfill, then the facility will last longer and a new one will not have to be built.

Kenilworth Dump

The conversion of dumps to landfills in the 1960s and 1970s was not an easy task. In most cases the landfill owners who had essentially no overhead to pay in the operation resisted having to increase their costs by daily burial of the refuse, which would control both fires and runoff. The problem was even more difficult when the owner of the dump was the United States Government.

The Kenilworth dump, on the other side of the Anacostia River from Washington DC and only blocks from the White House, was owned by the Department of the Interior that ran it for all of the governmental agencies in the District. It was a classic dump in that the waste, which in this case was mostly paper, was intentionally set on fire to reduce the volume. Landfill fires are not tidy things, and burn and smolder for a very long time, and often the fires go deep into the refuse pile, hollowing out burning crevasses, and leaping back to the surface when new fuel or oxygen is available. Pressure had been put on the Department of the Interior to stop the burning and to convert the dump to a maintained landfill, but the Department insisted that this would be too expensive.

One February afternoon in 1968 young Kelvin Mock was picking his way over the piles of paper, broken chairs, and discarded tires on top of the dump. Three of Kelvin's older friends, wiser and quicker than the seven-year-old, scampered ahead of him, stopping now and then to examine a discarded piece of metal or glass or a scrap of cloth. Suddenly the wind shifted. Flames raced across a river of computer paper and carbon, and burning debris swirled in all directions. Kelvin, cut off from his playmates, died in a pyre of trash.

The public reaction was immediate and overwhelming. After years of procrastination and promises, the Department of the Interior suddenly had the money to put the fires out and to construct a proper landfill. It took the death of a seven-year-old boy to get the U. S. Government to do the right thing and to close this dangerous eyesore.

8.9 SUSTAINABILITY

Without doubt, human use of resources as presently practiced is unsustainable. That is, present generations are using resources and polluting the earth at a rate that could make life more difficult (if not impossible) for future generations. The *precautionary principle* states that if a problem is sufficiently severe and the consequences sufficiently serious, one would not need proof before action is taken to alleviate the potential damage. This recognition led the World Commission on Environment and Development, sponsored by the United Nations, to argue for what they called "sustainable development." Also known as the Brundtland Commission, their published report in 1987 introduced the term *sustainable development* and defined it as "development that meets the needs of the present without compromising the ability of future generations to meet their own needs." The underlying purpose of the study was to help developing nations manage their resources, such as rain forests, without depleting these resources and making them unusable for future generations, and also to prevent the collapse of the global ecosystem. The report presumes that we have a core ethic of intergenerational equity, and that future generations have an equal opportunity to achieve a high quality of life. The report is silent, however, on just why we should embrace the ideal of intergenerational equity, or why one should be concerned about the survival of the human species.

Since the publication of the Brundtland Report, sustainable development has metamorphosized into *green engineering*, a term that recognizes not only the need to practice sustainable development, but also acknowledges that engineers are central to the practical application of these principles to everyday life. Sustainable living is related to the idea of sustainable development. That is, the concept of sustainability leads to a sustainable environment through green engineering.

But defining principles and a goal is not enough. Before one would be convinced to practice green engineering, there are two questions that demand answers: (1) Is sustainability, which is the goal of green engineering, even attainable? and (2) If sustainability is possible, what reasons are there for practicing green engineering?

One of the requirements of sustainable existence would be the complete recycling of all materials, just as natural ecosystems are able to do. Unfortunately, this is not an easy task. Some materials, such as aluminum, are easy to recycle. The material is essentially unaltered, and a recycled aluminum can has as much strength as a can made out of virgin aluminum. Some materials are possible to recycle, but the economics make this prohibitive. Examples are solvents, refrigerants, plastics, automobiles, etc. And finally there are materials that are not only economically prohibitive to recycle but for which the technology does not exist, such as paints, pesticides, fuels, detergents, fertilizers, etc.

As discussed in Chapter 8, level of recycling increases, the cost in energy and production of pollutants increases (for example, from the burning of fossil fuels). Eventually this will be too great and will prevent the recycling of many materials. Thus the use of many nonrenewable resources is inherently unsustainable, and attempts to close this loop will result in far greater damage than good.

The notion of full recycling/recovery of materials is essentially impossible given our present economic system. Even if we devote our limited energy resources to greatly enhancing recycling, it is likely that the effects will not be what we desire. The use of energy for materials recycling could result in a higher rate of the extinction of species, thus destroying part of what we are trying to save.

The collection and recycling of 100% of all materials and resources utilized by humans would be a requirement for what we can call *hard sustainability*. Hard sustainability is the complete elimination of human effect in the dispersion of materials or use of non-replenishable resources. But this is impossible, and any attempt to use enough energy to achieve such recycling will result in far greater harm than good to the global ecosystem. We can conclude therefore that given our present state of technology and our human resources, achieving hard sustainability is essentially impossible.

There is, however, another option, called *soft sustainability*, or a level of activity that is possible and feasible, and which will, as our societies develop, move increasingly toward the goal of hard sustainability. Where materials cannot be recovered without excessive expenditures or energy, then those materials should at least have less inherent hazard when dissipated.

The reasons for embracing the principles of green engineering are often self-serving, resulting in material gains for the firm or for the individual. But there are those who do the right thing, with or without others knowing of their work, simply because it gives them pleasure to do so. Even with the knowledge that hard sustainability is nearly impossible, many people are optimists and recognize that soft sustainability is both possible and feasible.

Kermit the Frog says it best. He concludes his song, "It's Not Easy Being Green," with these lines:

> I'm green,
> and it'll do fine;
> It's beautiful,
> and I think its what I want to be.

8.10 SYMBOLS

R = recycling rate, %
D = refuse diverted to materials recovery, tons/week

L = litter rating, percent of squares containing litter
G = refuse generated, tons/week
VOC = volatile organic compound

8.11 PROBLEMS

8.1 The objective of this assignment is to evaluate and report on the solid waste you personally generate. For one week, selected at random, collect all of the waste you would have normally discarded. This includes food, newspapers, beverage containers, etc. Using a scale, weigh your solid waste and report it as follows:

Component	Weight	Percent of Total Weight
Paper		
Plastics		
Aluminum		
Steel		
Glass		
Food		
Other		_____
		100

Your report should include a data sheet and a discussion. Answer the following questions.

   a. How does your percentage and total generation compare to national averages?

   b. What in your refuse might have been reusable (as distinct from recoverable), and if you had reused it, how much would this have reduced the refuse?

   c. What in your refuse is recoverable? How might this be done?

8.2 One of the costliest aspects of municipal refuse collection is the movement of the refuse from the household to the truck. Suggest a method by which this can be improved. Originality counts heavily here, practicality far less.

8.3 Plastic bags at food stores have become ubiquitous. Often recycling advocates point to plastic bags as the prototype of wastage and pollution, as stuff that clogs up our landfills. In retaliation, plastic bag manufacturers have begun a public relations campaign to promote their product. On one of the flyers (printed on paper) they say:

"The (plastic) bag does not emit toxic fumes when properly incinerated. When burned in waste-to-energy plants, the resulting by-products from combustion are carbon dioxide and water vapor, the very same by-products that you and I produce when we breathe. The bag is inert in landfills where it does not contribute to leaching, bacterial, or explosive gas problems. The bag photodegrades in sunlight to the point that normal environmental factors of wind and rain will cause it to break into very small pieces, thereby addressing the unsightly litter problem."

Critique this statement. Is all of it true? What part is not? Is anything misleading? Do you agree with their evaluation? Write a one-page response.

8.4 What is the content of recycled fiber in the paper used by your school/college/department? What is the goal established by the U. S. EPA and your state? (You can find this for the U. S. EPA at www.epa.gov.)

8.5 Suppose you are told by the town council to design a "recycling" program that will achieve at least 50% diversion from the landfill. What would be your response to the town council? If you agree to try to achieve such a diversion, how would you do it? For this problem, you are to prepare a formal response to the town council, including a plan of action and what would be required for its success. Numbers are mandatory. Make up data as needed. This response would include a cover letter addressed to the council and a report of several pages, all bound in a report format with title page and cover.

8.6 Same scenario as Problem 8.5, but now you are asked to develop a zero waste program. That is, the community is to produce no solid waste whatsoever. Respond with a letter to the town council.

8.7 How might you use the principles of waste reduction in doing your food shopping at the food store? Name at least three specific ways in which you might contribute to waste reduction in the purchase of your groceries.

8.8 We often see packaging labeled "Made from 100% recycled materials" or "Made from 50% recycled materials." The objective is to make the consumer believe that the company is environmentally conscious and caring, and that this will make you buy more of their products.

a. Why is a statement like the one above, "Made from, . . ." potentially misleading? What questions would you want to ask the company to determine if they truly are helping with materials recycling?

b. All things considered, if the statements are true, why ought you to buy their products in preference to products with no recycled material? Be specific.

c. If you were working for the company, how would you write the statement ("Made from . . .") to be more accurate?

8.9 Give three examples of waste reduction that you might be able to implement in your everyday life.

8.10 If you were responsible for marketing the paper collected by your city's curbside collection program, what would you tell the purchasing department if they asked you whether the city should buy copy paper that was (1) 75 percent recycled content (55 percent pre-consumer and 20 percent post-consumer) or (2) 50 percent recycled content paper (10 percent pre-consumer and 40 percent post-consumer)? Why?

8.11 Using either the Internet or the library, find some pictures of urban street scenes during the 19th century. Be sure to document the place and time. What are the conditions of the streets with regard to solid waste?

CHAPTER 9

Air Quality

WHAT is clean air? One definition is to list the gases that make up air;

	Percent by volume
Nitrogen	78.0
Oxygen	20.9
Argon	9.3
Carbon dioxide	0.31
Neon	0.18
Helium	0.05

and so on, but this "air" never occurs in nature. Natural air without exception has in it solid particles, liquid droplets, and gases that occur naturally (that is, without human intervention). These naturally occurring air pollutants are ubiquitous and it can become very difficult to define just what is meant by clean air. The reason the Appalachian Mountains in North Carolina are called the Smoky Mountains, for example, is that they are perpetually engulfed in a cloud of terpines given off by the pine trees. This cannot be defined as polluted air, even though the visibility is hindered and the volatile organic compounds can make eyes water. The sulfur odor given off by the geysers in Yellowstone Park is not pleasant, and yet we would not classify it as air pollution. So air pollution has to be limited to emissions into the atmosphere that are produced by humans. In this chapter we look at how these emissions move through the air to get from the source to the recipient, then review some of the most important air pollutants produced by humans, and finally discuss the effect of these on human health.

9.1 METEOROLOGY AND AIR MOVEMENT

The earth's atmosphere can be divided into easily recognizable strata,

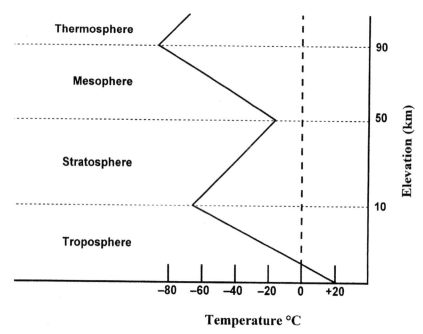

Temperature °C

Figure 9.1. Temperature profiles around the Earth.

depending on the temperature profile. Figure 9.1 shows a typical temperature profile for four major layers. The *troposphere*, where most of our weather occurs, ranges from about 5 km at the poles to about 18 km at the equator. The temperature here decreases with altitude. Over 80% of the air is within this well-mixed layer. On top of the troposphere is a layer of air where the temperature profile is inverted, and in this *stratosphere* little mixing takes place. Pollutants that migrate up to the stratosphere can stay here for many years. The stratosphere has a high ozone concentration, and the ozone adsorbs the sun's short wave ultraviolet radiation. Above the stratosphere are two more layers, the *mesosphere* and the *thermosphere*, which contain only about 0.1% of the air.

Other than the problems of global warming and stratospheric ozone depletion (Chapter 10), air pollution problems occur in the troposphere. Pollutants in the troposphere, whether produced naturally (such as terpenes in pine forests) or emitted from human activities (such as smoke from power plants) are moved by air currents, which we commonly call wind. Meteorologists identify many different kinds of winds, ranging from global wind patterns caused by the differential warming and cooling of the earth as it rotates around the sun, to local winds caused by differential temperatures between land and water masses. A sea breeze, for example, is a wind caused by the progressive warming of the land during a sunny day. The temperature of a large water body such as an ocean or large lake does not change as rapidly during the day, and the air over the warm land mass rises, creating a low pressure area towards which air flows, coming horizontally over the cooler large water body.

Wind not only moves the pollutants horizontally, but it causes the pollutants to disperse, reducing the concentration of the pollutant with distance from the source. The amount of dispersion is directly related to the stability of the air, or how much vertical air movement is taking place. The stability of the atmosphere is best explained using an ideal parcel of air.

As an imaginary parcel of air rises in the earth's atmosphere, it experiences lower and lower pressure from surrounding air molecules, and thus it expands. This expansion lowers the temperature of the air parcel. Ideally, a rising parcel of air cools at about 1°C/100 m or 5.4°F/1000 ft (or warms at 1°C/100 m if it is coming down.). This warming or cooling is termed the *dry adiabatic lapse rate* and is independent of prevailing atmospheric temperatures. The 1°C/100 m *always* holds (for dry air), regardless of what the actual temperature at various elevations might be. When there is moisture in the air the lapse rate becomes the *wet adiabatic lapse rate* because the evaporation and condensation of water influences the temperature of the air parcel. This is an unnecessary complication for the purpose of the following argument, where a moisture-free atmosphere is assumed.

The actual temperature-elevation measurements are called *prevailing lapse rates* and can be classified as shown in Figure 9.2. A *superadiabatic lapse rate*, also called a *strong lapse rate*, occurs when the atmospheric temperature drops more than 1°C/100 m. A *subadiabatic lapse rate*, also called a weak lapse rate,

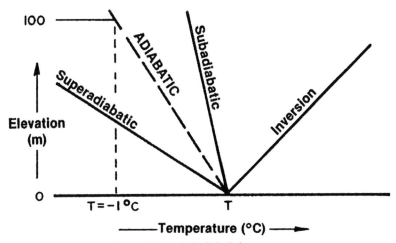

Figure 9.2. Typical adiabatic lapse rates.

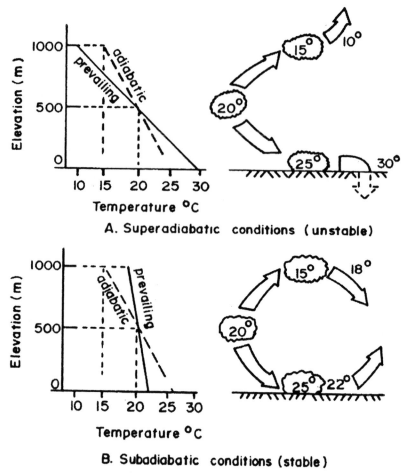

A. Superadiabatic conditions (unstable)

B. Subadiabatic conditions (stable)

Figure 9.3. Movement of air parcels in the atmosphere.

is characterized by drop of less than 1°C/100 m. A special case of the *weak lapse rate* is the *inversion*, a condition which has warmer air above colder air.

During a superadiabatic lapse rate the atmospheric conditions are unstable; a subadiabatic and especially an inversion characterizes a stable atmosphere. This can be demonstrated by depicting a parcel of air at 500 m (see Figure 9.3). If the temperature at 500 m is 20°C, during a superadiabatic condition the temperature at ground level might be 30°C and at 1000 m it might be 10°C. Note that this represents a change of more than 1°C/100 m.

If the parcel of air at 500 m and 20°C is moved upward to 1000 m, what would be its temperature? Remember that assuming adiabatic conditions, the parcel would cool 1°C/100 m. The temperature of the parcel at 1000 m is thus

5°C less than 20°C or 15°C. The prevailing temperature (the air surrounding the parcel) however, is 10°C, and the air parcel finds itself surrounded by cooler air. Will it rise or fall? Obviously, it will rise, since warm air is light and rises. We then conclude that once a parcel of air under superadiabatic conditions is displaced upward, it keeps right on going.

Similarly, if a parcel of air under superadiabatic conditions is displaced downward, say to ground level, the air parcel is 20°C + (500 m × [1/100 m]) = 25°C. It finds the air around it a warm 30°C and thus the cooler air parcel would just as soon keep going down if it could. Superadiabatic conditions also promote the downward air movement, adding to the instability. Superadiabatic atmospheres are characterized by a great deal of vertical air movement and turbulence, since any upward or downward movement tends to continue and not be dampened out.

The subadiabatic prevailing lapse rate is by contrast a very stable system. Consider again, as in Figure 9.3, a parcel of air at 500 m and at 20°C. A typical subadiabatic system has a ground level temperature of 21°C and 19°C at 1000 m. If the parcel is displaced to 1000 m, it will cool by 5°C to 15°C. But, finding the air around it a warmer 10°C, it will fall right back to its point of origin. Similarly, if the air parcel is brought to ground level, it would be at 25°C, and finding itself surrounded by 21°C air, it would rise back to 500 m. Thus the subadiabatic system would tend to dampen out vertical movement and is characterized by a limited vertical mixing.

The stability of the atmosphere is evident by the time of day and the shape of plumes being emitted from stacks. Typically, an inversion condition exists early in the morning and the atmosphere is stable. As the sun warms up the ground, the temperature increases and eventually the atmospheric condition becomes unstable (strong lapse rate). This is particularly evident on sunny summer afternoons, as anyone who has flown in an airplane will attest. The smoothest flights will occur in the morning, and the most turbulent in the afternoon of a sunny day. The shape of the plumes is an indicator of atmospheric stability. As shown in Figure 9.4, the superadiabatic or strong lapse rate produces a *looping* plume that carries vertically up and down, often reaching the ground. The very stable atmosphere in an inversion condition exhibits a *fanning* plume, which has little or no vertical movement. A *coning* plume would occur in a neutral atmosphere, and the worst case is the *fumigation* plume which has an inversion cap on top, and unstable condition below.

Certain inversions, called *subsidence inversions*, are due to the movement of a large warm air mass over cooler air. Such inversions, typical in Los Angeles, last for several days and are responsible for serious air pollution episodes. A more common type of inversion is the *radiation inversion*, caused by the radiation of heat to the atmosphere from the earth. During the night, as the earth cools, the air close to the ground loses heat, thus causing an inversion, which is not broken until the sun warms up the ground (Figure 9.5). The pollution emitted during the night is caught under this inversion lid and does not escape until the earth warms sufficiently to break the inversion.

In addition to inversions, serious air pollution episodes are almost always accompanied by fogs. These tiny droplets of water are detrimental in two ways. In the first place, fog makes it possible to convert SO_3 to H_2SO_4. Secondly, fog sits in valleys and prevents the sun from warming the valley floor and breaking inversions, often prolonging air pollution episodes.

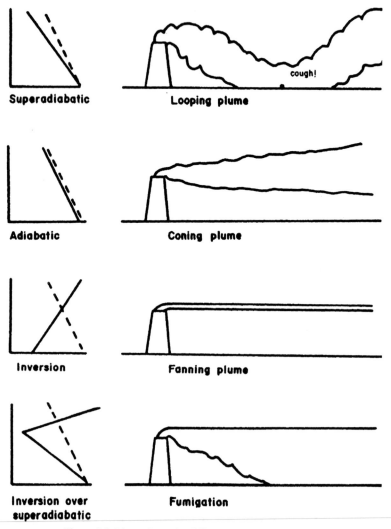

Superadiabatic   **Looping plume**

Adiabatic   **Coning plume**

Inversion   **Fanning plume**

Inversion over superadiabatic   **Fumigation**

Figure 9.4. Plume shapes for different atmospheric stabilities.

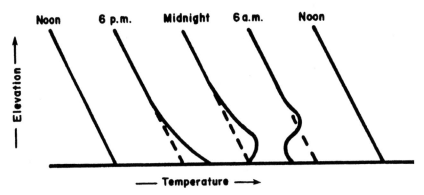

Figure 9.5. Formation and break-up of a radiation inversion.

9.2 MAJOR AIR POLLUTANTS AND THEIR MEASUREMENT

9.2.1 Particulates

Particulates in air pollution are everything in the air that is not a gas. A suspension of particulates in air is called an aerosol. Naturally occurring aerosols include sea spray, dust suspended by wind, smoke from forest fires, and plant spores. To this mix is added human-manufactured particulates like tobacco smoke, fly ash from coal combustion, cement dust from concrete plants, condensed paint pigments, and insecticides. Particulates in air can be divided technically into four categories:

* Fumes—condensations of volatile organics such as those produced from the drying of paint
* Dust—solid particles such as the ones in smoke
* Mist—tiny liquid droplets such as naturally occurring fog
* Spray— larger liquid droplets.

9.2.2 Measurement of Particulates

Particulate pollutants can be classified by their particle size, usually expressed in micrometers, μm. In air pollution control parlance, a micrometer is often referred to as a micron (μ) and this usage is adopted here.

The measurement of particulates is historically done using the high-volume sampler (or "hi-vol"). The high-volume sampler, (Figure 9.6), operates much like a vacuum cleaner by simply forcing over 2000 cubic meters of air through a filter in 24 hours. The analysis is gravimetric; the filter is weighed before and after, and the difference is the particulates collected. The particulate

The Donora Episode

In 1948, western Pennsylvania was the steel making capital of the world. Many small towns on the Monongahela and Ohio Rivers were there because of steel making. At the turn of the century, immigrants from central Europe poured into the area to provide the cheap but dedicated labor. The working conditions in the mills were abominable by today's standards. Some jobs were worse than others, such as the coke ovens, where it was known that for some jobs the life expectancy was only a few years. But all in all, the mills prospered, and the towns grew into proud communities.

Donora, Pennsylvania, was one of these towns. Nestled in a bend of the Monongahela River, it boasted a steel mill, a wire mill and a plating plant. The product was galvanized wire, and in 1948, the mills were humming.

Air pollution was accepted as part of life, the price paid for having a job. It was too bad if the curtains had to be washed every month, or that collars of white shirts would be soiled within a few hours, or that interior walls would get so dirty that they had to be literally erased with putty. Smoke meant jobs, and any intimation of controls on production were always met with threats by the company to move the plant elsewhere.

In October of 1948 an inversion occurred and effectively put a lid on the river valley and concentrated all of the pollutants directly above Donora. Within hours, animals began to be in distress, particularly parakeets. As time when on, and the pollutant concentrations under the inversion increased, people began to feel the effects. Many sought medical help, but the atmosphere was so impenetrable that the doctors had difficulty finding their way to the homes of the sick people. After three days, the meteorological lid finally lifted, and the pollution abated, leaving 27 people dead, and hundreds permanently affected by the episode.

The steel companies insisted they were not at fault, and indeed there never was any fault implied by the special inquiry into the incident. The companies were operating within the law, and were not coercing any of the workers to work there, or anyone to live in Donora. In the absence of legislation, the companies felt no obligation to pay for air pollution equipment or to change processes to reduce air pollution. They felt that if only their company was required to pay for and install air pollution control equipment, they would be at a competitive disadvantage, and would eventually go out of business.

Soon after the "Donora episode" it became evident that national legislation was needed in order to prevent acute health problems. Eventually state legislation began to appear, but it wasn't until 1972 that effective federal legislation was passed.

(continued)

(continued)

U. S. Steel moved out of Donora in 1966, and the plants were demolished, but not because of air pollution problems. Today Donora is a poor town with little industry, and some of the older people still believe that the 1948 episode gave Donora a reputation that has prevented new industry from moving into town. They often deny such an episode even occurred. The editor of the *Herald American*, Donora's local newspaper, wrote: "Except for the lesson to be learned by other communities, this small Pennsylvania town could wish for nothing better than to have the world forget 'the Donora episode.'"

concentration measured in this manner is often referred to as *total suspended particulates* to differentiate it from other measurements of particulates.

The air flow is measured by a small flow meter, usually calibrated in cubic feet of air per minute. Because the filter gets dirty during the 24 hours of operation, less air goes through the filter during the latter part of the test than in

Example 9.1

A clean filter is found to weigh 10.00 grams. After 24 hours in a hi-vol, the filter plus dust weighs 10.10 grams. The air flow at the start and end of the test is 2.0 and 1.6 m^3/minute, respectively. What is the particulate concentration?

Weight of the particulates (dust) = $(10.10 - 10.00)$g $\times 10^6$ µg/g = 0.1×10^6 µg
Average air flow = $(2.0 + 1.6)/2 = 1.8$ m^3/min
Total air through the filter = 1.8 m^3/min $\times$ 60 min/hr $\times$ 24 hr/day $\times$1 day = 2736 m^3
Total suspended particulates = $(0.1 \times 10^6$ µg$)/2736$ m^3 = 36 µg/m^3.

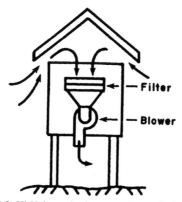

Figure 9.6. Hi-Vol sampler for measuring particulates in air.

The Laporte Weather Anomaly

Air pollution from cities can cause climatic changes far beyond their borders. Probably the best documented such case occurred in LaPorte, Indiana.

LaPorte is 30 miles east of a large complex of industrial plants in Chicago, and since 1925 it experienced an anomalous increase in total precipitation, number of rainy days, and number of thunderstorms. The discrepancy between LaPorte weather and that of surrounding towns first suggested that the meteorologists at LaPorte had been doctoring the data by being overzealous in reporting adverse weather. This, however, was statistically ruled out. Finally it was recognized that the LaPorte's lousy weather was due to the presence of nuclei, particulates emitted by the factories in the Chicago-Gary industrial complex. As these particulates rose into the atmosphere they acted as nuclei for the formation of rain droplets. It actually did rain more in LaPorte, and the cause was not overzealous weather forecasters but air pollution.

---

the beginning and the air flow must therefore be measured at both the start and end of the test period and the values averaged.

Another widely used measure of particulates in the environmental health area is of *respirable particulates*, or those particulates that would be respired into lungs. These are generally defined as being less than 0.3μ in size, and the measurements are with stacked filters. The first filter removes only particulates larger than 0.3μ, and the second filter, having smaller spaces, removes the small respirable particulates.

The EPA has also recognized that the measurement of particulates can be badly skewed if a few really large particles happen to fall into the sampler. To get around this problems, they now measure particles smaller than 10 microns only. Designated symbolically as PM_{10}, for "particulate matter less than 10μ", this measure is used in ambient air quality standards, as discussed later.

Woodburing stoves are apparently particularly effective emitters of small particulates, of the PM_{10} variety. In one Oregon city the PM_{10} concentrations reached above 700 μg/m³, whereas the national ambient air quality standard is only 150 μg/m³. Cities where the atmospheric conditions prevent the dispersal of wood smoke have been forced to pass woodburning bans during nights when the conditions could lead to high particulate levels in the atmosphere.

No continuous particulate measuring devices have yet been developed and accepted. The problem, of course, is that the measurement must be gravimetric, and it is difficult to construct a device that continuously weighs minute quantities of dust.

9.2.3 Gaseous Air Pollutants

In the context of air pollution control, gaseous pollutants include substances that are gases at normal temperature and pressure as well as vapors of substances that are liquid or solid at normal temperature and pressure. Among the gaseous pollutants of greatest importance in terms of present knowledge are: carbon monoxide, hydrocarbons, hydrogen sulfide, nitrogen oxides, ozone (and other oxidants), and sulfur oxides. Carbon dioxide should be added to this list because of its potential effect on climate. These and other gaseous air pollutants are listed in Table 9.1. Pollutant concentrations are commonly expressed as micrograms per cubic meter ($\mu g/m^3$).

TABLE 9.1. Some Gaseous Air Pollutants.

Name	Formula	Importance and effect
Sulfur Dioxide	SO_2	Colorless gas, damaging to property and health, at high concentrations causes intense choking, odiferous, highly soluble in water.
Sulfur Trioxide	SO_3	Soluble in water to form sulfuric acid, H_2SO_4. Highly corrosive.
Hydrogen Sulfide	H_2S	Rotten egg odor at low concentrations, odorless at higher (highly toxic) concentrations.
Nitrous Oxide	N_2O	Colorless gas, relatively inert, used as carrier gas in aerosol bottles, not produced in combustion.
Nitric Oxide	NO	Colorless gas, produced during high-temperature, high-pressure, combustion. Oxidizes to NO_2.
Nitrogen Dioxide	NO_2	Brown to orange gas, major component in the formation of photochemical smog.
Carbon Monoxide	CO	Colorless, odorless, and poisonous gas, produced as product of incomplete combustion.
Carbon Dioxide	CO_2	Colorless and odorless, formed during complete combustion. Greenhouse gas.
Ozone	O_3	Highly reactive, causes damage to vegetation and property. Produced mainly during the formation of photochemical smog.
Hydrocarbons	C_xH_y	Many hydrocarbons are emitted from automobiles and industries, others are formed in the atmosphere, especially during the formation of photochemical smog.
Methane	CH_4	Combustible, odorless. Major greenhouse gas.
Hydrogen fluoride	HF	Highly corrosive. Causes fluorosis, or mottling of bones.
Chlorofluorocarbons	CFC	Non-reactive, excellent thermal properties, depletes ozone in upper atmosphere.

Chlorine in New York City

In the first week of June, 1944, a truck carrying a cargo of compressed chlorine passed over the Manhattan Bridge into jam-packed downtown Brooklyn. Stopped at a light, the driver detected the unmistakable odor of his cargo. He pulled over and, on checking, found that one tank was leaking. He removed the tank, set it down, and then hurried off to call his dispatcher to ask for instructions. Inadvertently, he had placed the tank on a sidewalk grating that covered the ventilation intake of the Brooklyn-Manhattan Transit subway station at Myrtle Avenue and Flatbush Avenue Extension, one of the busiest stations of the BMT system.

Chlorine is heaver than air and it fell through the grating as it seeped out of the tank. Down below, the subway trains rushing through the tunnel acted much like the pistons of a vacuum pump, sucking green clouds of chlorine into the station mezzanine. People waiting for change were knocked over like bowling pins, passengers coming up the stairs from the platform beneath found themselves walking into a gas chamber. Blinded, many stumbled down the stairs, colliding with the passengers who were coming up. Meanwhile, the chlorine began to roll down the stairs as well and engulfed the passengers at the platform level as they were wating for the incoming train.

Within minutes panic took over the station. Scores of people groped about, trying to avoid falling off the narrow platform yet unable to see or breathe. Many fainted on the stairways, to be trampled on by the choking hordes that collided on the stairs while trying to escape the gas from above and below. As soon as the alarm was sounded, the city put into operation its wartime procedure for poison-gas attacks and police and emergency services rushed to the scene. Over a hundred people required hospital care, although nobody died.

9.2.4 Measurement of Gaseous Air Pollutants

While the units of particulate measurement are consistently in terms of micrograms per cubic meter, the expression of the concentration of gases can be as either parts per million (ppm) on volume to volume basis, or as micrograms per cubic meter. As first introduced in Chapter 2, the conversion from one to the other is

$$\mu g/m^3 = 40.9 \ (ppm) \ M \qquad (9.1)$$

where

M = molecular weight of the gas.

This equation is applicable for conditions of 1 atmosphere and 25°C. For 1 atmosphere and 0°C (273°K), the constant becomes 22.4.

Example 9.2

A stack gas contains carbon monoxide (CO) at a concentration of 10% by volume. What is the concentration of CO in $\mu g/m^3$? (Assume 25°C and 1 atmosphere pressure.)

Since 1% by volume is 10,000 ppm, 10% by volume is 100,000 ppm. Since the molecular weight of CO is 28, the concentration in micrograms per cubic meter is

$$40.9(100,000)28 = 114 \times 10^6 \ \mu g/m3$$

The earliest gas measurement techniques almost all involve the use of a *bubbler*, shown in Figure 9.7. The gas is literally bubbled through the liquid, which either reacts chemically with the gas of interest, or into which the gas is dissolved. Wet chemical techniques are then used to measure the concentration of the gas.

A simple (but now seldom used) bubbler technique for measuring SO_2 is to bubble air through hydrogen peroxide, so that the following reaction occurs:

$$SO_2 + H_2O_2 \rightarrow H_2SO_4$$

The amount of sulfuric acid formed can be determined by titrating the solution with a base of known strength.

A more accurate method of measuring SO_2 is the colorimetric pararosaniline method, in which SO_2 is bubbled into a liquid containing tetrachloromercurate (TCM). The SO_2 and TCM combine to form a stable complex. Pararosaniline is then added to this complex, with which it forms a colored solution. The amount

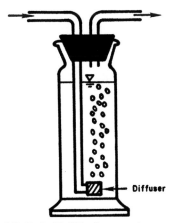

Figure 9.7. Typical bubbler for measuring gases in air.

The Poza Rica Accident

In 1950, the city of Poza Rica was in the center of Mexico's largest natural gas field. The gas, however, had a high concentration of hydrogen sulfide, H_2S, and this had to be stripped out, producing commercially valuable sulfur with the excess being incinerated in an odorless flare. Flaring is effective only if the amount of hydrogen sulfide fed to the flare is carefully controlled. In the early morning of November 24, the gas feeding system malfunctioned, and a large quantity of H_2S was fed to the flare. This excess was vented directly into the air where the denser-than-air gas flowed down to the ground and started to move toward the buildings surrounding the plant.

Hydrogen sulfide has a strong odor at low concentrations, but at higher concentrations the gas overwhelms the olfactory senses and does not have an apparent odor. Many of the people who awakened at 4 o'clock in the morning were aware of the strong odor, and saved their lives by evacuating the area. Other less fortunate people did not awaken and were found dead in their beds. In all, 320 people were taken ill and hospitalized, and 22 died from inhaling the H_2S.

---

of color is proportional to the SO_2 in the solution, and the color is measured with a spectrophotometer. (See Chapter 6 for ammonia measurement—another example of a colorimetric technique.)

9.2.5 Stack Sampling

One of the problems in air pollution control is the measurement of emissions coming out of smokestacks. The technique for obtaining a sample of an emission is called stack sampling. A hole is first drilled into the stack and a horizontal sampling tube is inserted through the hole into the stack (see Figure 9.8). At the end of the tube is a sampling port which detects temperature and pressure, and is able to draw a small sample into a tube leading out of the hole

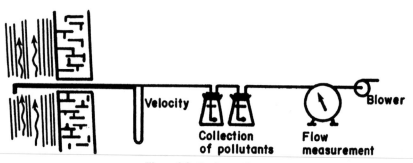

Figure 9.8. Stack sampling.

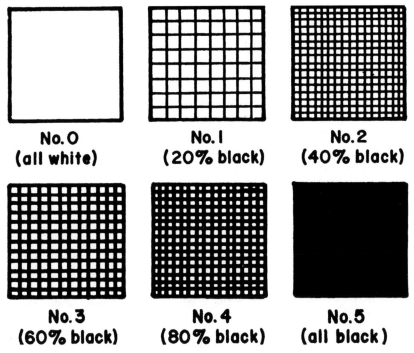

No. 0
(all white)

No. I
(20% black)

No. 2
(40% black)

No. 3
(60% black)

No. 4
(80% black)

No. 5
(all black)

Figure 9.9. Ringlmann scale for measuring smoke opacity.

and then into various instruments. The sample flows through sampling devices such as bubblers and then into a flow meter that measure the flow rate of the sample. The rod in the stack is moved from one side of the stack to the other in order to obtain representative and accurate measurements of pollutant concentrations in the stack emissions.

9.2.6 Measurement of Smoke

Air pollution has historically been associated with smoke—the darker the smoke, the more pollution. We now of course know that this isn't necessarily true, but many regulations (e.g., for municipal incinerators) are still written on the basis of smoke density. The density of smoke has for many years been measured on the *Ringlemann scale*, devised in the late 1800s by Maxmilian Ringlemann, a French professor of engineering. The scale runs from 0 for white or transparent smoke to 5 for totally black opaque smoke. The test is conducted by holding a card such as shown in Figure 9.9 and comparing the blackness of the card to the smoke. A fairly dark smoke is thus said to be "Ringlemann 4," for example.

9.2.7 Visibility

One of the obvious effects of air pollutants is reduction in visibility. Loss of visibility is often defined as the condition when, in bright daylight, it is just possible to identify a large object such as a building, or at night, not to be able to see a moderately bright light. This is of course a vague definition of visibility, but it is useful especially for defining the limits of visibility.

Reductions in visibility can occur due to natural air "pollutants," such as terpenes from pine trees or from human-produced emissions. Many naturally occurring constituents can cause attenuated visibility, such as water droplets (fog). However, the most effective reduction in visibility is from small

---

The Meuse Valley Episode

The Meuse River flows northward on the edge of Belgium. During the 1930s this area was the center of heavy industry in the region. With its coal mines and steel plants, and the transportation provided by the Meuse River, towns like Liege became wealthy and economically important. There was no control of air pollution, and the people in the region learned to live with the dirty atmosphere. Smoke meant jobs and prosperity.

The river valley from Huy to Liege is narrow with steep valley walls. Usually a wind sweeps the pollution out of the valley, but at the beginning of December 1930 there was no wind, and a fog had settled over the valley. A radiation inversion condition had occurred and did not break due to the fog. This condition lasted for only two days, but during this time the people knew that something was wrong. Eventually 63 people lost their lives, most of whom were elderly and already in poor health. But even healthy athletes were coughing and wheezing, and many workers at the mills sought medical assistance. The most telling was the effect of the smog on livestock and other animals, which showed various signs of stress. One farmer lost 48 cattle during the two days of the episode. Residents of Liege reported seeing birds falling out of the trees.

The cause of the disaster was never investigated. There were no instruments at the time that would have provided any help, and autopsies of the dead people did not provide many clues. The most likely cause was the emission of fluorides from the plating plants. In moist atmospheric conditions, hydrogen fluoride becomes a deadly acid capable of corroding vegetation and animal tissues, such as those in the lung. Corroded plants were reported by the farmers, supporting the assumption that fluoride was to blame for the health disaster.

Nothing was done after the episode to assist the people living in the Meuse Valley, but the episode received international attention, becoming the first documented air pollution episode involving human fatalities. The experience in the Meuse Valley helped investigators better understand the cause and effect of air pollution during the Donora episode 18 years later.

The Boston Black Rain

A memorable fallout of air pollutants occurred in South Boston on May 13, 1960. For one hour that morning the city was drenched in a downpour of oily soot that has become known as "black rain." The authorities blamed the Metropolitan Transit Authority power plant that burned either pulverized coal or oil. Apparently some engineers had attempted a tricky form of burning that combined the two fuels, but they did so without having the proper instruments or equipment. The combustion was inefficient, resulting in large quantities of soot being emitted from the stack. Instruments that were supposed to detect such upsets were either broken or not usable, and even the mirror attached to the top of the stack, which would have told the operators that something was wrong, was not usable because of low-hanging clouds. Thus the operators continued to use the combustion chamber, and the furnace kept pumping out huge quantities of oily soot.

Although the workers inside the plant had no idea what was happening, the people outside did, as roughly 1000 pounds of soot fell on the area surrounding the power plant. In the midst of intermittent rain, the soot came down like black snowfall, soaking all who did not seek cover. The soot was too oily to dissolve in the water, but when mixed with rain it produced an black ink that would not wash off surfaces and foamed when it hit the street. Some streets became ankle-deep in the foamy soot.

---

particulates. Particulates reduce visibility both by adsorbing as well as scattering the light. In the first case, the light does not enter the eye of the viewer, and in the second case the scattering reduces the contrast between light and dark objects.

9.3 HEALTH EFFECTS OF AIR POLLUTION

Much of the knowledge of the effects of air pollution on people comes from the study of acute air pollution episodes such as Donora, Pennsylvania and the Meuse Valley. In both episodes the pollutants affected a specific segment of the public—those individuals already suffering from diseases of the cardio-respiratory system. Another observation of great importance is that it was not possible to blame the adverse effects on any one pollutant. This observation puzzled the investigators (industrial hygiene experts) who were accustomed to studying industrial problems where one could usually relate health effects to a specific pollutant. The investigations following the episodes suggested that the health problems were due to the combined action of a particulate matter (solid or liquid particles) and sulfur dioxide, a gas. No one pollutant by itself, however, could have been responsible. Today there seems to be strong evidence that the problem at both Donora and Meuse Valley was actually hydrogen fluoride, which attacked the cardiovascular system.

Except for these episodes, scientists have little information from which to evaluate the health effects of air pollution. Laboratory studies with animals are of some help, but the step from a rat to a person (anatomically speaking) is quite large.

Four of the most difficult problems in relating air pollution to health are unanswered questions concerning: (1) the existence of thresholds, (2) the total body burden of pollutants, (3) the time-versus-dosage problem, and (4) the synergistic effects of various combinations of pollutants.

9.3.1 Threshold

The existence of a threshold in health effects of pollutants has been debated for many years. There are several dose-response curves possible for a dose of a specific pollutant (e.g., carbon monoxide) and the response (e.g., reduction in the blood's oxygen-carrying capacity). One possibility is that there will be no effect on human metabolism until a critical concentration (the threshold) is reached. On the other hand, some pollutants can produce a detectable response for any finite concentration. Nor do these curves have to be linear. In air pollution, the most likely dose-response relationship for many pollutants is non-linear without an identifiable threshold, but also a minimal response up to a higher concentration, at which point the response becomes severe. The problem is that for most pollutants the shapes of these curves are unknown.

9.3.2 Total Body Burden

Not all of the dose of pollutants comes from air. For example, although a person breathes in about 50 μg/day of lead, the daily intake of lead from water and food is about 300 μg/day. In the setting of air quality standards for lead, it must therefore be recognized that most of the lead intake is from food and water.

9.3.3 Time Versus Dosage

Most pollutants require time to react, and the time of contact is as important as the level. The best example of this is the effect of carbon monoxide which reduces the oxygen-carrying capacity of the blood by combining with the hemoglobin and forming carboxyhemoglobin. At about 60% carboxyhemoglobin concentration, death results from lack of oxygen. The effects of CO at sublethal concentrations are usually reversible. Because of the time-response problem, ambient air quality standards are set at maximum allowable concentrations for a given time.

9.3.4 Synergism

Synergism is defined as an effect that is greater than the sum of the parts. For example, black lung disease in coal miners occurs only when the miner is also a cigarette smoker. Coal mining by itself, or cigarette smoking by itself will not cause black lung, but the synergistic action of the two puts miners who smoke at high risk.

The major target of air pollutants is the respiratory system, pictured in Figure 9.10. Air (and entrained pollutants) enter the body through the throat and nasal cavities and pass to the lungs through the trachea. In the lungs, the air moves

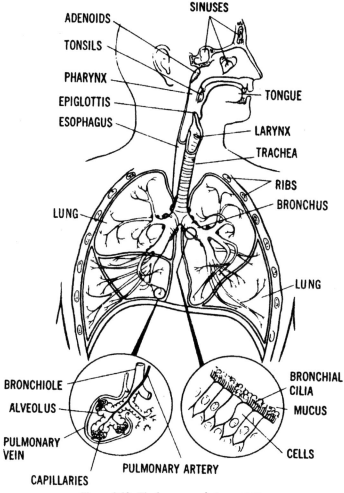

Figure 9.10. The human respiratory system.

Adverse Effect of Polluted Air on Things

One of the most destructive effects of automobile emission is the deterioration of buildings, statuary and other materials. Athens, Greece, for example, has one of the highest levels of photochemical smog, given its plethora of sunlight and incredible number of non-regulated automobiles. As a result, the buildings of the Acropolis are rapidly deteriorating, as are other remnants of ancient Greek civilization. Ironically, the theft of many of the most valuable pieces by the British around the turn of the century resulted in the saving of these treasures. The clean, controlled air in the British Museum is far better than the putrid atmosphere in Athens. But does this make the theft any less reprehensible?

Do we, in fact, have a duty to preserve things? What is worth preserving? And who decides? The things that may or may not be worth preserving are either constructed by humans, such as buildings, or natural things and places such as the Grand Canyon or the Gettysburg Battlefield. We may decide to preserve such landmarks for one of two reasons: to keep the structure or thing in existence purely for its own sake, or to not destroy it because we enjoy looking at it. There is a distinction of course between just enjoying the landmark and using it, as one would a bridge or building. If there is a use for the landmark, based on economics or other human desires, then there is no question that the landmark has some value. But some things do not have utility, such as old buildings and art. Is it possible to develop an argument for the preservation of mere things, irrespective of their utility to humans?

Recall the argument advanced by Christopher Stone in his provocative book *Should Trees Have Standing*. If corporations, municipalities and ships are considered legal entities, with rights and responsibilities, is it not reasonable to grant similar rights to trees, forests and mountains? Stone does not suggest that a tree should have the same rights as humans, but that a tree should be able to be represented in court (to have standing) just as corporations are. If this argument is valid, then buildings and other inanimate objects can similarly have legal rights, and their "interests" can be represented in court. It should then be possible to sue the city of Athens on behalf of the Acropolis.

But the argument by Stone for granting legal rights to trees and other objects is not very strong. Perhaps the only reason we would prefer not to destroy landmarks such as historic buildings and natural wonders is that these things are components of our physical environment and are necessary to a historical grounding of our civilization. The Acropolis in Athens is a bunch of rocks. The land would be worth millions if developed for condominiums. Yet we strive to protect it, and an international project has been hard at work treating the stone to prevent further deterioration. We do this because it is important to our heritage, and we want to preserve it for future generations. While it is often uncertain how we should act toward these future generations, it is quite obvious that if we allowed landmarks such as the Acropolis to be destroyed by air pollution from automobiles, future generations would condemn our inaction.

through bronchial tubes to the *alveoli*, small air sacks in which gas transfer takes place. Pollutants are either absorbed into the bloodstream, or moved out of the lungs by tiny hair cells called *cilia*, which are continually sweeping mucus up into the throat. The respiratory system can be damaged by both particulate and gaseous pollutants.

Particles greater than 0.1μ will usually be caught in the upper respiratory system and swept out by the cilia. However, particles less than 0.1μ in diameter can move into the alveoli, where there are no cilia, and they can stay there for a

Fluoride Poisoning in Garrison, Montana

One of the unwanted byproducts of phosphate fertilizer production is hydrogen fluoride. When the Rocky Mountain Phosphate Company in Garrison MT began operations in the early 1960s, it had great difficulty with pollution control and emitted high concentrations of sulfur oxides and fluorides. Health problems soon began to appear in Garrison, and the town went to court to stop the pollution. But every time the plant was ordered to control its pollution, it was up and running again a few days later claiming to have met the requirements of the court order. In 1964 the State of Montana Health Department charged Rocky Mountain Phosphate with ignoring the court orders and tried to shut them down, but the order was thrown out by another court. In 1965 Montana's Governor Babcock ordered a study to be conducted, but the study concluded (much to the amazement of the people of Garrison) that the pollutants did not show any adverse effect. Governor Babcock was known for his pro-industrial bias and he had won the job by pledging to bring more industries to Montana. He wasn't about to shut down a plant he had worked so hard to bring to the state.

By the summer of 1965 the vegetation in and around Garrison began to die, and soon all the vegetable gardens wilted. The land was barren of wildlife, with most animals dying off because of their inability to walk. Cattle in the farms around Garrison similarly were unable to stand up and soon died. Some of the ranchers brought a civil suit against the Rocky Mountain Phosphate company but almost unbelievably, given the overwhelming evidence, the judge ruled in favor or the company. Finally in 1967 the federal government intervened and forced the company to cease operations until it had installed emission equipment that would remove 99.9% of the fluoride. This order forced the plant to cease operations.

Fluoride emissions cause cattle (and people) to have "fluorosis," a disease in which their joints are swollen and they are unable to walk. In effect, the fluoride etches the bones and makes them brittle. The same effect occurs when people drink water that has too much fluoride in it, and their teeth become soft and mottled.

Tokyo-Yokohama Asthma

During the American occupation of Japan in 1946, servicemen stationed in Yokohama suddenly began to suffer attacks of wheezing and coughing. Shortness of breath incapacitated the men, and because they constantly needed to fight to breathe, they endured days at a time without sleep. Many required long term hospitalization. Virtually none of the sufferers had a history of allergy or asthma, yet all now were plagued by a severe asthmatic condition that was becoming progressively worse.

Army doctors were baffled. But they did observe that most patient's symptoms cleared up the instant they left the Yokohama area. Some patients, however, were transferred to hospitals in Tokyo, and the disease followed them there. Similar incidents occurred in a third city, Zama. Finally the common thread became clear—all of these cities are in the Kanto Plain, a large bowl ringed by high mountains much like Los Angeles. This area has weak winds and long periods of thermal inversions, and breathing the highly polluted air clearly brought on the asthma attacks.

Unfortunately, the sufferers became highly sensitized to such pollution, and often suffered asthma attacks when they returned to American cities with polluted air. Others who had sought relief back in the United States found that the symptoms returned when the flew back to the Tokyo area. No specific causative agent was ever identified.

---

long time and cause damage to the lung. Droplets of H_2SO_4 and lead particles represent the most serious forms of particulate matter in air.

The effects of carbon monoxide are potentially deadly. Human hemoglobin (Hb) has a carbon monoxide affinity 210 times greater than its affinity for oxygen, and thus CO combines, as noted above, with hemoglobin to form *carboxyhemoglobin* (COHb). The formation of COHb reduces the hemoglobin available to carry oxygen, and this can cause death by asphyxiation.

Ozone (O_3) in the atmosphere is an eye irritant and can cause damage to materials such as rubber in automobile tires. Ozone is usually a part of urban air and photochemical smog.

A number of diseases are thought to be causally related to air pollution, including lung cancer, emphysema and asthma. There is overwhelming evidence that air pollution can increase the risk of these diseases, specially when tied synergistically to cigarette smoking, yet the actual cause-and-effect is not medically proven. It is incorrect to assert that air pollution, no matter how bad, (or cigarettes, for that matter) cause lung cancer or other respiratory diseases.

Perhaps the single most important gas with reference to health is sulphur dioxide. It acts as an irritant, restricts air flow, and slows the action of the cilia. Since it is highly soluble, it should be readily removed by the mucus membrane, but it has been shown that it can extend to the deep reaches of the lung by first adsorbing onto tiny particles, then using this transport to enter the deep lung, a classic case of synergistic action.

9.4 INDOOR AIR

Indoor air quality is of importance to health simply because we spend so much time indoors, and the quality of the air we breathe is seldom monitored. Contaminated indoor air can cause any number of health problems, including eye irritation, headache, nausea, sneezing, dermatitis, heartburn, drowsiness and many other symptoms. These problems can arise as a result of breathing harmful pollutants such as:

- Carbon monoxide—from smoking, space heaters, stoves
- Formaldehyde—from carpets, ceiling tile, paneling
- Particulates—from smoking, fireplaces, dusting
- Nitrogen oxides—from kerosene stoves, gas stoves
- Ozone—from photocopying machines
- Radon—diffusion from the soil
- Sulfur dioxide—kerosene heaters
- Volatile organics—smoking, paints, solvents, cooking

While the air exchange and cleaning in office buildings and many apartments are controlled, private residence usually depend on natural ventilation to provide air exchange. Most houses are in fact poorly sealed, and air leakage takes place at many locations including through doors and windows, exhaust vents, and chimneys. In warmer weather some homes have forced ventilation, with whole-house fans or individual window fans.

One of the most insidious indoor air pollutants is secondary smoke from cigarettes and pipes. Most of the particulates in a cigarette are emitted into the room without being inhaled by the smoker, and these are then inhaled by everyone else in the room. Cigarette smoke also contains high amounts of CO, and when several persons are smoking in a room, the CO level can become high enough to affect performance. One of the smallest rooms we commonly live in is our car, and cigarette smokers can significantly affect the CO level in a car. Smokers exhale high levels of CO even when they are not smoking, and this has been blamed for the *sleepy driver syndrome* in commuter cars.

Another important and troublesome indoor air pollutant is *radon*, a naturally emitted gas entering homes through the basement, well water, and even the building materials. Radon and its radioactive daughters are part of the natural decay chain beginning with uranium and ending with lead. The decay products of radon—polonium, lead and bismuth—are easily inhaled and can readily find their way into lungs. The most important health effect of radon therefore is lung cancer.

9.5 SYMBOLS

M = molecular weight on the gas
ppm = parts per million on a volume basis
PM_{10} = particulate matter less than 10 μ

9.6 PROBLEMS

9.1 A car is driven an average of 1000 miles/month and emits 3.4 g/mi of hydrocarbons (HC) and 30 g/mi of carbon monoxide (CO).

a. How much CO and HC would be emitted during the year?

b. If there are 20 million cars in New York City, how much total HC and CO is emitted every year by automobiles?

c. Where does all this carbon monoxide go? (This is a thought question.)

9.2 Give three examples of synergism in air pollution.

9.3 If SO_2 is so soluble in water, how can it get to the deeper reaches of the lung without first dissolving in the mucus?

9.4 A hi-vol clean filter weighs 20.0 g and the dirty filter weighs 20.5 g. The initial and final air flows are 2.2 and 2.0 m³/min. What volume of air went through the filter in 24 hours?

9.5 A high-volume sampler draws air in at an average rate of 2.0 m³/min. If the particulate reading is 200 μg/m³, what was the weight of the dust on the filter?

9.6 Research and report on the famous 1952 air pollution episode in London.

9.7 What does Ringlemann No. 5 tell you about a smoke being emitted from a chimney?

9.8 The data for a hi-vol are as follows:

Clean filter:	20.0 g
Dirty filter:	20.5 g
Initial air flow:	2.3 m³/min.
Final air flow:	1.97 m³/min.
Time:	24 hours

a. What volume of air was put through the filter? (Answer in cubic meters.)

b. How is the air flow in a hi-vol measured?

c. What is the weight of the particulates collected?

d. What is the condition of the atmosphere?

9.9 If the primary ambient air quality standard for nitrogen oxides (as NO_2) is 100 µg/m^3, what is this in ppm? (Assume 25°C and 1 atmosphere pressure.)

9.10 Given the following temperature soundings:

Elevation (m)	Temperature (°C)
0	20
50	15
100	10
150	15
200	20
250	15
300	20

What type of plume would you expect if the exit temperature of the plume is 10°C and the smoke stack is

a. 50 m tall?

b. 150 m tall?

c. 250 m tall?

9.11 Consider a prevailing lapse rate which has these temperatures: ground = 21°C, 500 m = 20°C, 600 m = 19°C, 1000 m = 20°C. If a parcel of air is released at 500 m and at 20°C, would it tend to sink, rise or remain where it is? If a stack is 500 m tall, what type of plume would you expect to see?

9.12 A temperature sounding balloon feeds back the following data:

Elevation (m above ground level)	Temperature °C
0	20
20	20
40	20
60	21
80	21
100	20
120	17
140	16
160	14
200	12

What type of plume would you expect to see out of a stack 70 m high? Why?

CHAPTER 10

Air Pollution Control

CHANGING or eliminating a process that produces a polluting air emission is often easier than trying to trap the effluent. A process or product may be needed or necessary, but could be changed to control emissions. For example, automobile exhaust has caused high lead levels in urban air and elimination of lead from gasoline has reduced lead in urban air. Similarly, removal of sulfur from coal and oil before the fuel is burned reduces the amount of SO_2 emitted into the air. In these cases, the source of air pollution has been corrected. In many cases, however, the emissions have to be cleaned before being discharged into the atmosphere. In this chapter the control of emissions from both stationary and moving sources is discussed.

10.1 CONTROLLING AIR POLLUTION FROM STATIONARY SOURCES

10.1.1 Particulate Air Pollution

The effectiveness of an air pollution control device is measured the same way as the effectiveness of a device used in water pollution control. The removal of a pollutant from the emission is calculated as before in Equation 5.5,

$$R = \frac{C_o - C}{C_o} \times 100 \qquad (10.1)$$

where

R = removal of a pollutant, percent
C = concentration of the pollutant in emissions, $\mu g/m^3$
C_o = concentration of the pollutant in the dirty air stream, $\mu g/m^3$

The mass of pollutant removed is then

$$M = (C_o - C)Q \qquad (10.2)$$

where

M = mass of pollutant removed per time, $\mu g/min$
Q = flow rate, m^3/min

Removal can also be expressed in mass terms as

$$R = \frac{M}{M_o} \times 100$$

where

M_o = mass of pollutant going into the removal device, $\mu g/min$

Example 10.1

An air pollution control device receives an air flow of 40 m^3/min that contains 2000 $\mu g/m^3$ of particulates. The emission from the control device has a particulate concentration of 80 $\mu g/m^3$. What is the removal, expressed as a percent of particulates removed?

$$R = \frac{2000 - 80}{2000} \times 100 = 96\%$$

This can also be expressed in mass terms. The mass of particulates coming in to control device is:

$$2000 \ \mu g/m^3 \times 40 \ m^3/min = 80{,}000 \ \mu g/min$$

The mass of particulates that escapes the control device is

$$80 \ \mu g/m^3 \times 40 \ m^3/min = 3{,}200 \ \mu g/min$$

so the removal is 80,000 – 3,200 = 76,800 $\mu g/min$, giving a percent removal of

$$R = \frac{76{,}000}{80{,}000} \times 100 = 96\%$$

Cyclones

The cyclone is a popular, economical, and effective means of controlling particulates. Cyclones alone are generally not adequate to meet stringent air pollution control regulations, but serve as pre-cleaners for control devices like fabric filters or electrostatic precipitators. Figure 10.1 shows a simple diagram

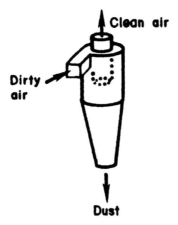

Figure 10.1. Cyclone used for removal of larger particulates.

of a cyclone. Dirty air enters the cyclone off-center creating a violent swirl of air in the cone of the cyclone. Particles are accelerated centrifugally outward toward the cyclone wall and friction at the wall slows the particles and they slide to the bottom, where they can be collected. Clean air exits at the center of the top of the cone. Cyclones are reasonably efficient for large particle collection, and are widely used as the first stage of dust removal.

The effectiveness of a cyclone for removing particulate matter is often expressed as

$$R = \frac{\pi N \rho d^2 v}{9 \mu W} \times 100$$

where

R = removal efficiency, percent of particulates removed by the cyclone
π = 3.14
N = number of turns the air makes in the tube section of the cyclone
ρ = density of the particles, kg/m^3
d = diameter of the particles, m
v = velocity of the air flow into the cyclone, m^3/hr
μ = viscosity of the fluid, usually air, kg/m-hr
W = inlet width, m

The value of this equation is not so much to solve for cyclone efficiency but to understand how the variables in this equation affect the removal efficiency.

If we want R to be as big as possible, then we want

1. N, the number of turns the air makes in the cyclone, to be high. The more

time the air spends in the cyclone the more chance is there for the particles to move to the inside wall and to drop out.

2. ρ, the density of the particles, to be high. Heavy particles will move toward the wall with greater velocity.

3. d, the diameter of the particles, to be large. Note also that this term is squared, so even a small difference in particle diameter will have a large effect on the efficiency of removal.

4. v, the velocity of the air flow into the cyclone, to be high. The higher this velocity, the more centrifugal force will be created to move the particles to the inside wall.

If we want to achieve the best performance from a cyclone for a given aerosol, we want to have a cyclone with a high inlet velocity and to have the cyclone be long enough for the air to make many turns inside the cyclone. If we are trying to apply the cyclone for the removal of particulates, we want a feed that has particles that are of high density and most importantly, big diameters. This is one reason why cyclones are seldom applied for applications where the particle diameters are less than 10 μm.

Fabric Filters

Fabric filters or *bag filters* used for controlling particulate matter (Figures 10.2) operate like a vacuum cleaner. Dirty gas is blown or sucked through a fabric filter bag. The fabric bag collects the dust, that is removed periodically by shaking the bag. Fabric filters can be very efficient collectors for even sub-micron-sized particles and are widely used in industrial applications and are especially effective in the removal of particulates from coal-fired power plant emissions. The basic mechanism of dust removal in fabric filters is thought to be similar to the action of sand filters in water quality management. Dust particles adhere to the fabric because of surface force that result in entrapment.

Figure 10.3 shows the theoretical capture mechanisms. The individual strands and the spaces between the strands in the fabric are much larger than the particles that are captured, and thus the removal mechanism is not simple sieving. We believe that three different mechanisms are actually going on. The first is *impaction* in which the small particles to be removed approach the fabric strands at a high velocity and simply are not able to negotiate the curve of the air flow around the strand, and crash into it. This is like taking a curve too fast in a car. The car will leave the road and crash into the side of the road, and in like manner the particle will leave the air stream in which it is suspended and crash into the strand. The second mechanism is *interception*, where the particle is able to stay in the air stream, but it is too large to sneak around the strand and gets caught on the surface. Finally, the third mechanism is *diffusion*, where the

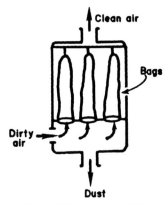

Figure 10.2. Fabric (bag) filter.

particles are small enough to be floating in irregular paths and by simply Brownian motion will collide with the strand.

If these are the three primary mechanisms for removal, then note that the first two favor large particles. The larger the particle the more likely it is to not be able to negotiate the curve around the strand, or to get caught as it tries to sneak

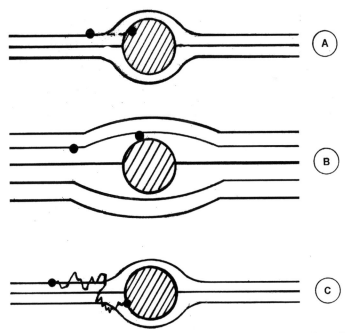

Figure 10.3. Removal mechanisms for a strand of fabric in a bag filter. A = impaction, B = interception, C = diffusion.

past. Thus we should conclude that the fabric filter should be most efficient for larger particles. But the third removal mechanism favors smaller particles which are so small as to be in random motion, bouncing off each other as they meander around the air stream. If this is a legitimate removal mechanism, then the fabric filter ought to be effective in removing very small particles as well. The actual effectiveness of a fabric filter is shown in Figure 10.4. Note that there is a point at around 1 μm where the efficiency is lowest. These particles are too small to be effectively removed by the first two removal mechanisms, and too large to be effectively removed by the third mechanism.

One of the problems in the application of fabric filters some years ago was that they were too sensitive to high temperatures. Filters made out of cotton or wool simply cannot withstand the temperature present in coal-fired power plant emissions, for example. But newer fabrics such as Teflon and fiberglass have solved that problem. The bags are still sensitive to humidity, however, and industrial emissions that have high moisture content are generally not candidates for bag filters.

As the particles adhere to the bag filter fabric, it becomes increasingly difficult to push air through the fabric. Just like in deep bed sand filtration, the

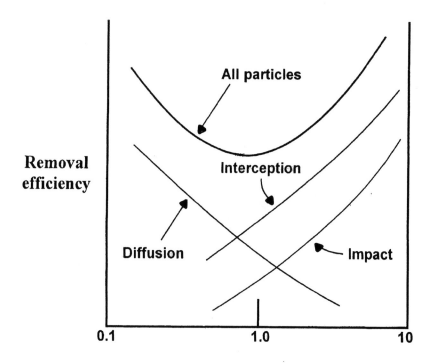

Particle diameter, μm

Figure 10.4. Efficiency of removal in a fabric filter.

fabrics have to be cleaned. The most used technique is to stop the air flow and to simply bang on the supports holding the bags. This sets up vibrations, which loosen the dust inside the bag and this falls down into the hopper below.

Wet Collectors

The *spray tower* or *scrubber* pictured in Figure 10.5 can remove larger particles effectively. More efficient scrubbers promote the contact between air and water by violent action in a narrow throat section into which the water is introduced. Generally, the more violent the encounter, hence the smaller the gas bubbles or water droplets, the more effective the scrubbing. The scrubber is essentially 100% efficient in removing particles greater than about 5 μ in diameter. Wet scrubbers can also trap gaseous pollutants, but they use a great deal of water that either itself requires further treatment or has limited use after being used to scrub dirty gas. Finally, scrubbers usually produce a visible plume of water vapor, which may be a public relations disadvantage.

A highly effective form of wet scrubber is called the *Venturi scrubber*, shown in Figure 10.6. In the Venturi scrubber the air flow is channeled through a duct which gets suddenly smaller. Recall the continuity equation first introduced in Chapter 2. This equation states that

$$Q = Av$$

where

Q = flow rate, m³/min
A = area through which this flow occurs, m²
v = the velocity of the flow, m/min

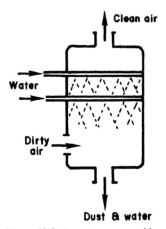

Figure 10.5. Spray tower or scrubber.

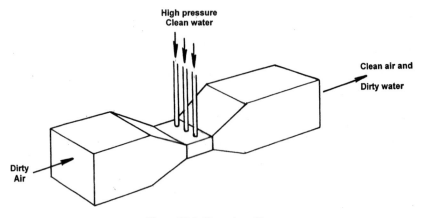

Figure 10.6. Venturi scrubber.

If some flow Q is sent through a progressively smaller opening, the area through which the flow travels gets smaller and this has to be compensated by a higher velocity. In the Venturi scrubber, the flow velocity is increased dramatically as it is pushed through the small opening. And just as the flow is in the smallest section at its highest velocity, a high-pressure spray of water is injected at right angles to the flow. This creates severe turbulence and results in the wetting and capturing of the particles suspended in the flow. Venturi scrubbers are highly efficient for a wide range of particle sizes, and they have the added advantage of being able to remove gases that might be dissolved in the water. The disadvantage, as with all scrubbers, is that a dirty effluent is created and must be disposed of.

Electrostatic Precipitators

Electrostatic precipitators are widely used to trap fine particulate matter in applications where large amount of gas needs treatment and where use of a wet scrubber is not appropriate. Coal-burning electric generating plants and incinerators often use electrostatic precipitators. In an electrostatic precipitator particles are removed when the dirty gas stream passes across high-voltage wires carrying a large negative voltage. The particles are electrically charged on passage past these electrodes and then migrate through the electrostatic field to a grounded positively charged collection electrode. The collection electrode can be either a cylindrical pipe surrounding the high-voltage charging wire or a flat plate, like that shown in Figure 10.7. In either case, the collection electrode must be periodically rapped with small hammerheads to loosen the collected particles from its surface. These trapped particles then fall into a hopper.

As the dust layer builds up on the collecting electrode, the collection efficiency may decrease, particularly if the collection electrode is the inside of a cylindrical pipe. Moreover, some dust has a highly resistive surface, does not discharge against the collection electrode, and sticks to the electrode. Heated or water-flushed electrodes may solve this difficulty. Electrostatic precipitators are efficient collectors and the amount of dust collected is directly proportional to the voltage, so that the electrical energy used by an electrostatic precipitator can be substantial, with resulting high operating cost. Their use in electrical power plants is a natural application, since the cost of electric power is lowest at these facilities.

The efficiency of collection depends on a number of variables, including the voltage used, the space between the electrodes, the residence time in the collector, and the nature of the particles. Particle size is also important. Large particles require a lot of charge before they start to move to the opposite electrode and thus they may not have time enough to pick up the charge before being discharged. Very small particles have difficulty taking on the charge, and they will continue to move in random paths, unaffected by the charge field. The electrostatic precipitator is best for particles in the 1 μm range, and the efficiency decreases for both larger and smaller particles. This is shown in Figure 10.8. Compare this figure with Figure 10.4, the efficiency plot for fabric filters. The electrostatic precipitator appears to be most efficient at precisely the range of particle sizes where the fabric filter is least efficient.

Comparison of Particulate Control Devices

Figure 10.9 shows the approximate collection efficiencies, as a function of particle size, for the devices discussed. Note the wide variation in removal efficiencies. The application of just the right removal device for a specific emission is a difficult job and can sometimes lead to disaster.

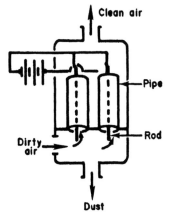

Figure 10.7. Electrostatic precipitator.

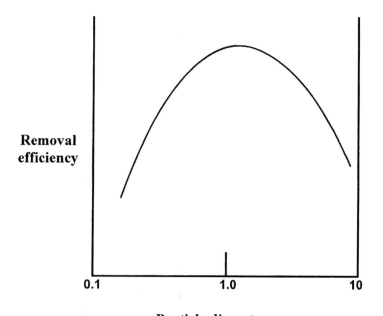

Particle diameter, μm

Figure 10.8. Collection efficiency for the electrostatic precipitator.

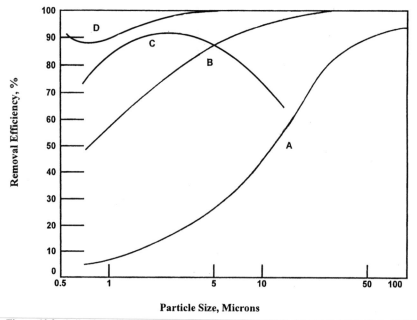

Particle Size, Microns

Figure 10.9. Collection efficiency of several particulate removal devices. A = cyclone, B = wet collector (scrubber), C = electrostatic precipitator, and D = bag filter.

354

10.1.2 Gaseous Air Pollutants

Gaseous pollutants may be removed from the effluent stream by trapping them from the stream, by changing them chemically, or by changing the process that produces the pollutants.

Wet scrubbers can remove pollutants by dissolving them in the scrubber solution. SO_2 and NO_2 in power plant off-gases are often controlled in this way. *Packed scrubbers*, spray towers packed with glass platelets or glass frit, carry out such solution processes more efficiently than ordinary wet scrubbers. Removal of fluoride from aluminum smelter exhaust gases is an example of packed scrubber use. *Adsorption*, or chemisorption, is the removal of organic compounds with an adsorbent like activated charcoal. *Incineration*, or *flaring*, is used when an organic pollutant can be oxidized to CO_2 and water, or in oxidizing H_2S to SO_2. *Catalytic combustion* is a variant of incineration in which the reaction is facilitated energetically and carried out at a lower temperature by surface catalysis. Catalytic converters are discussed later in this chapter.

Control of Sulfur Dioxide

Sulfur dioxide (SO_2) is both ubiquitous and a serious pollution hazard. The largest single source in the United States, and probably in the industrialized world, is generation of electricity by burning oil or coal, both of which contain sulfur. Increasingly strict standards for SO_2 control have prompted the development of a number of options and techniques for reducing SO_2 emissions, including changing to a low-sulfur fuel, removing the sulfur from the fuel before combustion (fuel desulfurization), and cleaning the emissions (flue-gas desulfurization).

Many homes, industries, and power companies in the United States have converted from burning high-sulfur coal or oil to using low-sulfur natural gas. Natural gas is low in sulfur, while oil and coal burned for industrial heat and electric power generation contains between 0.5% and 4% sulfur. The steady increase in natural gas prices, however, may soon eliminate this option.

Sometimes it is economical to treat the fuel before it is combusted, a process called fuel-desulfurization. Sulfur may be removed from heavy industrial oil by a number of chemical methods similar to those used to lower the H_2S content of crude oil. In coal, sulfur may be either inorganically bound, as pyrite (FeS_2), or organically bound. Pyrite can be removed by pulverizing the coal and washing with a detergent solution. Organically bound sulfur can be removed by washing with concentrated acid. Preferred methods are coal gasification, which produces pipeline-quality gas, or solvent extraction, which produces low-sulfur liquid fuel.

The off-gases from combustion or other SO_2-producing processes are called *flue gases*, and the removal of sulfur compounds from these gases is called

The Nashville RDF Combustion Plant

In 1974 the city of Nashville decided to embark on an imaginative and revolutionary scheme for heating and cooling the buildings in downtown Nashville. They decided to build a solid waste incinerator that would combust most of the refuse from the city, about 720 tons per day. This facility was a short distance from the city center and was intended to produce steam that could then by piped back to the buildings from whence the refuse came. In effect, the plant was to be an energy converter, using the energy in the office waste and burning it to produce useful steam for cooling and heating.

Because the plant was in the center of the city, the engineers realized that they needed excellent air pollution control efficiency. The trouble was that the project was running short on construction funds. The selected air pollution equipment was electrostatic precipitators because the required level of particulate removal was high. A typical solid waste combustion unit emits an uncontrolled 2.8 g/m³ of particulates, while the EPA requirement was 0.18 g/m³. This represents a recovery rate for the particulate control devices of about 96%, a very strict requirement. The electrostatic precipitators would have achieved this, but their installation would have resulted in large unacceptable cost overruns. So the engineers suggested using wet scrubbers instead.

When the plant first went into operation, the scrubbers were found to be woefully inadequate, and during the first day of operation the particulate emissions from the stack were impressive. The entire downtown area of Nashville was covered in a layer of fine dust from the combustion unit, and it was impossible to breathe the air. The plant was immediately shut down, and subsequent start-ups did not result in any improved air pollution control performance. Finally the city decided that since they had invested such a large sum of money in the plant they needed to protect their investment and had to purchase the electrostatic precipitators. These were installed and the plant has been operating for over 20 years with few upsets. The cost of the plant has been recovered many times over.

flue-gas desulfurization. SO₂ may be cleaned from flue gas by either wet scrubbing or dry scrubbing.

In *dry scrubbing*, lime, $Ca(OH)_2$, is sprayed into the hot gas stream and $CaSO_3$ (calcium sulfite) is formed.

$$Ca(OH)_2 + SO_2 \rightarrow CaSO_3 + H_2O$$

The calcium sulfite is a solid airborne particulate that must be removed using

---

Effective Disposal of Particulate Air Pollutants

If the objective of air pollution control is to produce a clean emission, the fate of the particulates removed from the dirty gases cannot at the same time be ignored. Something has to be done with the dust, and often this material is taken to the local landfill. Sometimes industrial plants contract out their waste disposal. Using contractors for the disposal of collected dust from air pollution control is common. But contracting out the disposal does not absolve the industrial firm of responsibility, either legal or moral, for the proper disposal of the waste.

Several years ago a large Kentucky textile firm installed bag filters and was pleased to be in compliance with all air quality standards. The dust collected in the filters was carried away on a weekly basis by a local farmer. After several months of such trouble-free operation, the plant engineers got curious as to what the farmer actually did with all this dust, and decided to follow him. It was then that the plant manager discovered that the farmer drove around town in the loaded, uncovered truck until the dust had all whipped off the back, and then he drove home, ready for a new load next week.

---

fabric filters. When the lime is sprayed into the hot gas, the water evaporates and a dry powder is produced.

In *wet scrubbing,* water or a mixture of water and a reagent are used to remove the sulfur oxides. If only water is used in the scrubbing, the SO_2 in the presence of excess oxygen is oxidized to sulfur trioxide, SO_3, which then dissolves in the water, producing dilute sulfuric acid.

$$2SO_2 + O_2 \rightarrow 2SO_3$$

$$SO_3 + H_2O \rightarrow H_2SO_4$$

The acid is, however, too dilute to make it commercially viable, and the removal efficiency is not high enough to make this process effective.

A more efficient form of wet scrubbing is to use limestone as the main removal agent. A wet slurry of limestone, $CaCO_3$, is sprayed into the hot gas. The first reaction that occurs is that the heat produces quicklime, CaO, which then reacts with the sulfur oxides to form $CaSO_3$. If there is excess oxygen available (which is normally the case) the calcium sulfite is converted to calcium sulfate, $CaSO_4$, or gypsum. The gypsum is removed using a wet scrubber and this goes to a settling tank where its solids concentration is increased. When the water is removed, the gypsum becomes a commercially valuable by-product used in making drywall for construction.

Control of Nitrogen Oxides

Wet scrubbers absorb NO_2 as well as SO_2, but are usually not installed primarily for NO_2 control. An effective method often used on fossil-fuel burning power plants is off-stoichiometric burning. This method controls NO formation by limiting the amount of air (or oxygen) in the combustion process to just a bit more than is needed to burn the hydrocarbon fuel in question. For example, the reaction for burning methane:

$$CH_4 + 2O_2 \rightarrow 2H_2O + CO_2$$

competes favorably with the high-temperature combination of nitrogen in the air with oxygen in the air to form NO (that eventually is oxidized to NO_2)

$$N_2 + O_2 \rightarrow 2NO$$

The stoichiometric ratio of oxygen needed in natural gas combustion is

$$32 \text{ g of } O_2 \rightarrow 16 \text{ g of } CH_4$$

A slight excess of oxygen in the combustion air will cause virtually all of the oxygen to combine with fuel rather than with nitrogen. In practice, off-stoichiometric combustion is achieved by adjusting the air flow to the combustion chamber until any visible plume disappears.

Control of Volatile Organic Compounds and Odors

Volatile organic compounds and odors are controlled by thorough destructive oxidation, either by incineration or by catalytic combustion, since they are only slightly soluble in aqueous scrubbing media. Catalytic incinerators that use a metal oxide or mixed metal oxide catalyst and operate at 450°C and sometimes higher temperatures, can achieve 95% to 99% destruction of volatile organic compounds like chlorinated hydrocarbons. These very efficient incinerators depend on catalysts that can withstand and function at high temperatures.

Control of Carbon Dioxide

Carbon dioxide is produced by oxidation of carbon compounds; that is, all combustion, all respiration, and all slow oxidative decay of vegetable matter produces CO_2. The world's oceans absorb CO_2 as carbonate, and plant photosynthesis removes CO_2 from the air. However, these natural phenomena have not kept pace with the steadily increasing concentration of CO_2 in the air, even though increasing CO_2 concentrations increases the rate of photosynthesis somewhat. Fossil fuel combustion for electrical production and for transportation appear to be the greatest contributors to increased CO_2 concentration.

CO_2 could be scrubbed from power plant effluent gas by alkaline solution and fixed as carbonate, but this process would require large quantities of scrubber solution and would produce large amounts of carbonate that would have to be disposed of. The most feasible alternative would seem to be the generation of electrical energy from prime energy sources other than fossil fuels. Although they do not produce CO_2, like all energy conversion methods, they all have some limitations and adverse environmental impacts. Hydroelectric and wind generation are limited by the finite number of physical locations where they can be constructed, and often run into local opposition. Nuclear power generation produces radioactive waste, has been shown to be susceptible to potentially catastrophic accidents, and offers targets for terrorists. Finally, solar power is relatively inefficient and requires a large land area. Non-fossil fuels sources such as wood chips and other biofuels would of course still produce CO_2. There seems to be little alternative, at least for the foreseeable future, to the combustion of fossil fuels.

10.2 CONTROLLING AIR POLLUTION FROM MOVING SOURCES

10.2.1 The Gasoline-Fueled Internal Combustion Engine

Mobile sources pose special pollution control problems, and one such source, the automobile, has received particular attention in air pollution control. Pollution control for other mobile sources, such as light-duty trucks, heavy trucks, and diesel engine-driven vehicles requires devices similar to those used for control of automobile emissions. The important pollution control points in an automobile are:

- evaporation of hydrocarbons (HC) from the fuel tank
- evaporation of HC from the carburetor
- emission of unburned gasoline and partly oxidized HC from the crankcase
- CO, HC and NO/NO_2 from the exhaust

Evaporative losses from the gas tank and carburetor often occur when the engine has been turned off and hot gasoline in the carburetor evaporates. These vapors may be trapped in an activated-carbon canister, and can be purged periodically with air, and then burned in the engine. The crankcase vent can be closed off from the atmosphere, and the blowby gases recycled into the intake manifold. The *positive crankcase ventilation* (PCV) *valve* is a small check valve that prevents buildup of pressure in the crankcase.

The exhaust accounts for about 60% of the emitted hydrocarbons and almost all the NO, CO and lead, and poses the most difficult problem of controlling emissions from mobile sources. Exhaust emissions depend on the engine operation. During acceleration, the combustion is efficient, CO and HC are low, and high compression produces a lot of NO/NO_2. On the other hand, deceleration results in low NO/NO_2 and high HC because of the presence of unburned fuel in the exhaust. This variation in emissions has prompted EPA to institute a standard acceleration-deceleration cycle for measuring emissions. Testing proceeds from a cold start through acceleration, cruising at constant speeds (on a dynamometer in order to load the engine), and deceleration, and a hot start.

The gasoline-fueled internal combustion engine can be tuned by altering the *air to fuel ratio*. The gasoline (fuel) is mixed with air in the carburetor and this mixture is injected into the cylinders. The ratio of air to fuel determines what the emissions will be. If the mixture is rich, that is, there is a lot of fuel for a little air (air: fuel ratio is low), much of the fuel is poorly combusted and a lot of CO and HC is produced. As the fraction of air is increased (air: fuel ratio increases) the additional oxygen makes the fuel burn hotter, producing less CO and HC, but resulting in higher concentrations of nitrogen oxides, NO/NO_2. If the air: fuel ratio is increased too far, the engine will stall because it does not get enough fuel.

Figure 10.10 shows that no single carburetor setting will result in the lowest values of the three primary pollutants. What is usually done is to run the engine at the leanest mixture possible (highest air: fuel ratio, almost at the stall point) to gain the best efficiency from the fuel, and then use emission controls to remove the pollutants.

A wide range of acceptable engine modifications is possible. Injection of water can reduce NO emissions, and fuel injection (bypassing or eliminating the carburetor) can reduce CO and HC emissions. Fuel injection is not compatible with water injection, however, since water may clog the fuel injectors. The stratified charge engine operates on a very lean air/fuel mixture, thus reducing CO and HC, but does not increase NO appreciably. The two compartments of the engine (the "stratification") accomplish this result: the first compartment receives and ignites the air/fuel mixture, and the second compartment provides a broad flame for an efficient burn. Better than 90% CO reduction can be achieved by this engine.

Recirculating the exhaust gas through the engine can achieve about 60% reduction of CO and hydrocarbons. The only major modification to an ordinary engine required by exhaust gas recirculation, in addition to the necessary fittings, is a system for cooling the exhaust gas before recirculating it, to avoid heat deformation of the piston surfaces. Exhaust gas recirculation, although it increased the rate of engine wear, was a popular and acceptable emission control method until 1980, but present-day exhaust emission standards require 90% CO control, which cannot be realized by this method.

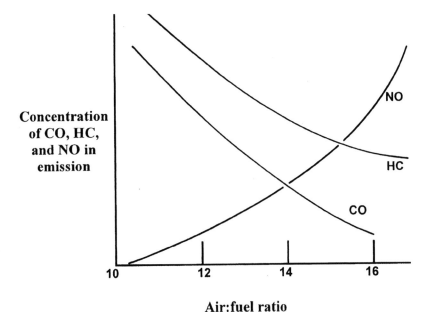

Air:fuel ratio

Figure 10.10. CO, HC, and NO emissions from an internal combustion engine as the air: fuel ratio is altered.

New cars sold in the United States since 1983 have required the use of a *catalytic reactor* ("catalytic converter") to meet exhaust emission standards, and the device is now standard equipment on new cars. The modern three stage catalytic converter performs two functions: *oxidation* of CO and hydrocarbons to CO_2 and water, and *reduction* of NO to N_2. A platinum-rhodium catalyst is used, and reduction of NO is accomplished in the first stage by burning a fuel-rich mixture, thereby depleting the oxygen at the catalyst. Air is then introduced in the second stage and CO and hydrocarbons are oxidized at a lower temperature. The reduction of NO to nitrogen gas N_2 occurs in the third reactor.

Catalytic converters are rendered inoperable by inorganic lead compounds, so that cars using catalytic converters require the use of unleaded gasoline, and this requirement prompted that change from leaded fuel to non-leaded gasoline.

10.2.2 Alternatives to the Gasoline-Fueled Internal Combustion Engine

Diesel engines produce the same three major pollutants as gasoline engines, although in somewhat different proportions. In addition, diesel-powered heavy duty vehicles produce annoying black soot, essentially unburned carbon.

Control of diesel exhaust was not required in the U.S. until passage of the 1990 Clean Air Act (nor is it required anywhere else in the world).

Natural gas may be used to fuel cars, but the limited supply of natural gas serves a number of competing uses. A changeover is not feasible at present. In addition to a steady supply, a changeover would require a different refueling system from the one used for gasoline. Electric cars are clean, but can store only limited power and have limited range. Generation of the electricity to power such cars also generates pollution, and the world's supply of battery materials would be strained to provide for a changeover to electric cars.

Hybrid cars have two powerplants, an electric motor and a standard gasoline engine. The advantage comes from using the kinetic energy of the car when coasting or running downhill to charge the electric batteries, thus eliminating the need to plug the car into an electric outlet for charging and simultaneously allowing for higher speeds on highways.

The 1990 Clean Air Act requires that cities in violation of the National Ambient Air Quality Standards sell oxygenated fuel during the winter months. Oxygenated fuel is gasoline containing 10% ethanol (CH_3CH_2OH) and is intended to bring about somewhat more efficient conversion of CO to CO_2. Its efficacy in cleaning up urban air remains to be seen.

10.3 AIR POLLUTION LAW

The mood of the United States Congress in the 1960s was to pass national legislation for promoting clean air, and it decided to write "technology driving" legislation restricting auto emissions. There were no cars that could meet these strict standards, but the Congress argued that there will not be any such cars unless emissions were controlled. Predictably, the American auto industry bemoaned its inability to meet these new standards and insisted that cars would become so complex and expensive as to be essentially undriveable. Much of their bombast was blunted when the very next year Honda came out with a vehicle that not only met all of the stringent exhaust requirements, but also got 40 miles to a gallon of gas!

The 1963 Clean Air Act, a major piece of legislation that for the first time anywhere set both emission and ambient air quality limits, was passed in great part due to the efforts of Senator Edmund Muskie of Maine. The Act required the federal government to set ambient air quality standards for seven major air pollutants. These standards were not implemented until 1970, and the states were allowed until 1975 to meet them. Many states have yet to meet these standards, and the city of Los Angeles may eventually have to go to extreme measures to reduce its photochemical smog.

The 1990 amendments to the Clean Air Act added over 180 hazardous materials (now know as air toxics) to the ambient air quality standards and requires significant cuts in the emission of sulfur and nitrogen oxides, precursors of acid rain. The act also extends the deadline for meeting the ambient ozone standard until the year 2007—which makes it 44 years since the original 1963 Clean Air Act!

Under Section 111 of the Clean Air Act, U. S. EPA has the authority to set emission standards (called *performance standards* in the Clean Air Act) only for new or markedly modified sources of the major pollutants. The states may set performance standards for existing sources and have the authority to enforce EPA's new source performance standards. New source performance standards for larger coal burning power plants, for example are:

- 1.0 lb particulates/million Btu
- 1.2 lb sulfur oxides/million Btu
- 0.7 lb nitrogen oxides/million Btu
- 20% opacity of the plume based on the Ringlemann scale.

Categories of industrial facilities that emit these listed pollutants must use removal technologies if the facility emits 10 tons per year of any single hazardous substance or 25 tons per year of any combination of hazardous substances. Typical categories of industrial sources include:

- Industrial external combustion boilers
- Printing and publishing
- Waste oil combustion
- Gasoline marketing
- Glass manufacturing

Each major source, defined by the 10 ton/yr or 25 ton/yr limit, must achieve a maximum achievable control technology for emissions. On a case by case basis, a state could require a facility to install control equipment, change an industrial or commercial process, substitute materials in the production process, change work practices, and train and certify operators and workers.

EPA also has the authority to set national emission, or performance, standards for hazardous air pollutants for all sources of those pollutants. The new Clean Air Act requires that EPA develop a list of substances to be regulated, and substances can be added to the list and removed from the list as a result of ongoing research. A substance that is so identified is said to be *listed*. The list currently contains about 170 pollutants, some of the compounds included commonly used substances such as ethylene glycol (anti-freeze), styrene (a plastic constituent), and methanol.

The emission standards are regulated (and for the most part set) by individual state air quality offices. As an example of an emission standard, an incinerator might not be allowed to exceed emissions of 0.01 g/m^3 of particulates at a specified CO_2 minimum level. The latter is necessary since particulate concentrations can be reduced by simply diluting with excess air, but since air contains negligible CO_2, the minimum CO_2 level prevents emission standard attainment by simple dilution.

The emissions from cars, trucks, and buses are regulated under the 1990 Clean Air Act. The present standards are:

- 1.7 g CO/mile
- 0.2 g NO_x/mile
- 0.125 g non-methane hydrocarbons/mile

Standards allowing slightly higher emissions for these pollutants, as well as a standard of 0.08 grams of particulate matter per mile, are set for light-duty trucks. Hazardous air pollutants emitted from vehicles, particularly benzene and formaldehyde, are also regulated.

The second type of air quality standards parallel the stream standards in water quality, and specify the minimum quality of ambient air. These are the *National Ambient Air Quality Standards* (NAAQS) and historically have been the focus of the nationwide strategy to protect air quality. The primary NAAQS are intended to protect human health; the secondary NAAQS, to protect welfare. The latter levels are actually determined as those needed to protect vegetation. Some of these standards are listed in Table 10.1.

NAAQS are set on the basis of extensive collections of information and data on the effects of these air pollutants and human health, ecosystems, vegetation and materials. These documents are called *criteria documents* by the EPA, and the pollutants for that NAAQS exist are sometimes referred to as *criteria pollutants*. Available data suggest that all criteria pollutants have some threshold below which there is no damage.

Under the Clean Air act, most enforcement power is delegated to the states by the EPA, and the states must show the EPA that they can clean up the air to the levels of the NAAQS.

Areas in the United States where the national ambient air quality standards are exceeded more than once or twice a year for any of the pollutants are known as a *non-attainment areas* for those pollutants. In such areas, air pollution control programs must be initiated to bring the area back into compliance, and industries contemplating expansion must show how they can improve the air quality by reducing other emissions. In some areas, automobiles will be required to either change fuels to reduce emissions or travel restrictions on the use of private automobiles will be initiated. Since the internal combustion engine is the single largest contributor of air pollution, a replacement that uses hydrogen or electricity would be a significant contribution to cleaner air.

The way the law was written, the U. S. EPA would step in only if the NAAQS are exceeded. This means that industries could locate where the air is

TABLE 10.1. Selected National Ambient Air Quality Standards.

Pollutant	Primary Standard ppm	Primary Standard $\mu g/m^3$	Secondary Standard ppm	Secondary Standard $\mu g/m^3$
Particulate matter less than 10 μm in diameter: PM_{10}				
24-hr average		150		150
annual geometric mean		50		50
Sulfur Dioxide				
3-hr maximum	0.5	1300	0.02	60
24-hr average	0.14	365	0.1	260
annual arithmetic mean	0.03	80		
Nitrogen Oxides				
annual arithmetic mean	0.053	100	same as primary	
Carbon Monoxide				
1 hr maximum	35	40,000	same as primary	
8 hr maximum	9	10,000		
Ozone				
1 hr maximum	0.12	210	same as primary	
Lead				
quarterly arithmetic mean		1.5	same as primary	

clean and be free to pollute until the NAAQS limits were reached, resulting in a situation where there would be no clean air left in the USA. In 1973, the Sierra Club sued the EPA for failing to protect the cleanliness of the air in those parts of the United States where the air was cleaner than the NAAQS, and won. In response, EPA is now mandated to prevent significant deterioration in locations where the air quality is better than the minimums set by the NAAQS. For this purpose, the United States is divided into class I and class II areas, with the possibility of class III designation for some areas. Class I includes the so-called "mandatory Class I" areas—all national wilderness areas larger than 5,000 acres and all national parks and monuments larger than 6,000 acres—and any area that a state or Native American tribe wishes to designate Class I. The rest of the United States is Class II, except that a state or tribe may petition the EPA for redesignation of a Class II area to Class III. To date, the only pollutants covered by this program are sulfur dioxide and particulate matter.

10.4 SYMBOLS

A = area through which this flow occurs
C = concentration of the pollutant in the clean air stream
C_o = concentration of the pollutant in the dirty air stream
d = diameter of the particles

N = number of turns the air makes in the tube section of the cyclone
Q = air flow rate
R = removal of a pollutant, percent
v = air velocity
W = inlet width
μ = viscosity of the fluid, usually air, kg/m-hr
π = 3.14
ρ = density of the particles, usually as kg/m^3

10.5 PROBLEMS

10.1 Taking into account cost, ease of operation, and ultimate disposal of residuals, what type of air pollution control device would you suggest for the following emissions?

a. A dust with particle range of 5 to 10μ?

b. A gas containing 20% SO_2 and 80% N_2?

c. A gas containing 90% HC and 10% O_2?

10.2 A stack emission has the following characteristics: 90% SO_2, 10% N_2, no particulates. What treatment device would you suggest and why?

10.3 An industrial emission has the following characteristics: 80% N_2, 19% O_2, and 1% CO_2 and no particulates. You are called in as a consultant to advise on the type of air pollution control equipment required. What would be your recommendation?

10.4 A new coal-fired power plant is to burn a coal of 3 percent sulfur, and a heating value of 11,000 Btu/lb. What is the minimum efficiency that a SO_2 scrubbing device will need to have in order to meet the new source emission standard of 1.2 lb SO_2/10^6 Btu?

10.5 Why can't the gasoline engine be tuned so that it results in the minimum production of all three pollutants, CO, HC and NO_x simultaneously?

10.6 What is the particulate removal efficiency of a cyclone if it has to remove particles of diameter

a. 100 microns

b. 10 microns

c. 1 micron

d. 0.1 microns?

CHAPTER 11

Pollution-Induced Atmospheric Changes

A N important approach to classification of air pollutants is that of primary and secondary pollutants. A primary pollutant is defined as one that is emitted as such to the atmosphere, whereas secondary pollutants are actually produced in the atmosphere by chemical reactions. In this chapter four examples of secondary pollutants and their effect on regional as well as global atmospheres is discussed.

11.1 PHOTOCHEMICAL SMOG

The components of automobile exhaust are particularly important in the formation of secondary pollutants. The well known and much discussed Los Angeles smog is a case of secondary pollutant formation. Photochemical smog starts with a high-temperature, high-pressure combustion such as occurs within the cylinders of an internal combustion (gasoline) engine. The first reaction is

$$N_2 + O_2 \xrightarrow{\text{high temperature and pressure}} 2NO$$

which immediately oxidizes to NO_2, as

$$2NO + O_2 \rightarrow 2NO_2$$

Nitrogen dioxide, NO_2, is very active photochemically, and breaks apart as the first reaction in the sequence of photochemical smog formation, as

$$NO_2 + hv \rightarrow NO + O$$

and the atomic oxygen molecule, being quite unstable, reacts as

$$O + O_2 + M \rightarrow O_3 + M$$

where M is some radical that catalyzes the reaction and hv is a symbol for sunlight. The ozone thus created is a strong oxidant, and it will react with whatever it finds to oxidize. From the previous reaction, NO is available, and thus

$$O_3 + NO \rightarrow NO_2 + O_2$$

and we are back where we started, with NO_2. Since all of these reactions are fast, there is no way that high levels of ozone could build up in the atmosphere. This puzzled early air pollution researchers, until Arie Haagen-Smit of CalTech came up with the answer. The answer is that one of the other constituents of polluted urban air is hydrocarbons, and some of these can oxidize the NO to NO_2, leaving ozone in excess. For example,

$$HCO_3^o + NO \rightarrow HCO_2^o + NO_2$$

where HCO_3^o and HCO_2^o denote general hydrocarbon radicals. This explanation allowed for the possibility of ozone buildup as well as the formation of other components of photochemical smog.

Ozone is an eye irritant and can cause severe damage to plants and to materials such as rubber. One of the major types of hydrocarbons produced in photochemical smog is peroxyacetyl nitrates (PAN) which are also effective in damaging crops and materials. Other hydrocarbons can cause breathing problems and also damage plants. And finally, NO_2 itself is an irritant and has a brown/orange color. Anyone who has ever flown into the Los Angeles airport on a sunny day would have seen the dramatic contrast between the blue sky above and the orange crud below.

Photochemical smog is possible if a series of conditions exist:

- high-temperature/high-pressure combustion of fossil fuels such as gasoline
- plentiful sunlight
- stable atmospheric conditions which allow time for the reactions to occur
- hydrocarbon emissions which scavenge the NO from the mix and allow for the buildup of ozone.

Cities in the United States that have these conditions include some of the largest urban areas such as Los Angeles, Denver, and Houston, and even rapidly growing areas such as Raleigh-Durham, North Carolina.

Reducing photochemical smog has been a challenge to environmental scientists and managers for many years. Different schemes for reducing the smog in Los Angeles, for example, have been suggested, such as boring a huge tunnel through the surrounding mountains and blowing the polluted air into the valleys on the other side. Never mind what the people on the other side of the

Los Angeles as a Cupola for Smog Formation

The unfortunate terrestrial and meteorological conditions in Los Angles are a major factor in the dismal air quality there. The potential problem was first recognized and recorded during the era of the Spanish explorers.

In October, 1542, the Spanish explorer and adventurer Juan Rodriquez Cabrillo dropped anchor in what is now San Pedro Bay, the harbor of present day Los Angeles. Cabrillo almost certainly was impressed by the spectacular setting, with mountains on three sides. What really astonished him, though, according to his diary, was a ghostly sight on shore: smoke from Indian campfires rising perpendicularly for a few hundred feet, then, as if striking an invisible ceiling, spreading horizontally over the valley. Cabrillo thus named the harbor the Bay of Smoke.

mountain might say. Also, rough calculations show that a tunnel large enough to be able to exchange the air over Los Angeles needs to be huge, so other solutions have been sought, the most reasonable being not producing NO_2 in the first place, by reducing the number of cars in the city and/or clearing up their emissions. But the cars are already very clean, and the people of LA are not keen on giving up their driving privilege, so this solution is not useful. Another is to replace all the gasoline-powered vehicles with electric cars, also a non-starter at least for the next few decades. As of this writing, there is no solution in sight.

11.2 ACID RAIN

Normal, uncontaminated rain has a pH of about 5.6. This low pH can be explained by its adsorption of carbon dioxide, CO_2. As the water droplets fall through the air, the CO_2 in the atmosphere dissolves in the water, setting up an equilibrium:

$$CO_2(\text{gas in air}) \rightarrow CO_2 \text{ (dissolved in the water)}$$

The CO_2 in the water reacts to produce hydrogen ions, as

$$CO_2 + H_2O \rightarrow H_2CO_3 \rightarrow H^+ + HCO_3^-$$

ARIE HAAGEN-SMIT

In the 1940s, as the number of clear days in Southern California became fewer and fewer, concern was being expressed as to what the cause of this bad visibility might be. But during the Second World War, little was done and the situation became progressively worse. When a butadiene plant was built in downtown Los Angeles, its severe upsets produced noxious fumes that caused office buildings to be evacuated. Although we now know that the butadiene plant had little to do with the poor visibility, the event caused a public outcry and a demand to do something about the air pollution. Experts were sent for, and they came with their instruments for measuring SO_2 and smoke, and these soon proved worthless since the levels of sulfur oxides were very low and certainly these could not be the cause of the bad air quality. In 1947 the County of Los Angeles got the police power to do something about the air pollution, and the LA County Air Pollution District was formed. Early efforts to reduce air pollution in LA all still concentrated on the reduction of SO_2 from such sources as backyard burning, but all to no avail. Finally, it was decided that research was necessary.

Arie Haagen-Smit (1900–1977), a Dutch-born biologist working at CalTech on the fumes emitted by pineapples, decided to distill the contents of the Los Angeles air, and discovered peroxy-organic substances, which were no doubt the cause of the eye irritation. The source of these pollutants, if this was true, had to have been the gasoline-powered automobile. When Haagen-Smit published this research, a fire storm of protest erupted, led by the automobile and oil industries. Scientists at the Stanford Research Institute, which had been doing smog research on behalf of the transportation industry, presented a paper at CalTech essentially accusing Haagen-Smit of bad science. This made him so angry that he abandoned his pineapple work and he began to work full time on the smog problem.

Some of his previous research had been on the damage caused by ozone on plants, and he discovered that the effect of automobile exhaust on plants produced a similar injury, suggesting that the exhaust contained ozone. But ozone was not emitted by automobiles, so where did it come from? He finally hit on the idea of mixing automobile exhaust with hydrocarbons in a large air chamber and subjecting the mixture to strong light, in effect modeling the atmosphere over the city. With this experiment he was able to demonstrate that ozone is not emitted directly from any source, but is formed by reactions in the atmosphere, and the ozone in the LA smog was created by reactions that began with the oxides of nitrogen in automobile exhaust.

The powerful automobile and gasoline industries denied that this could be the cause of the smog, and suggested instead that the ozone had to have entered the smog from some other source. They even speculated that the ozone had somehow descended from the stratosphere, which of course was nonsense. Haagen-Smit's courageous work, taking on the most powerful political forces in California, paved the way for an eventual (but not yet present!) solution to the smog problem in Los Angeles.

$$HCO_3^- \rightarrow 2H^+ + CO_3^{2-}$$

Thus the more CO_2 dissolves in the rain, the lower the pH will be. But this reaction will, at equilibrium, produce rainwater with a pH of 5.6, and no lower. How is it then that some rain can have a pH of 2 or even lower?

This pH reduction occurs as the result of interactions with air contaminants. For example, sulfur oxides produced in the burning of fossil fuels (especially coal) is a major contributor to low pH in rain. In its simplest terms, SO_2 is emitted from the combustion of fuels containing sulfur, the reaction being

$$S + O_2 \xrightarrow{\text{heat}} SO_2$$

$$SO_2 + O \xrightarrow{\text{sunlight}} SO_3$$

$$SO_3 + H_2O \rightarrow H_2SO_4 \rightarrow 2H^+ + SO_4^{2-}$$

Sulfur oxides do not literally produce sulfuric acid in the clouds, but the idea is the same. The precipitation from air containing high concentrations of sulfur oxides is poorly buffered and readily drops its pH.

Nitrogen oxides, emitted mostly from automobile exhaust but also from any other high temperature combustion sources, contribute to the acid mix in the atmosphere. The chemical reactions that apparently occur with nitrogen are

$$N_2 + O_2 \rightarrow 2NO$$

$$NO + O_3 \rightarrow NO_2 + O_2$$

$$2NO_2 + O_3 + H_2O \rightarrow 2HNO_3 + O_2 \rightarrow 2H^+ + 2NO_3^- + O_2$$

where HNO_3 is of course nitric acid.

The effect of acid rain has been devastating. Hundreds of lakes in North America and Scandinavia have become so acidic that they no longer can support fish life. In a recent study of Norwegian lakes, more than 70% of the lakes having a pH of less than 4.5 contained no fish, and nearly all lakes with a pH of 5.5 and above contained fish. The low pH not only affects fish directly, but contributes to the release of potentially toxic metals such as aluminum, thus magnifying the problem.

In North America, acid rain has already wiped out all fish and many plants in 50% of the high mountain lakes in the Adirondacks. The pH in many of these lakes has reached such levels of acidity as to replace the trout and native plants with acidtolerant mats of algae.

The deposition of atmospheric acid on freshwater aquatic systems prompted EPA to suggest a limit of from 10 to 20 kg SO_4^- per hectare per year of deposition from the atmosphere. If "Newton's law of air pollution" is used

The Mystery of the Disappearing Fish

In the 1960s, the mountain lakes of Norway were full of cold-water game fish, but the populations of these fish seemed to be decreasing. Studies conducted by the Norwegian Institute for Water Research showed a steady decline in fish. The lakes were in remote areas, far away from sources of pollution, and various non-pollution causes were considered for the declining fish populations. Using water quality samples, the Institute found that the pH in the lakes had been steadily decreasing. More importantly, rain gauges set up around the lakes showed a wide variation in the pH of the rain. But where was this acid rain coming from? The mystery was solved by tracking the paths of storms that produced particularly low pH rainwater and determining where these storms had been prior to dumping the rainwater in Norway. The map plots the paths of three of these storms providing incontrovertible evidence that the pollution originated in the industrial regions of middle Europe.

Normal rainwater pH in Kristiansand is about 5.8. The storm of 25 August had a pH of 4.2, the storm of 26 August had a pH of 4.0, and the storm of 27 August had a pH of 4.1. All storms passed over highly industrialized areas.

Knowing what was happening to the fish in Norwegian lakes was one thing, of course, and quite another was trying to do something about it. Pollution across political boundaries is a particularly difficult regulatory problem. The big stick of police power is no longer available. Why *should* the U.K. worry about acid rain in Scandinavia? Why *should* the Germans clean up the Rhine before it flows through The Netherlands? Why *should* Israel stop taking water out of the Dead Sea, which it shares with Jordan? Laws are no longer useful, and threats of retaliation are unlikely. What forces are there to encourage these countries to do the right thing? Is there such a thing as "international ethics"?

(what goes up must come down), it is easy to see that the amount of sulfuric and nitric oxides emitted is vastly greater than this limit. For example, just for the state of Ohio alone, the total annual emissions are 2.4×10^6 metric tons of SO_2 per year. If all of this is converted to SO_4^{2-} and is deposited on the State of Ohio, the total would be 360 kg per hectare per year.

But not all of this sulfur falls on the folks in Ohio, and much of it is exported by the atmosphere to places far away. Similar calculations for the sulfur emissions for northeastern United States indicates that the rate of sulfur emission is 4 to 5 times greater than the rate of deposition. Where does it all go?

The Canadians have a ready and compelling answer. They have for many years blamed the United States for the formation of most of the acid rain that crosses their border. Similarly, much of the problem in Scandinavia can be traced to the use of tall stacks in Great Britain and the lowland countries of continental Europe. For years British industry simply built taller and taller stacks as a method of air pollution control, reducing the immediate ground level concentration, but emitting the same pollutants into the higher atmosphere. The air quality in the United Kingdom improved, but at the expense of acid rain in other parts of Europe.

11.3 OZONE DEPLETION

The atmosphere surrounding the Earth is conveniently divided into identifiable layers, as shown in Chapter 8, Figure 8.1. More than 80% of the air mass is in the lowest layer, the troposphere which is between 10 and 12 km deep. The fastest airliners fly at around 38,000 feet, which is about 11.6 km. The mass of air in the *troposphere* varies as the distance from the equator. At the equator, the troposphere is about 18 km deep, while at the poles it is only about 6 km deep. The temperature decreases with elevation, and the air within the troposphere is in motion with significant mixing.

Above the troposphere is the *stratosphere* which is stable and reaches to about 50 km. There is little movement in or out of the layer and anything that gets to that layer can be expected to stay there for a long time. Taken together, the troposphere and the stratosphere account for about 99.9% of all the air, and above this layer is considered "space".

One of the major gases in the stratosphere is ozone. Ozone (O_3) is an eye irritant at usual urban levels, but urban ozone should not be confused with stratospheric ozone. The latter ozone acts as an ultraviolet radiation shield, and there is concern that this protective shield is being destroyed, leaving the earth and its inhabitants vulnerable to the effects of ultraviolet radiation which can result in an increased risk of skin cancer as well as producing changes in the global ecology in unpredictable ways.

Ozone in the stratosphere is created by the reaction of oxygen with light energy, first splitting the oxygen molecule into free oxygen atoms,

$$O_2 + hv \rightarrow O + O$$

where

hv = symbol for light energy.

The important point is that this reaction can take place only if the light energy is at a low wavelength, in the ultraviolet region, below about 400 mn.

Because the oxygen atoms are highly unstable, they will seek out oxygen molecules to form ozone,

$$O + O_2 + M \rightarrow O_3 + M$$

where M is some third body used to carry away the heat generated.

Ozone is in turn destroyed by light energy as ultraviolet radiation, as

$$O_3 + h\nu \rightarrow O_2 + O$$

and the atomic oxygen can then once again react with molecular oxygen to form ozone. The light energy responsible for this reaction has a wavelength between 200 and 320 nm, which is exactly in the middle of the ultraviolet part of the spectrum, thus adsorbing the UV and preventing it from reaching the Earth's surface. If the stratosphere does not contain ozone, the ultraviolet energy will pass through the stratosphere, causing damage to living tissues. In the stratosphere there is a balance with oxygen being broken up, forming ozone, and the ozone being destroyed back to oxygen molecules, all the time using up the energy from the ultraviolet radiation. The presence of the ozone at a steady state concentration is immensely important in protecting all life on Earth from damaging ultraviolet radiation.

The problem with the depletion of upper atmospheric ozone is due to the manufacture and discharge of a class of chemicals called chlorofluorocarbons (CFC). These useful chemicals find wide use in aerosols and refrigeration systems but may be responsible for both global warming as well as the depletion of the protective ozone layer in the stratosphere.

Before the invention of CFCs, refrigeration units used CO_2, isobutene, methyl chloride, or sulfur dioxide, all of which are either toxic, flammable, or inefficient. CFCs are non-toxic, not water soluble, non-reactive, non-biodegradable, and are easily liquefied under pressure. When they evaporate they produce very cold temperatures, making them ideal refrigerants. Two of the most important CFC's are trichlorofluoromethane, $CFCl_3$ (industrial designation CFC-11) and dichlorodifluoromethane, CF_2Cl_2 (industrial designation CFC-12, also known as Freon when manufactured by DuPont.). Other chlorinated refrigerants include hydrochlorfluorocarbon (HCFCs) such as CHF_2Cl (industrial designation HCFC-22) and hydrofluorocarbons (HFCs) that do not contain chlorine. One HFC of importance is CH_2FCF_3 (HFC 134a) which is becoming widely used in automobile air conditioners. Both the HCFCs and HFCs are still potentially ozone-depleting gases, but they can be broken down by sunlight and thus their lifetime in the atmosphere is significantly shorter than the CFCs which end up in the stratosphere, whirling around for hundreds of years.

CFCs are no longer manufactured, although there are many air-conditioning systems that still require CFC and thus the potential for damage by CFCs remains high. The problem is that coolants can escape from the refrigeration units and enter the atmosphere where they are inert and non-water soluble and do not wash out.

The effect of the CFCs on the delicate ozone balance can be devastating. Because the CFCs are so non-reactive and non-soluble, once they have drifted into the atmosphere they stay there for a long time. Eventually they will be broken down by ultraviolet radiation, forming other chlorinated fluorocarbons, and releasing atomic chlorine. For example, CFC-12, a leading refrigerant, breaks down as

$$CF_2Cl_2 + h\nu \rightarrow CF_2Cl + Cl$$

The atomic chlorine acts as a catalyst in breaking down ozone as

$$Cl + O_3 \rightarrow ClO + O_2$$

$$ClO + O \rightarrow Cl + O_2$$

with atomic chlorine again forming to continue to promote more destruction of ozone. The atomic oxygen, meanwhile, can also help in destroying ozone as

$$O + O_3 \rightarrow 2O_2$$

Thus a single Cl atom can make thousands of loops until it eventually reacts with something like methane and is chemically tied up. The reaction with methane produces HCl and ClO, which can react with nitrogen dioxide (another air pollutant) as

$$ClO + NO_2 \rightarrow ClONO_2 \text{ (chlorine nitrate)}$$

The chlorine in HCl and $ClONO_2$ is trapped and can no longer enter into the above reactions. The problem is that under certain conditions, both of these chemicals, HCl and $ClONO_2$, can break up and re-release the chlorine so it can continue to do damage. These reactions occur when there are surfaces available for the reaction to take place, and these surfaces are provided by molecules such as sulfate aerosols. The breakup of HCl and $ClONO_2$ plays an important role in understanding the "ozone hole" over the Antarctic.

During the winter months there exists a polar vortex, formed as a whirling mass of very cold air. This vortex is so strong that it isolates air above the pole from the rest of the atmosphere. This exceedingly cold (−90°C) air forms clouds of ice crystals in the stratosphere. On the surfaces of these crystals the "inert" chlorinated compounds react as

$$ClONO_2 + H_2O \rightarrow HOCl + HNO_3$$

$$HOCl + HCl \rightarrow Cl_2 + H_2O$$

$$ClONO_2 + HCl \rightarrow Cl_2 + HNO_3$$

These reactions tie up the chlorine on the ice crystals and it cannot enter into the reactions to break up ozone. But when spring comes, this large quantity of stored chlorine suddenly becomes available, and light energy breaks it apart, as

$$Cl_2 + h\nu \rightarrow 2Cl$$

The sudden dumping of all that chlorine into the stratosphere results in a high rate of ozone destruction, producing the "hole" in the ozone layer. In later spring (November) air from other parts of the globe rushes in, replenishing the ozone and reducing the size of the "hole". The effect of the annual depletion is strongly felt in countries such as Australia and New Zealand, which are close to the Antarctic continent.

The size of the ozone hole has been increasing over the years, prompting increasing concern about the effect of ultraviolet radiation on the Earth. Calculations show that a 1% increase in UV radiation can result in a 0.5% increase in melanoma, a particularly aggressive form of skin cancer, and a 2.5% increase in other forms on skin cancer. Epidemiological statistics show that melanoma is in fact increasing at a rate of 2 to 3% annually in the United States.

Believe Thine Own Data: The Story of the Ozone Hole

Ozone levels in the stratosphere have been measured since the 1970s using satellites. In addition to satellite data, a British team of scientist in the Antarctic began to measure the concentration of ozone in the stratosphere using sophisticated instruments at the base near the pole. In 1985 they discovered that there had been a dramatic drop in the ozone levels in the stratosphere, creating a large hole over the south pole. The satellite data however did not show such a hole, and the scientists concluded that their instruments must be faulty. They had a new set of instruments flown in, and even with careful calibration, they had to conclude that this huge hole had indeed developed. They studied the satellite data and found that the software was designed to ignore very low levels of ozone, and the software reported this simply as "no data". The satellite had seen this hole for many years, but the data reduction did not recognize the presence of the hole. When the scientists adjusted the program to report low levels of ozone, the satellite data agreed with the data obtained on the ground. There indeed was a huge hole right in the middle of the South Pole's atmosphere.

F. SHERWOOD ROWLAND   **MARIO J. MOLINA**

Mario Molina (1943—) was born and raised in Mexico. Following his graduation from university with a degree in physical chemistry, he went to Berkeley where he did research on photochemical reactions. In 1973 he joined the research group headed by Sherwood Rowland (1927—) at the University of California at Irvine. Rowland had come to Irvine after an undergraduate degree from Ohio Wesleyan and a PhD in chemistry from the University of Chicago, and offered several research opportunities to Molina. Molina fortunately chose a little-known problem of understanding the fate of chlorofluorocarbons in the atmosphere. In 1972, Rowland had heard a talk by the British scientist James Lovelock about detecting these chemicals in the atmosphere, but not understanding their fate. At first the study was only one of scientific curiosity, but soon they began to realize that the presence of CFCs in the stratosphere would have profound environmental consequences. Other scientists had recognized the effect of chlorine on stratospheric ozone, but none had shown that the CFCs would have a dramatic impact on the ozone concentration. Molina and Rowland published their findings in *Nature* in 1974 and immediately became the targets of severe scientific criticism from industrial interests. The two scientists persevered, however, and went to great lengths to publicize their results and to testify at congressional hearings. Finally the scientists at DuPont, the largest manufacturer of CFCs, acknowledged that Molina and Rowland were correct, and pledged to cease the manufacture of these compounds.

In 1974, Rowland and Molina, along with Paul Crutzen of the Max Planck Institute were awarded the Nobel Prize in chemistry for their work in understanding the chemical processes involved in ozone depletion.

It must be pointed out, however, that much of this increase might be explained by a longer life span, more outdoor activities and recreation.

UV radiation can also cause eye damage and can suppress the immune system in humans. A very troublesome effect of UV radiation is the suppression of photosynthesis in aquatic plants. A slower rate of photosynthesis would result in a higher atmospheric concentration of CO_2, exacerbating the problems with global warming.

The story of the ozone layer and CFCs is one of two environmental researchers who published a courageous paper in 1974 suggesting that the depletion of ozone was possible as a direct result of escaped refrigerants. Industrial interests at first contested these findings, and it might never have become a global concern were it not for the discovery of the Antarctic ozone hole. From that point, the international concern forced action, culminating in the *Montreal Protocol on Substances that Deplete the Ozone Layer*, with 23 nations including the United States agreeing to cut the use of CFCs by 50% by 1999 and to eventually cease production totally. In 1988 DuPont, the largest manufacturer of CFCs, stopped making them, prompting a huge industry of smuggling these refrigerants into the United States, since many cooling systems depended on them. In 1990, the availability of new refrigerants, and the growing concern with the ozone hole in the Antarctic prompted a revised schedule, with a new timetable calling for a complete phaseout of CFCs by 2000, with a few exceptional chemicals needed for medical and other uses to be phased out by 2010. A convention in Copenhagen in 1992 again accelerated the phaseout, with the production and importation of CFCs banned as of 1994. The U. S. EPA has responded to these protocols by banning non-essential use of CFCs, such as noise horns, requiring automobile mechanics to save and reuse CFC in air conditioners, and mandating the removal of CFCs from cars headed for demolition.

The Montreal Protocol was mainly the product of both the United Kingdom's and the United States' leadership, and was possible because the two largest manufacturers of refrigerants, DuPont in the United States and ICI in the United Kingdom, already had developed alternative coolants. The Protocol was therefore welcomed by these industries since it represented a way of increasing their market share in these chemicals.

11.4 GLOBAL WARMING

The first question that comes to mind in discussing global warming is "is it happening?" Is the earth really getting warmer? Quite obviously, the average temperature of the earth is difficult to measure, but all techniques lead to the same answer: yes, the Earth is getting warmer. This is a very small overall change that would not be detectable to humans due to short-term and regional variations. Overall, however, there seems to be no doubt that the temperature of the Earth is increasing.

The second question, providing we have convinced ourselves that the earth is in fact getting warmer, is what might be causing this increase in temperature.

One explanation is that we are simply in a natural cycle of temperature fluctuations, such as have occurred on the Earth for hundreds of thousands of years. Studies of ice cores in Russia have shown that, amazingly enough, the mean temperature of the Earth has not changed in over 200,000 years! There have been wide fluctuations such as the ice ages, but on balance, the mean temperature has remained constant, prompting some scientists to suggest whimsical causes for such consistency.

There is another explanation for the sudden (relative to geological time) increase we are presently experiencing. It is possible that certain gases in the atmosphere are causing the Earth to not be able to reflect enough heat energy from the sun back into space. The earth acts as a reflector of the sun's rays, receiving the radiation from the sun, reflecting some of it into space (called *albedo*) and adsorbing the rest, only to reradiate this into space as heat. In effect the earth acts as a wave converter, receiving the high energy high frequency radiation from the sun and converting most of it into low-energy low-frequency heat to be radiated back into space. In this manner, the earth maintains a balance of temperature.

In order to better understand this balance, the light energy and the heat energy have to be defined in terms of their radiation patterns, as shown in Figure 11.1. The incoming radiation (light) wavelength has a maximum at around 0.5 nm and almost all of it is less than 3 nm. The heat energy spectrum, or that energy reflected back into space, has a maximum of about 10 nm and almost all or it at a wavelength higher than 3 nm.

As both the light and heat energy pass through the earth's atmosphere, they encounter the aerosols and gases surrounding the earth. These can either allow the energy to pass through, or can interrupt it by scattering or absorption. If the atoms in the gas molecules vibrate at the same frequency as the light energy, they will absorb the energy and not allow it to pass through. Aerosols will scatter the light and provide a "shade" for the earth.

The absorptive potential of several important gases is shown in Figure 11.2, along with the spectra for the incoming light (short-wavelength) radiation and the outgoing heat (long-wavelength) radiation. The incoming radiation is impeded by water vapor and oxygen and ozone, as discussed in the preceding section. Most of the light energy comes through unimpeded.

The heat energy, however, encounters several potential impediments. As it is trying to reach outer space, it finds that water vapor, CO_2. CH_4, O_3, and N_2O all have absorptive wavelengths right in the middle of the heat spectrum. Quite obviously, an increase in the concentration of any of these will greatly limit the amount of heat transmitted into space. These gases are appropriately called *greenhouse gases*, because their presence will limit the heat escaping into space, much like the glass of a greenhouse or even the glass in your car limits the amount of heat that can escape, thus building up the temperature under the glass cover.

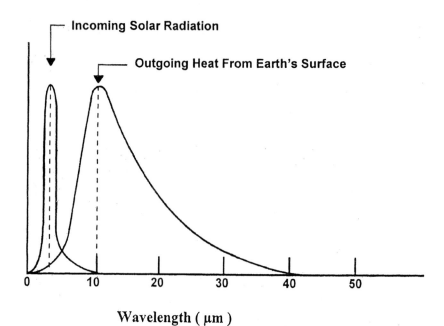

Wavelength (μm)

Figure 11.1. Radiation patterns for heat and light energy.

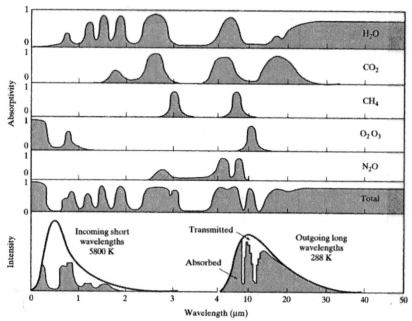

Figure 11.2. Adsorptive potential of several important gases in the atmosphere. Also shown are the spectra for the incoming solar energy and the outgoing thermal energy from the earth. Note that the wavelength scale changes at 4 μm. (Courtesy of Gilbert Masters, *Introduction to Environmental Engineering and Science* Prentice Hall, Englewood Cliffs NJ, 1998.)

380

The effectiveness of any one gas to promote global warming (or cooling, as is the case with aerosols) is known as *forcing*. The gases of most importance in forcing are listed in Table 11.1.

Carbon Dioxide

Carbon dioxide is the product of decomposition of organic material, whether biologically or through combustion. The effectiveness of CO_2 as a global warming gas has been known for over 100 years, but the first useful measurements of atmospheric CO_2 were not taken until 1957. The data from Mauna Loa in Hawaii are exceptionally useful since they show that even in the 1950s the CO_2 concentration had increased from the baseline 280 ppm to 315 ppm, and has continued to climb over the last 50 years at a constant rate of about 1.6 ppm per year. The most serious problem with CO_2 is that the effect on global temperature due to its greenhouse effect are delayed. What we have already done will, even if we stopped emitting any new CO_2 into the atmosphere, will increase CO_2 from our present 370 ppm to possibly higher than 600 ppm! That is, if we stopped everything now, we have already contaminated the global atmosphere to where we will have doubled the CO_2 concentration. The effect of this is discussed below.

Methane

Methane is the product of anaerobic decomposition and human food production. One of the highest producers of methane in the world is New Zealand which boasts 80 million sheep. Methane also is emitted during the combustion of fossil fuels and cutting and clearing of forests. The concentration of CH_4 in the atmosphere has been steady at about 0.75 ppm for over a thousand years, and then increased to 0.85 ppm in 1900. Since then, in the space of only a hundred years, has skyrocketed to 1.7 ppm. Methane is removed from the atmosphere by reaction with the hydroxil radical (OH) as

$$CH_4 + OH + 9O_2 \rightarrow CO_2 + 0.5H_2 + 2H_2O + 5O_3$$

But in so doing, it creates carbon dioxide, water vapor, and ozone, all of

TABLE 11.1. Relative Forcing of Increased Global Temperature.

Gas	Percent of Relative Radiative Forcing
Carbon dioxide, CO_2	64
Methane, CH_4	19
Halocarbons (mostly CFCs)	11
Nitrous Oxide, N_2O	6

The Gaia Hypothesis

The most remarkable thing about the global temperature might be that it has been so constant over millions of years. Somehow the Earth has been able to develop the thermal conditions suitable for the creation and support of life. Speculating on this idea, James Lovelock, a noted biologist and ecologist, suggested (tongue in cheek) that there must be an earth goddess, Gaia, who watches over the earth. Gaia has been able to compensate for changing conditions and has been able to fight off diseases in order to keep the Earth a healthy place to live. The name Gaia comes from the Greek name for a nurturing earth goddess.

Much to Lovelock's chagrin, many people took this to be the truth, and a new religion was born. Some Gaians have interpreted this notion in its broadest and most spiritual sense, taking the view that the earth really is one organism, albeit an unusual one; and it has many of the characteristics of other organisms. Others see the hypothesis as nothing but a feedback response model, and do not ascribe mysticism to it.

But suppose humans are simply one part of a whole living organism, the Earth, much like brain cells are a part of the animal. If this is true, then it makes no sense to destroy one's own body, and therefore it makes no sense for humans to destroy the rest of the creatures that co-inhabit the earth. To think of the earth as a living organism suggests a spiritual rather than a scientific approach, especially if it is taken literally.

If we accept this notion, some interesting implications could result. One could speculate (and this is pure whimsy!) that the earth (Gaia) is still developing, and is going through adolescent stages. Most notably, it has not settled on the carbon balance. Millions of years ago, most of the carbon on earth used to be in the atmosphere as carbon dioxide, and the preponderance of CO_2 promoted the development of plants. But as the plants grew, they soon started to rob the atmosphere of carbon dioxide, replacing it with oxygen. This change in atmospheric gases promoted the growth of animals which converted the oxygen back to carbon dioxide, once again seeking a balance. Unfortunately, Gaia made a mistake and did not count on the animals dying in such huge numbers, trapping the carbon in deep geologic deposits which eventually became coal, oil and natural gas.

How to get this carbon out? What was needed was a semi-intelligent being with reversible digits. So Gaia invented humans! It can therefore be argued that our sole purpose on earth is to dig up the carbon deposits as rapidly as possible and to liberate the carbon dioxide. This is the long-sought-for "meaning of life."

Unfortunately, Gaia again miscalculated, and did not count on the humans becoming so prolific and destructive, and especially did not count on humans inventing nuclear power, which eliminates their sole purpose for existing. The sheer numbers of humans (especially if they are not going to dig up the carbon) became a problem, much as a pathogenic bacterium causes an infection in an organism. The growth of the human population on Earth is like the growth of cancer cells in an organism.

So next, Gaia has to limit the number of humans, and she will do so by developing "antihumatics" (as in antibiotics) that will kill off a sufficient number

(continued)

(continued)

of humans and once again bring the earth to a proper balance. The increase in highly-resistant strains of bacteria and viruses may be the first indications of a general culling of the human population.

The application of the Gaia hypothesis to environmental ethics leads to existentialism—the notion that life is meaningless (except as a burner of fossil fuel) and that one should not concern oneself with the environment.

---

which are greenhouse gases, so the effect of one molecule of methane enhances the greenhouse effect.

Halocarbons

The same gang of suspects in the destruction of atmospheric ozone is also at work in promoting global warming. The most effective global warming gases are CFC-11 and CFC-12, both of which are no longer manufactured, and the banning of these substances has led to a leveling off of their presence in the stratosphere.

Nitrous Oxide

Nitrous oxide (laughing gas) is also in the atmosphere, mostly as a result of human activities. The greatest problem with nitrous oxide is that there appear to be no natural removal processes for this gas and so its residence time in the stratosphere is indefinitely long.

The net effect of these global pollutants is still being debated. Atmospheric models used to predict temperature change over the next hundred years vary widely. They nevertheless agree that some positive change will occur, even if we do something drastic today (which does not seem likely). By the year 2100, even if we do not increase our production of greenhouse gases and if the United States signs the Kyoto Accord, which encourages reduction in greenhouse gas production, is followed, the global temperature is likely to be between 0.5 and 1.5°C warmer. This does not seem like much, but the effect of this on natural systems and ocean currents will be devastating. The melting of glaciers and polar ice caps will result in higher sea levels and the extinction of species such as the polar bear.

11.5 PROBLEMS

11.1 Photochemical smog is a serious problem in many large cities.

a. Draw a graph showing the concentration of NO, NO_2, HC and O_3, in the Los Angeles area during a sunny, smoggy day. (concentration vs. time of day)

b. Draw another graph showing how the same curves appear on a cloudy day. Explain the difference.

c. The only feasible way of reducing the formation of photochemical smog in Los Angeles seems to be to prevent automobiles from entering the city. Draw the same curves as they might appear if *all* cars are banned from LA streets.

11.2 An earlier chapter tells the story of how the Spanish Conquistadors first saw the smoke from the Indian campfires in the Los Angeles basin rise and then level off. Why is this observation so important in the subsequent air pollution problems in Los Angeles? Explain using your knowledge of atmospheric stability.

11.3 Ozone is a pollutant in photochemical smog. Why is it that ozone is also so necessary in the stratosphere? Explain.

11.4 Why is rain that falls through unpolluted air acidic (has a low pH)?

11.5 Explain, using the precautionary principle, why global warming is such a concern?

11.6 How are some of the major petroleum companies reacting to global warming? Search the corporate Internet sites for several companies and see if you detect a corporate policy, and if there are differences of opinion.

11.7 One of the results of global warming is that more hurricanes would be generated. Is this happening? Search the U.S. Weather Service web site and produce a plot showing number of hurricanes vs. year.

CHAPTER 12

Noise Pollution

NOISE has been described as the perfect pollutant since it can be eliminated instantaneously with no residuals simply by ceasing the noisemaking. It is also, however, a subtle pollutant, since the effect of noise on our everyday life is far greater than we suspect.

12.1 DEFINING SOUND

We begin this discussion of noise pollution by defining sound. *Sound* is created when pressure waved in the atmosphere reach our ears and we detect these pressure waves as sound. *Noise* is simply unwanted sound. It is sound that irritates or even causes harm. But before we talk about noise we have to understand sound.

Pure sound is described by pressure waves traveling through a medium (air, in almost all cases). These pressure waves are described by their amplitude and their frequency. With reference to Figure 12.1, note that the pure soundwave can be described as a sinusoidal curve, having positive and negative pressures within one cycle. The number of these cycles per unit time is called the sound *frequency*, often expressed as cycles per second, or *Hertz* in honor of the German physicist, Heinrich Hertz. Typical sounds that healthy human ears hear range from about 15 Hz to about 20,000 Hz, a huge range. Low frequency noise is a deep sound, while the high frequency represents high pitched sounds. For example, a middle A on the piano is at a frequency of 440 Hz. Speech is usually in the range of 1000 to 4000 Hz.

The wide frequency spectrum is significantly reduced by age and environmental exposure to loud noise. The most significant sources of such damaging noise come from occupational sources and loud music, particularly rock concerts and earphones turned too high. High-quality sound reproduction equipment is designed to reproduce a full spectrum of frequencies, although

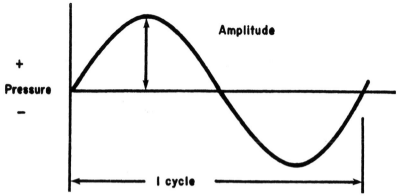

Figure 12.1. Defining sound amplitude and frequency.

advertising that the equipment is capable of less than 10 Hz reproduction is somewhat silly, since most people can only hear down to 20 Hz.

Young, healthy persons (particularly young women) can hear very high frequencies, often including such signals as automatic door openers. With age, and unhappily with the damage so many young people do to their ears, the ability to detect a wide frequency range drops. Older people especially tend to lose the high end of the hearing spectrum, and might start to complain that "everyone is mumbling."

The loudness of a noise is expressed by its *amplitude*. With reference again to Figure 12.1, the energy in a pressure wave is the total area under the curve. Because the first half is a positive pressure and the second is negative, adding these would produce zero net pressure. The trick is to use a *root mean square* analysis of a pressure wave, first multiplying the pressure by itself and then taking its square root. Since the product of two negative numbers is a positive number, the result is a positive pressure number. In the case of sound waves, the pressure is expressed as Newtons per square meter (N/m^2), although sometimes sound pressure is also expressed in bars or atmospheres. In this text we use the modern designation of N/m^2.

The human ear is a phenomenal instrument, being able to hear sound at both a huge frequency range, but at an even more spectacular range of pressures. Human hearing covers a pressure range of over 10^{18}, a fact that makes it difficult to express sound pressure in some meaningful and useful way. What has developed over the years is a convention to use ratios to express sound amplitude.

Psychologists have known for many years that human response to stimuli are not linear. In fact, the ability to detect an incremental change in any response such as heat, cold, odor, taste, or sound, depends entirely on the level of that stimulus. For example, suppose you are blindfolded and are holding a 20 pound weight in your hand, and 0.1 pounds is added to it. You will probably not detect

---

Urban Noise

Noise has been a part of urban life since the first cities. The noise of chariot wheels in Rome, for example, caused Julius Caesar to forbid chariot riding after sunset so he could sleep. In England, the birthplace of the industrial revolution, cities were incredibly noisy. As one American visitor described it:

> The noise surged like a mighty heart-beat in the central districts of London's life. It was a thing beyond all imaginings. The streets of workaday London were uniformly paved in "granite" sets... and the hammering of a multitude of iron-shod hairy heels, the deafening side-drum tattoo of tyred wheels jarring from the apex of one set (of cobblestones) to the next; the sticks dragging along a fence; the creaking and groaning a chirping and rattling of vehicles, light and heavy, thus maltreated, the jangling of chain harness; augmented by the shrieking and bellowings called for from those of God's creatures who desired to impart information or proffer a request vocally—raised a din that is beyond conception. It was not any such paltry thing as noise.

In the United States, cities grew increasingly noisy, and all manner of ordinances were contrived to hold down the racket. In Dayton, Ohio, the city proclaimed that "It is illegal for hawkers or peddlers to disturb the peace and quiet by shouting or crying their wares." Most of these ordinances were of little use and soon disappeared into oblivion, as the noise increased.

Quote from: Still, H. In *Quest of Quiet,* Stackpole Books, Harrisburg PA 1970

---

the change. On the other hand, if you are holding 0.1 pounds in your hand, and another 0.1 is added (a 100% increase), you would be able to detect the change. Thus the ability to detect stimuli can be described on a logarithmic scale.

Since there is no good way to express sound levels in absolute terms, a useful expression would be to compare the given sound level to something else. But what reference to use? It appears that the least pressure a human ear is able to detect is about 2×10^{-5} N/m^2 and this would make a convenient datum. Using 2×10^{-5} N/m^2 as a reference value, the *Sound Pressure Level* is then defined in terms of *decibels*, designated by the symbol dB, as

$$SPL = 20 \log_{10} \frac{P}{P_{ref}} \qquad (12.1)$$

where

SPL = Sound Pressure Level, dB
P = the sound pressure as measured in N/m^2
P_{ref} = the reference sound pressure, 2×10^{-5} N/m^2

Typical sound pressure levels are shown is Table 12.1. Note that the highest possible SPL, at which point the air molecules can no longer carry pressure waves, is 194 dB, while 0 dB is the threshold of hearing. A typical classroom might be at about 50 dB, while mowing your lawn could subject you to about 90 dB. Conversational speech is at about 62 dB. A ringing alarm clock (next to your head) is about 80 dB, while passenger jet on takeoff can produce over 110 dB. Rock concerts often average 110 dB, far above the threshold permitting conversation. The loudest sound recorded seems to be the Saturn rocket, at 140 dB.

Remember that the scale is logarithmic. One dB difference from 40 to 41 is considerably less energy than a one dB difference from 80 to 81. Every 3 dB increase produces a doubling of the energy level, and a doubling of the danger of damage from excessive noise.

Because sound levels are logarithmic ratios, they cannot be added. If two sources of sound are combined, the pressures have to be calculated from the SPL equation, these added, and a new SPL calculated. As a rule of thumb, two equal sounds added result in a 3 dB increase in overall sound level. If the difference between two sounds is greater than 10 dB, the lesser of the two does not contribute to the overall level of sound.

Example 12.1

A machine shop has two machines, one producing a sound pressure level of 70 dB and one producing 58 dB. A new machine producing 70 dB is brought into the room. What is the new sound pressure level in the room?

First, add 70 and 58 dB. Since the difference is greater than 10, the effect of the 58 dB sound is negligible, and the room would have a sound pressure level of 70 dB. If another 70 dB machine is brought in, the two sounds are equal, producing an increase of 3 dB. Thus the sound pressure level in the room would be 73 dB.

TABLE 12.1. Typical Sound Levels in the Workplace
and the Environment (dB).

Source of Sound	Typical dB
Saturn rocket at one mile	140
Pneumatic riveter	128
Jetliner at 500 feet overhead	110
Air hammer	100
Air compressor	92
Heavy city traffic	88
Average traffic	70
Conversational speech, 3 ft	62
Business office	50
Average home (windows closed)	45
Soft whisper, 5 ft	35
Room in a quiet house at night	30

Sound in the atmosphere travels uniformly in all directions, radiating out from its source. The sound intensity is reduced as the square of the distance away from the source of the sound according to the inverse square law. That is, the sound pressure level is proportional to $1/r^2$, where r is the radial distance from the source.

An approximate relationship can be developed if the sound power is expressed as a logarithmic ratio based on some standard reference power, such that

$$SPL_r = SPL_o - 10 \log(r^2) \qquad (12.2)$$

where

SPL_r = sound pressure level at some distance r from the sound source, dB

SPL_o = sound pressure level at the source, dB

r = distance away from the sound, m

Example 12.2

A sound source generates 80 dB. What would the SPL be 100 m from the source?

$$SPL_r = 80 - 10 \log (100^2) = 40 \text{ dB}$$

This is of course an approximation. We assume that the sound propagates in all directions evenly, but in the real world this is never true. If, for example, the sound occurs on a flat surface, so that the area through which it propagates is a hemisphere instead of a sphere, the approximate addition is 3 dB to the SPL calculated in Example 9.2. In enclosed spaces, reverberation can also greatly increase the sound pressure level since the energy is not dissipated. The most important point to remember is that the SPL is reduced approximately as the log of the square of the distance away from it.

The frequency in cycles per second (Hz) and the amplitude in decibels describe a pure sound at a specific frequency. All environmental sounds, however, are quite "dirty," with many frequencies. A true picture of such sounds is obtained when the sound pressure level is measured at a number of different frequencies. Using meters that filter out all sounds but the sound at a small range of frequencies. Figure 12.2 shows such a plot. The noise described by this plot has some low frequency sound, particularly at about 300 Hz, but there is a lot of sound at about 5000 Hz.

Such pictures of environmental noise are useful in noise control because the

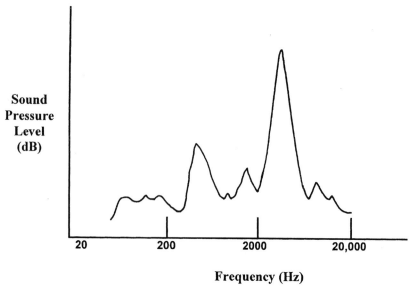

Frequency (Hz)

Figure 12.2. A typical frequency analysis, showing a "dirty" noise.

frequency of the loudest sound pressure levels can be identified and corrective measures taken. If the loudest sound is of a low frequency, then mitigating procedures for such low sounds might be used, while the same control measure might be totally ineffective for high-frequency sounds. This is discussed further below.

12.2 MEASUREMENT OF SOUND

Sound is measured with a *sound level meter* that converts the energy in the air pressure waves to an electrical signal. A microphone picks up the pressure waves and a meter reads the sound pressure level, directly calibrated into decibels. But this sound pressure level is not necessarily what the human ear hears. While we can detect frequencies over a wide range, this detection is not equally effective at all frequencies (our ear does not have a *flat response* in audio terms). If the sound level meter is to simulate the efficiency of the human ear in detecting sound, the signal has to be filtered.

Using many thousands of experiments, researches discovered that on average the human ear has an efficiency over the audible frequency range that resembles Figure 12.3. At very low frequencies, our ears are less efficient than at middle frequencies, say between 1000 and 2000 Hz. At higher frequencies, the ear becomes increasingly inefficient, finally petering out at some high frequency where no sound can be detected. This curve is called the *A-weighed* filtering curve, since there are other filtering curves for different purposes. The band of greatest efficiency for the human ear is very close to human speech range.

Using the curve shown in Figure 12.3, instrument designers have constructed a meter that filters out some of the very low frequency sound and much of the very high frequency sound, so that the sound measured represents somewhat the hearing of a human ear. Such a measurement is called a *sound level*, and is designated dB(A), since it represents modified dB value, with the A filter.

Such analyses are useful in measuring hearing ability. Using the hearing ability of an average young ear as the standard, an *audiometer* measures the hearing ability at various frequencies producing an *audiogram*. Such audiograms are then used to identify the frequencies where the hearing aids must be able to boost the signal.

Figure 12.4, for example, shows three audiograms. Person A has excellent hearing with a basically flat response. Person B has lost hearing at a specific frequency range, in this case around 4000 Hz. Such hearing losses are often due to industrial noise which destroys the ear's ability to hear sounds at a specific frequency. Since speech is near that range, this person already has difficulty

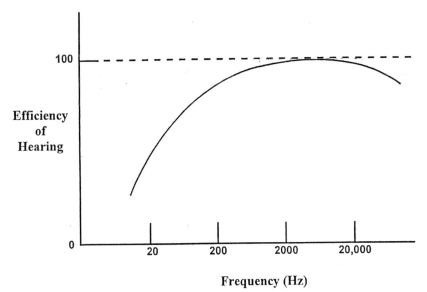

Frequency (Hz)

Figure 12.3. Approximate efficiency of the human ear to various frequencies of sound.

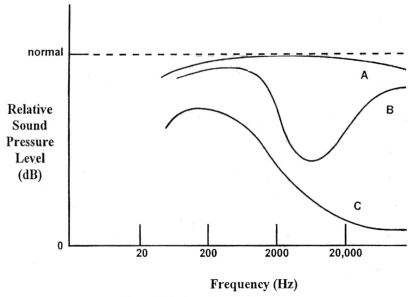

Frequency (Hz)

Figure 12.4. Three typical audiograms.

hearing. The third curve on Figure 12.4, Person C, shows a typical audiogram either for an older person who has lost much of the higher frequency range and probably is a candidate for a hearing aid or a young person who has sustained severe damage to his/her hearing by subjecting it to loud music.

The Power Mower

One of the ubiquitous irritants in any neighborhood is the power mower (or the person who starts mowing his lawn at 6 am on a Sunday morning.) Is it not possible, you ask, to build a quiet mower?

Yes, of course it is. A typical gasoline-powered lawn mower emits about 72 dB(A) at 50 feet, but for less than $15 additional cost, the mower can be quieted to 59 dB(A), representing a four-fold decrease in noise. There are two reasons why this is not done: 1) there is no government agency that would limit the noise in lawnmowers, and 2) there is no public pressure to do so. At the present time, the noise standards for lawnmowers are set by an industry group, the Society of Automotive Engineers, and this organization, run and supported by the manufacturers, has little incentive to reduce the noise of the lawnmower. They therefore set their "standards" according to what the manufacturers produce, and it is no wonder that the SAE noise standard for lawnmower noise is 72 dBA at 50 feet!

12.3 EFFECT OF NOISE ON HUMAN HEALTH

Figure 12.5 shows a schematic of a human ear. The air pressure waves first hit the ear drum (*tympanic membrane*), causing it to vibrate. The cavity leading to the tympanic membrane and the membrane itself are often called the *outer ear*. The tympanic membrane is attached physically to three small bones in the middle ear which start to move when the membrane vibrates. The purpose of these bones, called colloquially the *hammer, anvil* and *stirrup* due to their shapes, is to amplify the physical signal (to achieve some gain in audio terms). This air-filled cavity is called the *middle ear*. The amplified signal is then sent to the *inner ear* by first vibrating another membrane called the *round window membrane*, which is attached to a snail-shaped cavity called the *cochlea*.

Within this fluid-filled cochlea is another membrane, the basilar membrane, which is attached to the round window membrane. Attached to the basilar membrane are two sets of tiny *hair cells*, pointing in opposite directions. As the round window membrane vibrates, the fluid in the inner ear is set in motion and the thousands of hair cells in the cochlea shear past each other, setting up electrical impulses which are then sent to the brain through the *auditory nerves*. The frequency of the sound determines which ones of the hair cells will move. The hair cells close to the round window membrane are sensitive to high frequencies and those in the far end of the cochlea respond to low frequencies.

Damage to the human ear can occur in several ways. First, very loud impulse noise can burst the ear drum, causing mostly temporary loss of hearing, although frequently torn eardrums heal poorly and can result in permanent damage. The bones in the middle ear are not usually damaged by loud sounds,

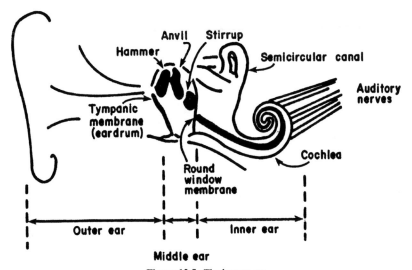

Figure 12.5. The human ear.

although they can be hurt by infections. Since our sense of balance depends very much on the middle ear, a middle ear infection can be debilitating. Finally, the most important and most permanent damage can occur to the hair cells in the inner ear. Very loud sounds will stun these hair cells and cause them to cease functioning. Most of the time this is a temporary condition, and time will heal the damage. Unfortunately, if the insult to the inner ear is prolonged, the damage can be permanent. This damage cannot be repaired by an operation or corrected by hearing aids. It is this permanent damage to young people, inflicted by loud music, that is most frequent, insidious, and sad. Is it really worth it to spend your times in front of huge loudspeakers in concerts, or turn up the volume on the headphones, when the result will be that you will not be able to hear *any* music by the time you are 40 years old?

But loud noise does more than cause permanent hearing damage. Noise to our ancestors meant danger and thus the human body reacts to loud noise so as to protect itself from imminent harm. The bodily reactions are amazing—the eyes dilate, the adrenaline flows, the blood vessels dilate, the senses are alerted, the heartbeat is altered, blood thickens—all to get the person "up." Apparently, such "up" state, if prolonged, is quite unhealthy. People who live and work in noisy environments have measurably greater general health problems, are

The Ill-Fated Comet

After the Second World War the British aircraft industry was the finest in the world and had great manufacturing capacity. It turned to the design and manufacture of jet passenger aircraft, and DeHavilland become the first company to fly a prototype of a jet airliner, called the Comet. The Comet went into service and was a huge success, being able to fly across the Atlantic at unheard of speeds. But then tragedy struck. First one, then another, and then a third Comet fell out of the skies. One minute they were cruising along, and the next minute they disappeared off the radar screen. There was no warning. It was as if the planes exploded in midair.

DeHavilland, in desperation, decided to instrument a full-sized Comet to the maximum and run it through the flight procedures while in the hangar, including pressurizing the cabin and running the engines at full throttle. In a few weeks they got their answer. Small cracks were beginning to form in the fuselage, and these cracks eventually became sufficiently large that the pressurized cabin simply blew apart. After much analysis, the engineers determined that it was the *noise* of the jet engines that had shaken the fuselage so much that the fatigue cracks formed. DeHavilland had its answer, but was unable to find a cure, and within a few years American companies such as Boeing introduced their jet airliners and established the United States as the world's center for manufacturing commercial jet airplanes.

grouchy and ill-tempered, and have trouble concentrating. Noise that reduces sleep carries with it an additional array of health problems. What is most important is that we cannot adapt to high noise levels. Industrial workers are often seen walking around with their ear protectors hanging around their necks, as if this is some macho thing to do. They think they can "take it." Perhaps they can take it because they already are deaf at the frequency ranges prevalent in that environment.

12.4 NOISE ABATEMENT

The most effective single legislation to pass Congress in the noise control area is the Occupational Health and Safety Act of 1970 in which OSHA was given authority to control industrial noise levels. One of the most important regulations promulgated by OSHA is the limits of industrial noise. Recognizing that both intensity and duration are important in preventing damage to hearing, the OSHA regulations limit noise as shown in Table 12.2.

Industrial noise is mostly constant over an 8-hour working day. Community noise, however, is intermittent noise. If an airplane goes overhead only a few time a day, on average the sound level is quite low, but the irritation factor is high. For this reason, there have been a plethora of noise indexes developed, all of which are intended to estimate the psychological effect of the noise. Most of these begin by using the cumulative distribution technique, similar to the analysis of natural events such as floods, as discussed previously.

Excellent work has also been done by acoustic engineers working to quiet commercial aircraft. Modern airliners are amazingly quiet compared to earlier, louder models such as the Boeing 727, one of the loudest large planes still in service. The control of noise in the urban environment or noise produced by machines and other devices has not received much governmental or public support. Little progress has been made in controlling noise during the past few decades.

One of the problems in measuring environmental noise such as the noise from highways is that environmental noise is intermittent, while industrial

TABLE 12.2. OSHA Industrial Noise Limits.

Duration (hours)	Sound Levels [dB(A)]
8	90
6	92
4	95
3	97
2	100
1	105
0.5	110
0.25	115

Sound levels greater than 115 dB(A) are clearly detrimental and should not be permitted.

The Kokomo Hum

Kokomo, Indiana, is a medium-sized industrial city that hums. Many residents have complained about a hum they hear, especially at night, and this hum is becoming more than just an annoyance. They claim that the hum is causing physical illness and a number of people have had to move out of town to escape the hum. Since 1999, over 100 complaints have been made, and the community has initiated a study to try to find the source of the hum.

But the acoustical engineers hired to find out what is going on are baffled. They can measure the hum in homes and therefore acknowledge that it exists, but they have no idea where it comes from. Apparently the hum is of very low frequency (less than 10 Hz), and these sounds can travel for long distances. When the sound enters rooms that are just the right size, they can set up resonance and bounce from wall to wall, increasing in amplitude. The source is unknown, and could be any number of factories, miles away. The sound they emit would be masked by the noise of the factories and there would be little chance of finding it. The high-frequency sounds would die off in a short distance, but the low-frequency sounds would continue on. Once this sound finds an appropriately sized room, it would start bouncing around and increasing in amplitude, creating the audible

---

noise is most often steady. It is easy to measure such industrial noise with a sound-level meter and report the results, but getting a handle on highway noise, with occasional very noisy trucks going by, is a different problem.

The solution once again is in frequency analysis, the technique first introduced in Chapter 2. To measure highway noise, a series of reading are made, at certain time intervals such as 10 seconds. These data are tabulated on a frequency analysis plot and the results interpreted as "percent of time a given level of sound is exceeded." The example below demonstrates this technique.

Example 12.3

Highway noise was measured intermittently, every 15 seconds. The results are as shown:

Time (sec)	dB(A)
0	80
15	90
30	85
45	78
60	65
75	70
90	88
105	69

This analysis is identical to the procedures previously used for grappling with natural events such as floods. The first step is to rank the data, the sound level measurements, from the loudest to the least loud. The total number of data points, n, in this case is 8. The probability of the level of noise returning is m/n.

Rank, m	dB(A)	m/n
1	90	0.125
2	88	0.25
3	85	0.375
4	80	0.5
5	78	0.625
6	70	0.75
7	69	0.875
8	65	1.00

The m/n values are plotted against the noise level as shown in Figure 12.6. The average noise level from this highway, or that time that the noise level that is exceeded 50% of the time is read off the plot as 81 dB(A).

The symbol L is often used in noise pollution engineering to signify the fraction (or commonly the percent) of time that a noise is statistically exceeded, so that L_{50} would mean that 50% of the time a certain sound level has been exceeded. In the above example, $L_{50} = 81$ dB(A). Similarly, L_{10} means that this sound level has been exceeded only 10% of the time. From Figure 12.6, L_{10} can be read as about 91 dB(A).

Such numbers have been used to estimate that effect of transient noise, such as traffic or aircraft noise on humans using various combinations and permutations of the L values. One such equation that has found wide use is the *noise pressure level* (NPL), defined as

$$NPL = L_{50} + (L_{10} - L_{90}) + (L_{10} - L_{90})^2/60 \qquad (12.3)$$

where

$$NPL = \text{noise pressure level in dB(A)}$$
L_{10}, L_{50}, and L_{90} = sound pressure level that is exceeded, 10, 50 and 90% of the time, respectively.

Note that the average noise level, L_{50}, is enhanced if the difference between the very loud noise (the one exceeded only 10% of the time) and the less loud noise (the one exceeded 90% of the time) is great. This means that very loud transient noises such as aircraft taking off overhead can be expected to be quite annoying, which they certainly are. Anyone who has tried to watch a baseball

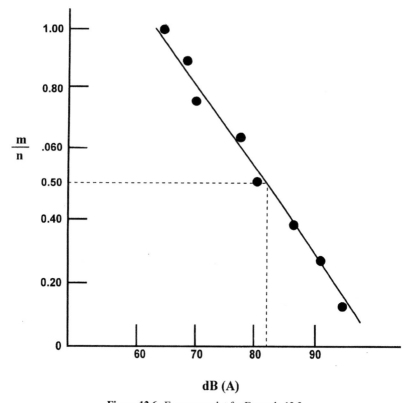

Figure 12.6. Frequency plot for Example 12.3.

game in Shea Stadium in New York, not far from the JFK airport, will attest to the fact that the noise of large planes taking off every few minutes can ruin an otherwise pleasant summer afternoon watching the Mets.

Example 12.4

Given the results of sound-pressure measurements shown in Figure 12.6, what is the noise pressure level?

From Figure 12.6, $L_{10} = 91$ dB(A), $L_{50} = 81$ dB(A), and $L_{90} = 67$ dB(A). Using Equations 12.3,

$$NPL = 81 + (91 - 67) + (91 - 67)^2/60 = 114 \text{ dB(A)}$$

One of the most ubiquitous transient noises in contemporary American society is traffic noise. The U. S. Department of Transportation has been concerned with traffic noise and has established maximum sound levels for

TABLE 12.3. Design Noise Levels for Highways.

Land category and use	Design noise level, L_{10} Exterior	Interior
A. Activites that require special quiet, like outdoor amphitheaters, memorial gardens, etc.	60 dB(A)	
B. Residences, motels, hospitals, schools, parks, libraries, and other public buildings.	70 dB(A)	55 dB(A)
C. Developed land not otherwise included in classifications A and B above.	75 dB(A)	
D. Undeveloped land	No limit	

different vehicles. Modern trucks and automobiles are considerably quieter than they were only a few years ago, but much of this has been forced not by governmental fiat but by public demand. The exception is renegade trucks that continue to be the major noisemakers on the highways, and have little chance of being caught by police who do not see noise as a serious problem compared to other highway safety concerns.

The Federal Highway Administration has established design levels for land close to highways, and these limits are shown in Table 12.3. Note that these levels have been set based on L_{10}, so that infrequent transitory noises such as the occasional very loud truck, are more than likely ignored in such an analysis.

It is not difficult to exceed these limits. Figure 12.7 shows some data for

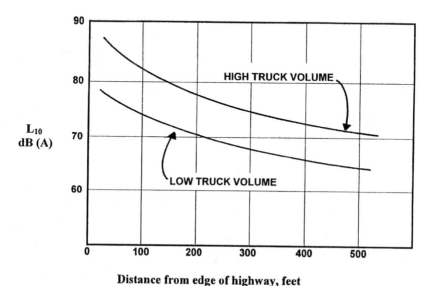

Distance from edge of highway, feet

Figure 12.7. Noise from highways is slowly attenuated, and the level depends on truck traffic.

The Red Flag Law

Shortly after the first steam-powered and very noisy horseless carriage appeared in England, the British Parliament in 1865 passed the famous Red Flag Law, which required a man to precede a horseless carriage on foot, carrying a red flag by day and a lantern by night. The primary aim of the law was to warn people about the loud noises produced by the vehicles. This law, in fact, was credited with slowing the development of automobiles until 1896, when it was repealed. Coincidentally, that year saw some of the early applications of the internal combustion engine to vehicles in England, France, Germany, and the United States. When the explosive noises of the automobile disturbed the New England countryside, Vermont in 1894 passed an adaptation of the English law (repealed in two years). To protect the nervous systems of horses, and perhaps people's ears as well, the Vermont law required a man to walk "several hundred feet" in advance of a moving car.

---

traffic noise as a function of distance away from the highway. Note that the 60 dB(A) level for quiet places is difficult to achieve even if an area is a considerable distance away from a highway.

A recent development has been the growth of noisewalls around highways. Noise produces a shadow, and the loudest noise comes when the recipient is in a direct line with the source. Placing walls between the trucks and people in homes is intended to take advantage of the noise shadow effect. Unfortunately, noise is not like light. A noise shadow is not perfect and noise can bend and bounce off air, depending on the atmospheric conditions. Sometimes noises miles away can be heard as the pressure waves bounce off inversion layers. The noise fences along highways are, for the most part, psychological, and do not provide much protection except for the homes immediately in the lee of the fence.

Highway sound can also be abated by heavy natural growth. Trees by themselves are not very effective, but a dense growth will reduce the sound pressure level by several decibels per 100 feet of dense forest. Cutting down the natural growth to widen a highway invariably will cause increased noise problems.

12.5 SYMBOLS

L_{10}, L_{50}, and L_{90} = sound pressure level that is exceeded, 10, 50 and 90% of the time, respectively.

NPL = noise pressure level in dB(A)

P = the sound pressure as measured in N/m^2

P_{ref} = the reference sound pressure, 2×10^{-5} N/m^2

r = distance away from the sound, m

SPL = Sound Pressure Level, dB

SPL_r = sound pressure level at some distance r from the sound source, dB

SPL_o = sound pressure level at the source, dB

12.6 PROBLEMS

12.1 A dewatering building in a wastewater treatment plant has a centrifuge that operates at 89 dB(A). Another similar machine is being installed in the same room. Will the operator be able to work in this room?

12.2 A lawn mower emits 80 dB(A). Suppose two other lawn mowers that produce 80 and 61 dB(A) respectively join the mowing operation.

a. What will be the cacophony, in dB(A)?

b. What will be the sound pressure level 200 m away in a neighbor's yard?

12.3 A 20-year-old rock star has an audiogram done. The results are shown as Curve B in Figure 12.4

a. Is he able to hear normal conversational speech?

b. If curve C shown in Figure 12.4 represents a normal loss of hearing due to aging for a 65 year old person, draw the audiogram for the rock star when he reaches 65 years old.

c. Will he be able to hear anything at all?

12.4 OSHA requirements for noisy work places often include protective devices such as earmuffs. Managers whose responsibility it is to make sure the workers wear the protection complain that the workers will take them off as soon as the managers are out of sight. They complain that the earmuffs are uncomfortable, and believe that they don't need them. Why might this latter be true? Why would persons who might have worked in a factory for many years believe that they don't need ear protection?

12.5 If the sound level in a factory is 98 dB(A), how many minutes out of an 8-hour working day should the workers be subjected to such sound?

12.6 A study was done on highway noise with the following results:

Location	Time (sec)	dB(A)
At edge of pavement	0	75
	10	68
	20	80
	30	83
	40	74
	50	78
	60	79
	70	80
50 meters from the pavement	0	75
	10	74
	20	73
	30	69
	40	78
	50	75
	60	63
	70	70
100 meters from the pavement	0	68
	10	61
	20	73
	30	67
	40	68
	50	71
	60	72
	70	73

Calculate the L_{50}, or that sound level exceeded 50% of the time for each of the sampling stations. Plot the L_{50} as distance from the road. Does this result agree with the Department of Transportation data?

Epilogue

FOLLOWING the destructive hurricane Isabella that brought devastation to the North Carolina shore, the Reverend Tom Fairley offered this prayer at the releasing of sea turtles from the turtle hospital in Topsail Beach NC.

God of all creatures, both great and small, if Isabel was a tiny demonstration of your power and might, then we stand dutifully in awe. But frankly, such demonstrations are getting very boring. If You truly hear our prayers, then we pray for You to re-evaluate Your own priorities. Unleash your fury on polluters who poison Your Oceans instead of Your lovely coastlines. Visit your wrath on profiteers who would make a buck by any means possible, even to the detriment of this tiny dust speck we call home. Strike down politicians bought by the highest bidder whose vote to clean the environment is silenced by big money.

In a similar manner, O God, infuse into us a mighty dose of Your indignation and courage. As we ask you to get Your priorities straight, so, too, do we pray that you will help us to get our priorities in order. Give us the might and the fury of Isabel to attack polluters, profiteers, and crooked politicians at every opportunity. Let them know that because of us, by comparison, Isabel was nothing but a puff. Let them learn to fear us more than they fear a hurricane, and give us the courage of our resolve to act as Your agents. Help us to be your hurricanes for good.

Once again, we ask You to pay attention to a handful of Your sea turtles. Keep them out of polluted waters, and away from all of our gadgets that cut them or drown them. But most importantly, keep them away from our kind who attacks them for money or sport. And may the calm that follows a hurricane become the peace we have in our hearts because we have done Your work, and have done it well. Amen.

ALBERT SCHWEITZER

Albert Schweitzer was born in Alsace, and following in the footsteps of his father and grandfather, entered into theological studies in 1893 at the University of Strasbourg where he obtained a doctorate in philosophy in 1899, with a dissertation on religious philosophy. He began preaching at St. Nicholas Church in Strasbourg in 1899 and served in various high ranking administrative posts. In 1906 he published *The Quest of the Historical Jesus,* a book on which much of his fame as a theological scholar rests.

Schweitzer had a parallel career as an organist. He had begun his studies in music at an early age and performed in his father's church when he was nine years old. He eventually became an internationally known interpreter of the organ works of Johann Sebastian Bach. From his professional engagements he earned funds for his education, particularly his later medical schooling.

He decided to embark on a third career, as a physician, and to go to Africa as a medical missionary. After obtaining his MD at Strasbourg in 1913 he founded his hospital at Lambaréné in French Equatorial Africa. In 1917, however, the war intervened and he and his wife spent 1917 in a French internment camp as prisoners of war. Returning to Europe after the war, Schweitzer spent the next six years preaching in his old church, and giving lectures and concerts to raise money for the hospital.

Schweitzer returned to Lambaréné in 1924 and except for relatively short periods of time, spent the remainder of his life there. With the funds earned from his own royalties and personal appearance fees and with those donated from all parts of the world, he expanded the hospital to seventy buildings which by the early 1960's could take care of over 500 patients in residence at any one time.

On one of his trips up the Congo to his hospital Schweitzer saw a group of hippopotamuses along the shore, and had a sudden inspiration for a new philosophical concept that he called "reverence for life" which has had wide influence in Western environmental thought.

He was awarded the Nobel Peace Prize in 1953.

Appendix

Mutiply	by	to obtain
acres	0.404	ha
acres	43,560	ft^2
acres	4047	m^2
acres	4840	yd^2
acre ft	1233	m^3
atmospheres	14.7	lb/in^2
atmospheres	29.95	in mercury
atmospheres	33.9	ft of water
atmospheres	10,330	kg/m^2
Btu	252	cal
Btu	1.053	kJ
Btu	1,053	J
Btu/ft^3	8,905	cal/m^3
Btu/lb	2.32	kJ/kg
Btu/lb	0.555	cal/g
Btu/s	1.05	kW
Btu/ton	278	cal/tonne
Btu/ton	0.00116	kJ/kg
calories	4.18	J
calories	0.0039	Btu
calories/g	1.80	Btu/lb
calories/m^3	0.000112	Btu/ft^3
calories/tonne	0.00360	Btu/ton
centimeters	0.393	in
cubic ft	1728	in^3
cubic ft	7.48	gal
cubic ft	0.0283	m^3
cubic ft	28.3	L
cubic ft/lb	0.0623	m^3/kg
cubic ft/s	0.646	million gal/day

Mutiply	by	to obtain
cubic ft/s	0.0283	m^3/s
cubic ft/s	449	gal/min
cubic ft of water	61.7	lb
cubic in of water	0.0361	lb
cubic m	35.3	ft^3
cubic m	264	gal
cubic m	1.31	yd^3
cubic m/day	264	gal/day
cubic m/hr	4.4	gal/min
cubic m/hr	0.00638	million gal/day
cubic m/s	1	cumec
cubic m/s	35.31	ft^3/s
cubic m/s	15,850	gal/min
cubic m/s	22.8	mil gal/day
cumec	1	m^3/s
cubic yards	0.765	m^3
cubic yards	202	gal
C-ration	100	rations
decacards	52	cards
feet	0.305	m
feet/min	0.00508	m/s
feet/s	0.305	m/s
feet/s	720	in/min
fish	10^{-6}	microfiche
foot lb (force)	1.357	J
foot lb (force)	1.357	Nm
gallon of water	8.34	lb
gallons	0.00378	m^3
gallons	3.78	L
gallons/day	43.8×10^{-6}	L/s
gallons/day/ft^2	0.0407	$m^3/day/m^2$
gallons/min	0.00223	ft^3/s
gallons/min	0.0631	L/s
gallons/min	0.227	m^3/hr
gallons/min	6.31×10^{-5}	m^3/s
gallons/min/ft^2	2.42	$m^3/hr/m^2$
gallons of water	8.34	lb water
grams	0.0022	lb
grams/cm^3	1,000	kg/m^3
hectares	2.47	acre
hectares	1.076×10^5	ft^2
horsepower	0.745	kW
horsepower	33,000	ft lb/min
inches	2.54	cm
inches/min	0.043	cm/s
inches of mercury	0.49	lb/in^2

Mutiply	by	to obtain
inches of mercury	0.0038	N/m^2
inches of water	249	N/m^2
joules	0.239	cal
joules	9.48×10^{-4}	Bru
joules	0.738	ft lb
joules	2.78×10^{-7}	kWh
joules	1	Nm
joules/g	0.430	Btu/lb
joules/g	1	W
kilocalories	3.968	Btu
kilocalories/kg	1.80	Btu/lb
kilograms	2.20	lb
kilograms	0.0011	tons
kilograms/ha	0.893	lb/acre
kilograms/hr	2.2	lb/hr
kilograms/m^3	0.0624	lb/ft^3
kilograms/m^3	1.68	lb/yd^3
kilograms/m^3	1.69	lb/yd^3
kilograms/tonne	2.0	lb/ton
kilojoules	9.49	Btu
kilojoules/kg	0.431	Btu/lb
kilometers	0.622	mile
kilometer/hr	0.622	miles/hr
kilowatts	1.341	horsepower
kilowatt-hr	3600	kJ
kilowatt-hr	3.410	Btu
liters	0.0353	ft^3
liters	0.264	gal
liters/s	15.8	gal/min
liters/s	0.0228	million gal/day
meters	3.28	ft
meters	1.094	yd
meters/s	3.28	ft/s
meters/s	196.8	ft/min
microscope	10^{-6}	mouthwash
miles	1.61	km
miles/hr	0.447	m/s
miles/hr	88	ft/min
miles/hr	1.609	km/hr
milligrams/L	0.001	kg/m^3
million gal	3,785	m^3
million gal/day	43.8	L/s
million gal/day	3785	m^3/day
million gal/day	157	m^3/hr
million gal/day	0.0438	m^3/s
million gal/day	0.0438	m^3/s
million gal/day	1.547	ft^3/s
million gal/day	694	gal/min

Mutiply	by	to obtain
newtons	0.225	lb (force)
newtons/m^2	2.94×10^{-4}	in mercury
newtons/m^2	1.4×10^{-4}	lb/in^2
newtons/m^2	10	poise
newton m	1	J
pounds (mass)	0.454	kg
pounds (mass)	454	g
pounds (mass)/acre	1.12	kg/ha
pounds (mass)/ft3	16.04	kg/m^3
pounds (mass)/ton	0.50	kg/tonne
pounds (mass/yd^3	0.593	kg/m^3
pounds (force)	4.45	N
pounds (force)/in^2	0.068	atmospheres
pounds (force)/in^2	2.04	in of mercury
pounds (force)/in^2	6895	N/m^2
pounds (force)/in^2	6.89	kPa
pound cake	454	Graham crackers
pounds of water	0.01602 ft^3	ft^3
pounds of water	27.68	in^3
pounds of water	0.1198	gal
square ft	0.0929	m^2
square m	10.74	ft^2
square m	1.196	yd^2
square m	2.471×10^{-4}	acres
square miles	2.59	km^2
tons (2000 lb)	0.907	tonnes (1000 kg)
tons	907	kg
tons/acre	2.24	tonnes/ha
tonnes (1000 kg)	1.10	ton (2000 lb)
tonnes/ha	0.446	tons/acre
two kilomockingbirds	2000	mockingbird
watts	1	J/s
yard	0.914	m

Index